Bayesian Nonparametrics

Bayesian nonparametrics works – theoretically, computationally. The theory provides highly flexible models whose complexity grows appropriately with the amount of data. Computational issues, though challenging, are no longer intractable. All that is needed is an entry point: this intelligent book is the perfect guide to what can seem a forbidding landscape.

Tutorial chapters by Ghosal, Lijoi and Prünster, Teh and Jordan, and Dunson advance from theory, to basic models and hierarchical modeling, to applications and implementation, particularly in computer science and biostatistics. These are complemented by companion chapters by the editors and Griffin and Quintana, providing additional models, examining computational issues, identifying future growth areas, and giving links to related topics.

This coherent text gives ready access both to underlying principles and to state-of-the-art practice. Specific examples are drawn from information retrieval, neuro-linguistic programming, machine vision, computational biology, biostatistics, and bioinformatics.

NILS LID HJORT is Professor of Mathematical Statistics in the Department of Mathematics at the University of Oslo.

CHRIS HOLMES is Professor of Biostatistics in the Department of Statistics at the University of Oxford.

PETER MÜLLER is Professor in the Department of Biostatistics at the University of Texas M. D. Anderson Cancer Center.

STEPHEN G. WALKER is Professor of Statistics in the Institute of Mathematics, Statistics and Actuarial Science at the University of Kent, Canterbury.

CAMBRIDGE SERIES IN STATISTICAL AND PROBABILISTIC MATHEMATICS

This series of high-quality upper-division textbooks and expository monographs covers all aspects of stochastic applicable mathematics. The topics range from pure and applied statistics to probability theory, operations research, optimization, and mathematical programming. The books contain clear presentations of new developments in the field and also of the state of the art in classical methods. While emphasizing rigorous treatment of theoretical methods, the books also contain applications and discussions of new techniques made possible by advances in computational practice.

A complete list of books in the series can be found at
http://www.cambridge.org/uk/series/sSeries.asp?code=CSPM
Recent titles include the following:

6. *Empirical Processes in M-Estimation,* by Sara van de Geer
7. *Numerical Methods of Statistics,* by John F. Monahan
8. *A User's Guide to Measure Theoretic Probability,* by David Pollard
9. *The Estimation and Tracking of Frequency,* by B. G. Quinn and E. J. Hannan
10. *Data Analysis and Graphics using R (2nd Edition),* by John Maindonald and John Braun
11. *Statistical Models,* by A. C. Davison
12. *Semiparametric Regression,* by David Ruppert, M. P. Wand and R. J. Carroll
13. *Exercises in Probability,* by Loïc Chaumont and Marc Yor
14. *Statistical Analysis of Stochastic Processes in Time,* by J. K. Lindsey
15. *Measure Theory and Filtering,* by Lakhdar Aggoun and Robert Elliott
16. *Essentials of Statistical Inference,* by G. A. Young and R. L. Smith
17. *Elements of Distribution Theory,* by Thomas A. Severini
18. *Statistical Mechanics of Disordered Systems,* by Anton Bovier
19. *The Coordinate-Free Approach to Linear Models,* by Michael J. Wichura
20. *Random Graph Dynamics,* by Rick Durrett
21. *Networks,* by Peter Whittle
22. *Saddlepoint Approximations with Applications,* by Ronald W. Butler
23. *Applied Asymptotics,* by A. R. Brazzale, A. C. Davison and N. Reid
24. *Random Networks for Communication,* by Massimo Franceschetti and Ronald Meester
25. *Design of Comparative Experiments,* by R. A. Bailey
26. *Symmetry Studies,* by Marlos A. G. Viana
27. *Model Selection and Model Averaging,* by Gerda Claeskens and Nils Lid Hjort
28. *Bayesian Nonparametrics,* edited by Nils Lid Hjort *et al.*
29. *Finite and Large Sample Statistical Theory (2nd Edition),* by Pranab K. Sen, Julio M. Singer and Antonio C. Pedrosa-de-Lima
30. *Brownian Motion*, by Peter Mörters and Yuval Peres

Bayesian Nonparametrics

Edited by

Nils Lid Hjort
University of Oslo

Chris Holmes
University of Oxford

Peter Müller
University of Texas
M.D. Anderson Cancer Center

Stephen G. Walker
University of Kent

CAMBRIDGE UNIVERSITY PRESS
Cambridge, New York, Melbourne, Madrid, Cape Town, Singapore,
São Paulo, Delhi, Dubai, Tokyo

Cambridge University Press
The Edinburgh Building, Cambridge CB2 8RU, UK

Published in the United States of America by Cambridge University Press, New York

www.cambridge.org
Information on this title: www.cambridge.org/9780521513463

First published 2010

Printed in the United States of America

A catalog record for this publication is available from the British Library

Library of Congress Cataloging in Publication data
Bayesian nonparametrics / edited by Nils Lid Hjort . . . [et al.].
p. cm. – (Cambridge series in statistical and probabilistic mathematics ; 28)
Includes index.
ISBN 978-0-521-51346-3 (hardback)
1. Nonparametric statistics. 2. Bayesian statistical decision theory. I. Hjort, Nils Lid. II. Title.
QA278.8.B39 2009
519.5′42 – dc22 2009037744

ISBN 978-0-521-51346-3 Hardback

Contents

Contributors

David B. Dunson
Institute of Statistics
and Decision Sciences
Duke University
Durham, NC 27708-0251, USA

Subhashis Ghosal
Department of Statistics
North Carolina State University
Raleigh, NC 27695, USA

Jim Griffin
Institute of Mathematics, Statistics
and Actuarial Science
University of Kent
Canterbury CT2 7NZ, UK

Nils Lid Hjort
Department of Mathematics
University of Oslo
N-0316 Oslo, Norway

Chris Holmes
Oxford Centre for Gene Function
University of Oxford
Oxford OX1 3QB, UK

Michael I. Jordan
Department of Electrical Engineering
and Computer Science
University of California, Berkeley
Berkeley, CA 94720-1776, USA

Antonio Lijoi
Department of Economics
and Quantitative Methods
University of Pavia
27100 Pavia, Italy

Peter Müller
Department of Biostatistics
M. D. Anderson Cancer Center
University of Texas
Houston, TX 77030-4009, USA

Igor Prünster
Department of Statistics
and Applied Mathematics
University of Turin
10122 Turin, Italy

Fernando Quintana
Department of Statistics
Pontifical Catholic University of Chile
3542000 Santiago, Chile

Yee Whye Teh
Gatsby Computational
Neuroscience Unit
University College London
London WC1N 3AR, UK

Stephen G. Walker
Institute of Mathematics, Statistics
and Actuarial Science
University of Kent
Canterbury CT2 7NZ, UK

An invitation to Bayesian nonparametrics

Nils Lid Hjort, Chris Holmes, Peter Müller and Stephen G. Walker

This introduction explains why you are right to be curious about Bayesian nonparametrics – why you may actually need it and how you can manage to understand it and use it. We also give an overview of the aims and contents of this book and how it came into existence, delve briefly into the history of the still relatively young field of Bayesian nonparametrics, and offer some concluding remarks about challenges and likely future developments in the area.

Bayesian nonparametrics

As modern statistics has developed in recent decades various dichotomies, where pairs of approaches are somehow contrasted, have become less sharp than they appeared to be in the past. That some border lines appear more blurred than a generation or two ago is also evident for the contrasting pairs "parametric versus nonparametric" and "frequentist versus Bayes." It appears to follow that "Bayesian nonparametrics" cannot be a very well-defined body of methods.

What is it all about?

It is nevertheless an interesting exercise to delineate the regions of statistical methodology and practice implied by constructing a two-by-two table of sorts, via the two "factors" parametric–nonparametric and frequentist–Bayes; Bayesian nonparametrics would then be whatever is not found inside the other three categories.

(i) *Frequentist parametrics* encompasses the core of classical statistics, involving methods associated primarily with maximum likelihood, developed in the 1920s and onwards. Such methods relate to various optimum tests, with calculation of p-values, optimal estimators, confidence intervals, multiple comparisons, and so forth. Some of the procedures stem from exact probability calculations for models that are sufficiently amenable to mathematical derivations, while others relate to the application of large-sample techniques (central limit theorems, delta methods, higher-order corrections involving expansions or saddlepoint approximations, etc.).

(ii) *Bayesian parametrics* correspondingly comprises classic methodology for prior and posterior distributions in models with a finite (and often low) number of parameters. Such methods, starting from the premise that uncertainty about model parameters may somehow be represented in terms of probability distributions, have arguably been in existence for more than a hundred years (since the basic theorem that drives the machinery simply says that the posterior density is proportional to the product of the prior density with the likelihood function, which again relates to the Bayes theorem of *c.* 1763), but they were naturally limited to a short list of sufficiently simple statistical models and priors. The applicability of Bayesian parametrics widened significantly with the advent and availability of modern computers, from about 1975, and then with the development of further numerical methods and software packages pertaining to numerical integration and Markov chain Monte Carlo (MCMC) simulations, from about 1990.

As for category (i) above, asymptotics is often useful for Bayesian parametrics, partly for giving practical and simple-to-use approximations to the exact posterior distributions and partly for proving results of interest about the performance of the methods, including aspects of similarity between methods arising from frequentist and Bayesian perspectives. Specifically, frequentists and Bayesians agree in most matters, to the first order of approximation, for inference from parametric models, as the sample size increases. The mathematical theorems that in various ways make such statements precise are sometimes collectively referred to as "Bernshteĭn–von Mises theorems"; see, for example, Le Cam and Yang (1990, Chapter 7) for a brief treatment of this theme, including historical references going back not only to Bernshteĭn (1917) and von Mises (1931) but all the way back to Laplace (1810). One such statement is that confidence intervals computed by the frequentists and the Bayesians (who frequently call them "credibility intervals"), with the same level of confidence (or credibility), become equal, to the first order of approximation, with probability tending to one as the sample size increases.

(iii) *Frequentist nonparametrics* is a somewhat mixed bag, covering various areas of statistics. The term has historically been associated with test procedures that are or asymptotically become "distribution free," leading also to nonparametric confidence intervals and bands, etc.; for methodology related to statistics based on ranks (see Lehmann, 1975); then progressively with estimation of probability densities, regression functions, link functions etc., without parametric assumptions; and also with specific computational techniques such as the bootstrap. Again, asymptotics plays an important role, both for developing fruitful approximations and for understanding and comparing properties of performance. A good reference book for learning about several classes of these methods is Wasserman (2006).

(iv) What ostensibly remains for our fourth category, then, that of *Bayesian nonparametrics*, are models and methods characterized by (a) big parameter spaces (unknown density and regression functions, link and response functions, etc.) and (b) construction of probability measures over these spaces. Typical examples include Bayesian setups for density estimation (in any dimension), nonparametric regression with a fixed error distribution, hazard rate and survival function estimation for survival analysis, without or with covariates, etc. The divisions between "small" and "moderate" and "big" for parameter spaces are not meant to be very sharp, and the scale is interpreted flexibly (see for example Green and Richardson, 2001, for some discussion of this).

It is clear that category (iv), which is the focus of our book, must meet challenges of a greater order than do the other three categories. The mathematical complexities are more demanding, since placing well-defined probability distributions on potentially infinite-dimensional spaces is inherently harder than for Euclidean spaces. Added to this is the challenge of "understanding the prior"; the ill-defined transformation from so-called "prior knowledge" to "prior distribution" is hard enough to elicit in lower dimensions and of course becomes even more challenging in bigger spaces. Furthermore, the resulting algorithms, for example for simulating unknown curves or surfaces from complicated posterior distributions, tend to be more difficult to set up and to test properly.

Finally, in this short list of important subtopics, we must note that the bigger world of nonparametric Bayes holds more surprises and occasionally exhibits more disturbing features than one encounters in the smaller and more comfortable world of parametric Bayes. It is a truth universally acknowledged that a statistician in possession of an infinity of data points must be in want of the truth – but some nonparametric Bayes constructions actually lead to inconsistent estimation procedures, where the truth is not properly uncovered when the data collection grows. Also, the Bernshteĭn–von Mises theorems alluded to above, which hold very generally for parametric Bayes problems, tend not to hold as easily and broadly in the infinite-dimensional cases. There are, for example, important problems where the nonparametric Bayes methods obey consistency (the posterior distribution properly accumulates its mass around the true model, with increased sample size), but with a different rate of convergence than that of the natural frequentist method for the same problem. Thus separate classes of situations typically need separate scrutiny, as opposed to theories and theorems that apply very grandly.

It seems clear to us that the potential list of good, worthwhile nonparametric Bayes procedures must be rather longer than the already enormously long lists of Bayes methods for parametric models, simply because bigger spaces contain more than smaller ones. A book on Bayesian nonparametrics must therefore limit itself to only some of these worthwhile procedures. A similar comment applies to the *study* of these methods, in terms of performance, comparisons with results from other approaches, and so forth (making the distinction between the construction of a method and the study of its performance characteristics).

Who needs it?

Most modern statisticians have become well acquainted with various nonparametric and semiparametric tools, on the one hand (nonparametric regression, smoothing methods, classification and pattern recognition, proportional hazards regression, copulae models, etc.), and with the most important simulation tools, on the other (rejection–acceptance methods, MCMC strategies like the Gibbs sampler and the Metropolis algorithm, etc.), particularly in the realm of Bayesian applications, where the task of drawing simulated realizations from the posterior distribution is the main operational job. The *combination* of these methods is becoming increasingly popular and important (in a growing number of ways), and each such combination may be said to carry the stamp of Bayesian nonparametrics.

One reason why combining nonparametrics with Bayesian posterior simulations is becoming more important is related to practical feasibility, in terms of software packages and implementation of algorithms. The other reason is that such solutions contribute to the solving of actual problems, in a steadily increasing range of applications, as indicated in this book and as seen at workshops and conferences dealing with Bayesian nonparametrics. The steady influx of good real-world application areas contributes both to the sharpening of tools and to the sociological fact that, not only hard-core and classically oriented statisticians, but also various schools of other researchers in quantitative disciplines, lend their hands to work in variations of nonparametric Bayes methods. Bayesian nonparametrics is used by researchers working in finance, geosciences, botanics, biology, epidemiology, forestry, paleontology, computer science, machine learning, recommender systems, to name only some examples.

By prefacing various methods and statements with the word "Bayesian" we are already acknowledging that there are different schools of thought in statistics – Bayesians place prior distributions over their parameter spaces while parameters are fixed unknowns for the frequentists. We should also realize that there are different trends of thought regarding how statistical methods are actually used (as partly opposed to how they are constructed). In an engaging discussion paper, Breiman (2001) argues that contemporary statistics lives with a Snowean "two cultures" problem. In some applications the careful study and interpretation of finer aspects of the model matter and are of primary concern, as in various substantive sciences – an ecologist or a climate researcher may place great emphasis on determining that a certain statistical coefficient parameter is positive, for example, as this might be tied to a scientifically relevant finding that a certain background factor really influences a phenomenon under study. In other applications such finer distinctions are largely irrelevant, as the primary goals of the methods are to make efficient predictions and classifications of a sufficient quality. This pragmatic goal, of making good enough "black boxes" without specific regard to the components of the box in question, is valid in many situations – one might be satisfied with a model that predicts climate parameters and the number of lynx in the forest, without always needing or aiming to understand the finer mechanisms involved in these phenomena.

This continuing debate is destined to play a role also for Bayesian nonparametrics, and the right answer to what is more appropriate, and to what is more important, will be largely context dependent. A statistician applying Bayesian nonparametrics may use one type of model for uncovering effects and another for making predictions or classifications, even when dealing with the same data. Using different models for different purposes, even with the very same data set, is not a contradiction in terms, and relates to different loss functions and to themes of interest-driven inference; cf. various focused information criteria for model selection (see Claeskens and Hjort, 2008, Chapter 6).

It is also empirically true that some statistics problems are easier to attack using Bayesian methods, with machineries available that make analysis and inference possible, in the partial absence of frequentist methods. This picture may of course shift with time, as better and more refined frequentist methods may be developed also, for example for complex hierarchical models, but the observation reminds us that there is a necessary element of pragmatism in modern statistics work; one uses what one has, rather than spending three extra months on developing alternative methods. An eclectic view of Bayesian methods, prevalent also among those statisticians hesitant to accept all of the underlying philosophy, is to use them nevertheless, as they are practical and have good performance. Indeed a broad research direction is concerned with reaching performance-related results about classes of nonparametric Bayesian methods, as partly distinct from the construction of the models and methods themselves (cf. Chapter 2 and its references). For some areas in statistics, then, including some surveyed in this book, there is an "advantage Bayes" situation. A useful reminder in this regard is the view expressed by Art Dempster: "a person cannot be Bayesian or frequentist; rather, a particular *analysis* can be Bayesian or frequentist" (see Wasserman, 2008). Another and perhaps humbling reminder is Good's (1959) lower bound for the number of different Bayesians (46 656, actually), a bound that may need to be revised upwards when the discussion concerns nonparametric Bayesians.

Why now?

Themes of Bayesian nonparametrics have engaged statisticians for about forty years, but now, that is around 2010, the time is ripe for further rich developments and applications of the field. This is due to a confluence of several different factors: the availability and convenience of computer programs and accessible software packages, downloaded to the laptops of modern scientists, along with methodology and machinery for finessing and finetuning these algorithms for new applications; the increasing accessibility of statistical models and associated methodological tools for taking on new problems (leading also to the development of further methods and algorithms); various developing application areas paralleling statistics that find use for these methods and sometimes develop them further; and the broadening meeting points for the two flowing rivers of nonparametrics (as such) and Bayesian methods (as such).

Evidence of the growing importance of Bayesian nonparametrics can also be traced in the archives of conferences and workshops devoted to such themes. In addition to having been on board in broader conferences over several decades, an identifiable subsequence of workshops and conferences set up for Bayesian nonparametrics per se has developed as follows, with a rapidly growing number of participants: Belgirate, Italy (1997), Reading, UK (1999), Ann Arbor, USA (2001), Rome, Italy (2004), Jeju, Korea (2006), Cambridge, UK (2007), Turin, Italy (2009). Monitoring the programs of these conferences one learns that development has been and remains steady, regarding both principles and practice.

Two more long-standing series of workshops are of interest to researchers and learners of nonparametric Bayesian statistics. The BISP series (Bayesian inference for stochastic processes) is focused on nonparametric Bayesian models related to stochastic processes. Its sequence up to the time of writing reads Madrid (1998), Varenna (2001), La Mange (2003), Varenna (2005), Valencia (2007), Brixen (2009), alternating between Spain and Italy. Another related research community is defined by the series of research meetings on objective Bayes methodology. The coordinates of the O'Bayes conference series history are Purdue, USA (1996), Valencia, Spain (1998), Ixtapa, Mexico (2000), Granada, Spain (2002), Aussois, France (2003), Branson, USA (2005), Rome, Italy (2007), Philadelphia, USA (2009).

The aims, purposes and contents of this book

This book has in a sense grown out of a certain event. It reflects this particular origin, but is very much meant to stand solidly and independently on its constructed feet, as a broad text on modern Bayesian nonparametrics and its theory and methods; in other words, readers do not need to know about or take into account the event that led to the book being written.

A background event

The event in question was a four-week program on Bayesian nonparametrics hosted by the Isaac Newton Institute of Mathematical Sciences at Cambridge, UK, in August 2007, and organized by the four volume editors. In addition to involving a core group of some twenty researchers from various countries, the program organized a one-week international conference with about a hundred participants. These represented an interesting modern spectrum of researchers whose work in different ways is related to Bayesian nonparametrics: those engaged in methodological statistics work, from university departments and elsewhere; statisticians involved in collaborations with researchers from substantive areas (like medicine and biostatistics, quantitative biology, mathematical geology, information sciences, paleontology); mathematicians; machine learning researchers; and computer scientists.

For the workshop, the organizers selected four experts to provide tutorial lectures representing four broad, identifiable themes pertaining to Bayesian nonparametrics. These were not merely four themes "of interest," but were closely associated with the core models, the core methods, and the core application areas of nonparametric Bayes. These tutorials were

- Dirichlet processes, related priors and posterior asymptotics (by S. Ghosal),
- models beyond the Dirichlet process (by A. Lijoi),
- applications to biostatistics (by D. B. Dunson),
- applications to machine learning (by Y. W. Teh).

The program and the workshop were evaluated (by the participants and other parties) as having been very successful, by having bound together different strands of work and by perhaps opening doors to promising future research. The experience made clear that nonparametric Bayes is an important growth area, but with side-streams that may risk evolving too much in isolation if they do not make connections with the core field. All of these considerations led to the idea of creating the present book.

What does this book do?

This book is structured around the four core themes represented by the tutorials described above, here appearing in the form of invited chapters. These core chapters are then complemented by chapters written by the four volume editors. The role of these complementary chapters is partly to discuss and extend the four core chapters, in suitably matched pairs. These complements also offer further developments and provide links to related areas. This editorial process hence led to the following list of chapters, where the pairs 1–2, 3–4, 5–6, 7–8 can be regarded as units.

1 S. G. Walker: Bayesian nonparametric methods: motivation and ideas
2 S. Ghosal: The Dirichlet process, related priors and posterior asymptotics
3 A. Lijoi and I. Prünster: Models beyond the Dirichlet process
4 N. L. Hjort: Further models and applications.
5 Y. W. Teh and M. I. Jordan: Hierarchical Bayesian nonparametric models with applications
6 J. Griffin and C. Holmes: Computational issues arising in Bayesian nonparametric hierarchical models
7 D. B. Dunson: Nonparametric Bayes applications to biostatistics
8 P. Müller and F. Quintana: More nonparametric Bayesian models for biostatistics

As explained at the end of the previous section, it would not be possible to have "everything important" inside a single book, in view of the size of the expanding topic. It is our hope and view, however, that the dimensions we have probed are sound, deep and relevant ones, and that different strands of readers will benefit from working their way through some or all of these.

The *first* core theme (Chapters 1 and 2) is partly concerned with some of the cornerstone classes of nonparametric priors, including the Dirichlet process and some of its relatives. General principles and ideas are introduced (in the setting of i.i.d. observations) in Chapter 1. Mathematical properties are further investigated, including characterizations of the posterior distribution, in Chapter 2. The theme also encompasses properties of the behavior of the implied posterior distributions, and, specifically, consistency and rates of convergence. Bayesian methodology is often presented as essentially a machinery for coming from the prior to the posterior distributions, but is at its most powerful when coupled with decision theory and loss functions. This is true in nonparametric situations as well, as also discussed inside this first theme.

The *second* main theme (Chapters 3 and 4) is mainly occupied with the development of the more useful nonparametric classes of priors beyond those related to the Dirichlet processes mentioned above. Chapter 3 treats completely random measures, neutral-to-the-right processes, the beta process, partition functions, clustering processes, and models for density estimation, with Chapter 4 providing further methodology for stationary time series with nonparametrically modeled covariance functions, models for random shapes, etc., along with pointers to various application areas, such as survival and event history analysis.

The third and fourth core themes are more application driven than the first two. The *third* core theme (Chapters 5 and 6) represents the important and growing area of both theory and applications of Bayesian nonparametric hierarchical modeling (an area related to what is often referred to as machine learning). Hierarchical modeling, again with Dirichlet processes as building blocks, leads to algorithms that solve problems in information retrieval, multipopulation haplotype phasing, word segmentation, speaker diarization, and so-called topic modeling, as demonstrated in Chapter 5. The models that help to accomplish these tasks include Chinese restaurant franchises and Indian buffet processes, in addition to extensive use of Gaussian processes, priors on function classes such as splines, free-knot basis expansions, MARS and CART, etc. These constructions are associated with various challenging computational issues, as discussed in some detail in Chapter 6.

Finally the *fourth* main theme (Chapters 7 and 8) focuses on biostatistics. Topics discussed and developed in Chapter 7 include personalized medicine (a growing trend in modern biomedicine), hierarchical modeling with Dirichlet processes, clustering strategies and partition models, and functional data analysis. Chapter 8 elaborates on these themes, and in particular discusses random partition priors and certain useful variations on dependent Dirichlet processes.

How do alternative models relate to each other?

Some comments seem in order to put the many alternative models in perspective. Many of the models are closely related mathematically, with some being a special case of others. For example, the Dirichlet process is a special case of the normalized random measure with independent increments introduced in Chapter 3. Many of the models introduced in later chapters are natural generalizations and extensions of earlier defined models. Several of the models introduced in Chapter 5 extend the random partition models described in the first four chapters, including, for example, a natural hierarchical extension of the Dirichlet process model. Finally, Chapters 7 and 8 introduce many models that generalize the basic Dirichlet process model to one for multiple related random probability measures. As a guideline for choosing a model for a specific application, we suggest considering the data format, the focus of the inference, and the desired level of computational complexity.

If the data format naturally includes multiple subpopulations then it is natural to use a model that reflects this structure in multiple submodels. In many applications the inference of interest is on random partitions and clustering, rather than on a random probability measure. It is natural then to use a model that focuses on the random partitions, such as a species sampling model. Often the choice will simply be driven by the availability of public domain software. This favors the more popular models such as Dirichlet process models, Pólya tree models, and various dependent Dirichlet process models.

The reader may notice a focus on biomedical applications. In part this is a reflection of the history of nonparametric Bayesian data analysis. Many early papers focused on models for event time data, leading naturally to biomedical applications. This focus is also a reflection of the research experience of the authors. There is no intention to give an exhaustive or even representative discussion of areas of application. An important result of focusing on models rather than applications is the lack of a separate chapter on hierarchical mixed effects models, although many of these feature in Chapters 7 and 8.

How to teach from this book

This book may be used as the basis for master's or Ph.D. level courses in Bayesian nonparametrics. Various options exist, for different audiences and for different levels of mathematical skill. One route, for perhaps a typical audience of statistics students, is to concentrate on core themes two (Chapters 3 and 4) and four (Chapters 7 and 8), supplemented with computer exercises (drawing on methods exhibited in these chapters, and using for example the software `DPpackage`, described in Jara, 2007). A course building upon the material in these chapters would focus on data analysis problems and typical data formats arising in biomedical research problems. Nonparametric Bayesian probability models would be introduced as and when needed to address the data analysis problems.

More mathematically advanced courses could include more of core theme one (Chapters 1 and 2). Such a course would naturally center more on a description of nonparametric Bayesian models and include applications as examples to illustrate the models. A third option is a course designed for an audience with an interest in machine learning, hierarchical modeling, and so forth. It would focus on core themes two (Chapters 2 and 3) and three (Chapters 5 and 6).

Natural prerequisites for such courses as briefly outlined here, and by association for working with this book, include a basic statistics course (regression methods associated with generalized linear models, density estimation, parametric Bayes), perhaps some survival analysis (hazard rate models, etc.), along with basic skills in simulation methods (MCMC strategies).

A brief history of Bayesian nonparametrics

Lindley (1972) noted in his review of general Bayesian methodology that Bayesians up to then had been "embarrassingly silent" in the area of nonparametric statistics. He pointed out that there were in principle no conceptual difficulties with combining "Bayesian" and "nonparametric" but indirectly acknowledged that the mathematical details in such constructions would have to be more complicated.

From the start to the present

Independently of and concurrently with Lindley's review, what can be considered the historical start of Bayesian nonparametrics occurred in California. The 1960s had been a period of vigorous methodological research in various nonparametric directions. David Blackwell, among the prominent members of the statistics department at Berkeley (and, arguably, belonging to the Bayesian minority there), suggested to his colleagues that there ought to be Bayesian parallels to the problems and solutions for some of these nonparametric situations. These conversations produced two noteworthy developments, both important in their own right and for what followed: (i) a 1970 UCLA technical report titled "A Bayesian analysis of some nonparametric problems," by T. S. Ferguson, and (ii) a 1971 UC Berkeley technical report called "Tailfree and neutral random probabilities and their posterior distributions," by K. A. Doksum. After review processes, these became the two seminal papers Ferguson (1973) in *Annals of Statistics*, where the Dirichlet process is introduced, and Doksum (1974) in *Annals of Probability*, featuring his neutral-to-the-right processes (see Chapters 2 and 3 for descriptions, interconnections and further developments of these classes of priors). The neutral-to-the-right processes are also foreshadowed in Doksum (1972). In this very first wave of genuine Bayesian nonparametrics work, Ferguson (1974) also stands out, an invited review paper for *Annals of Statistics*. Here he gives early descriptions of and results for Pólya trees, for example, and points to fruitful research problems.

We ought also to mention that there were earlier contributions to constructions of random probability measures and their probabilistic properties, such as Kraft and van Eeden (1964) and Dubins and Freedman (1966). More specific Bayesian connections, including matters of consistency and inconsistency, were made in Freedman (1963) and Fabius (1964), involving also the important notion of tailfree distributions; see also Schwartz (1965). Similarly, a density estimation method given in Good and Gaskins (1971) may be regarded as having a Bayesian nonparametric root, involving an implied prior on the set of densities. Nevertheless, to the extent that such finer historical distinctions are of interest, we would identify the start of Bayesian nonparametrics with the work by Ferguson and Doksum.

These early papers gave strong stimulus to many further developments, including research on various probabilistic properties of these new prior and posterior processes (probability measures on spaces of functions), procedures for density estimation based on mixtures of Dirichlet processes, applications to survival analysis (with suitable priors on the random survivor functions, or cumulative hazard functions, and with methodology developed to handle censoring), a more flexible machinery for Pólya trees and their cousins, etc. We point to Chapters 2 and 3 for further information, rather than detailing these developments here.

The emphasis in this early round of new papers was perhaps simply on the construction of new prior measures, for an increasing range of natural statistical models and problems, along with sufficiently clear results on how to characterize the consequent posterior distributions. Some of these developments were momentarily hampered or even stopped by the sheer computational complexity associated with handling the posterior distributions; sometimes exact results could be written down and proved mathematically, but algorithms could not always be constructed to evaluate these expressions. The situation improved around 1990, when simulation schemes of the MCMC variety became more widely known and implementable, at around the time when statisticians suddenly had real and easily programmable computers in their offices (the MCMC methods had in principle been known to the statistics community since around 1970, but it took two decades for the methods to become widely and flexibly used; see for example Gelfand and Smith, 1990). The MCMC methods were at the outset constructed for classes of finite-parameter problems, but it became apparent that their use could be extended to solve problems in Bayesian nonparametrics as well.

Another direction of research, in addition to the purely constructive and computational sides of the problems, is that of performance: how do the posterior distributions behave, in particular when the sample size increases, and are the implicit limits related to those reached in the frequentist camp? Some of these questions first surfaced in Diaconis and Freedman (1986a, 1986b), where situations were exhibited in which the Bayesian machine yielded asymptotically inconsistent answers; see also the many discussion contributions to these two papers. This and similar research made it clearer to researchers in the field that, even though asymptotics typically led to various mathematical statements of the comforting type "different Bayesians agree among themselves, and also with the frequentists, as the sample size tends to infinity" for *finite-dimensional* problems, results are rather more complicated in infinite-dimensional spaces; see Chapters 1 and 2 in this book and comments made above.

Applications

The history above deals in essence with theoretical developments. A reader sampling his or her way through the literature briefly surveyed there will make the anthropological observation that articles written after say 2000 have a different look to them than those written around 1980. This partly reflects a broader trend, a transition of sorts that has moved the primary emphases of statistics from more mathematically oriented articles to those nearer to actual applications – there are fewer sigma-algebras and less measure theoretic language, and more attention to motivation, algorithms, problem solving and illustrations.

The history of applications of Bayesian nonparametrics is perhaps more complicated and less well defined than that of the theoretical side. For natural reasons, including the general difficulty of transforming mathematics into efficient algorithms and the lack of good computers at the beginning of the nonparametric Bayes adventure, applications simply lagged behind. Ferguson's (1973, 1974) seminal papers are incidentally noteworthy also because they spell out interesting and nontrivial applications, for example to adaptive investment models and to adaptive sampling with recall, though without data illustrations. As indicated above, the first broad theoretical foundations stem from the early 1970s, while the first noteworthy real-data applications, primarily in the areas of survival analysis and biostatistics, started to emerge in the early 1990s (see for example the book by Dey, Müller and Sinha, 1998). At the same time rapidly growing application areas emerged inside machine learning (pattern recognition, bioinformatics, language processing, search engines; see Chapter 5). More information and further pointers to actual application areas for Bayesian nonparametrics may be found by browsing the programs for the Isaac Newton Institute workshop 2007 (`www.newton.ac.uk/programmes/BNR/index.html`) and the Carlo Alberto Programme in Bayesian Nonparametrics 2009 (`bnpprogramme.carloalberto.org/index.html`).

Where does this book fit in the broader picture?

We end this section with a short annotated list of books and articles that provide overviews of Bayesian nonparametrics (necessarily with different angles and emphases). The first and very early one of these is Ferguson (1974), mentioned above. Dey, Müller and Sinha (1998) is an edited collection of papers, with an emphasis on more practical concerns, and in particular containing various papers dealing with survival analysis. The book by Ibrahim, Chen and Sinha (2001) gives a comprehensive treatment of the by-then more prominently practical methods of nonparametric Bayes pertaining to survival analysis. Walker, Damien, Laud and Smith (1999) is a read discussion paper for the Royal Statistical Society, exploring among other issues that of more flexible methods for Pólya trees. Hjort (2003) is a later discussion paper, reviewing various topics and applications, pointing to research problems, and making connections to the broad "highly structured stochastic systems" theme that is the title of the book in question. Similarly Müller and Quintana (2004) provides another review of established results and some evolving research areas. Ghosh and Ramamoorthi (2003) is an important and quite detailed, mathematically oriented book on Bayesian nonparametrics, with a focus on precise probabilistic properties of priors and posteriors, including that of posterior consistency (cf. Chapters 1 and

2 of this book). Lee (2004) is a slim and elegant book dealing with neural networks via tools from Bayesian nonparametrics.

Further topics

Where might you want to go next (after having worked with this book)? Here we indicate some of the research directions inside Bayesian nonparametrics that nevertheless lie outside the natural boundaries of this book.

Gaussian processes Gaussian processes play an important role in several branches of probability theory and statistics, also for problems related to Bayesian nonparametrics. An illustration could be of regression data (x_i, y_i) where y_i is modeled as $m(x_i)+\epsilon_i$, with say Gaussian i.i.d. noise terms. If the unknown $m(\cdot)$ function is modeled as a Gaussian process with a known covariance function, then the posterior is another Gaussian process, and Bayesian inference may proceed. This simple scenario has many extensions, yielding Bayesian nonparametric solutions to different problems, ranging from prediction in spatial and spatial-temporal models (see e.g. Gelfand, Guindani and Petrone, 2008) to machine learning (e.g. Rasmussen and Williams, 2006). Gaussian process models are also a popular choice for inference with output from computer simulation experiments (see e.g. Oakley and O'Hagan (2002) and references there). An extensive annotated bibliography of the Gaussian process literature, including links to public domain software, is available at `www.gaussianprocess.org/`. Regression and classification methods using such processes are reviewed in Neal (1999). Extensions to treed Gaussian processes are developed in Gramacy (2007) and Gramacy and Lee (2008).

Spatial statistics We touched on spatial modeling in connection with Gaussian processes above, and indeed many such models may be handled, with appropriate care, as long as the prior processes involved have covariance functions determined by a low number of parameters. The situation is more complicated when one wishes to place nonparametric priors on the covariance functions as well; see some comments in Chapter 4.

Neural networks There are by necessity several versions of "neural networks," and some of these have reasonably clear Bayesian interpretations, and a subset of these is amenable to nonparametric variations. See Lee (2004) for a lucid overview, and for example Holmes and Mallick (2000) for a particular application. Similarly, flexible nonlinear regression models based on spline bases provide inference that avoids the restrictive assumptions of parametric models. Bayesian inference for penalized spline regression is summarized in Ruppert, Wand and Carroll (2003,

Chapter 16) and implementation details are discussed in Crainiceanu, Ruppert and Wand (2005). For inference using exact-knot selection see, for example, Smith and Kohn (1996) or Denison, Mallick and Smith (1998). In addition, there is more recent work on making the splines more adaptive to fit spatially heterogeneous functions, such as Baladandayuthapani, Mallick and Carroll (2005) and BARS by DiMatteo, Genovese and Kass (2001).

$p \gg n$ problems A steadily increasing range of statistical problems involve the "$p \gg n$" syndrome, in which there are many more covariates (and hence unknown regression coefficients) than individuals. Ordinary methods do not work, and alternatives must be devised. Various methods have been derived from frequentist perspectives, but there is clear scope for developing Bayesian techniques. The popular lasso method of Tibshirani (1996) may in fact be given a Bayesian interpretation, as the posterior mode solution (the Bayes decision under a sharp 0–1 loss function) with a prior for the large number of unknown regression coefficients being that of independent double exponentials with the same spread. Various extensions have been investigated, some also from this implied or explicit Bayesian nonparametric perspective.

Model selection and model averaging Some problems in statistics are attacked by working out the ostensibly best method for each of a list of candidate models, and then either selecting the tentatively best one, via some model selection criterion, or averaging over a subset of the best several ones. When the list of candidate models becomes large, as it easily does, the problems take on nonparametric Bayesian shapes; see for example Claeskens and Hjort (2008, Chapter 7). Further methodology needs to be developed for both the practical and theoretical sides.

Classification and regression trees A powerful and flexible methodology for building regression or classifiers via trees, with perhaps a binary option at each node of the tree, was first developed in the CART system of Breiman, Friedman, Olshen and Stone (1984). Several attempts have been made to produce Bayesian versions of such schemes, involving priors on large families of growing and pruned trees. Their performance has been demonstrated to be excellent in several classes of problems; see for example Chipman, George and McCulloch (2007). See in this connection also Neal (1999) mentioned above.

Performance Quite a few journal papers deal with issues of performance, comparisons between posterior distributions arising from different priors, etc.; for some references in that direction, see Chapters 1 and 2.

Computation and software

A critical issue in the practical use of nonparametric Bayesian prior models is the availability of efficient algorithms to implement posterior inference. Recalling the earlier definition of nonparametric Bayesian models as probability models on big parameter spaces, this might seem a serious challenge at first glance. But we run into some good luck. For many popular models it is possible to marginalize analytically with respect to some of the infinite-dimensional random quantities, leaving a probability model on some lower-dimensional manageable space. For example, under Gaussian process priors the joint probability model for the realization at any finite number of locations is simply a multivariate normal distribution. Similarly, various analysis schemes for survival and event history models feature posterior simulation of beta processes (Hjort, 1990), which may be accomplished by simulating and then adding independent beta-distributed increments over many small intervals. Under the popular Dirichlet process mixture-of-normals model for density estimation, the joint distribution of the observed data can be characterized as a probability model on the partition of the observed data points and independent priors for a few cluster-specific parameters. Also, under a Pólya tree prior, or under quantile-pyramid-type priors (see Hjort and Walker, 2009), posterior predictive inference can be implemented considering only finitely many levels of the nested partition sequence.

Increased availability of public domain software greatly simplifies the practical use of nonparametric Bayesian models for data analysis. Perhaps the most widely used software is the R package `DPpackage` (Jara, 2007, exploiting the R platform of the R Development Core Team, 2006). Functions in the package implement inference for Dirichlet process mixture density estimation, Pólya tree priors for density estimation, density estimation using Bernshteĭn–Dirichlet priors, nonparametric random effects models, including generalized linear models, semiparametric item-response type models, nonparametric survival models, inference for ROC (relative operating characteristic) curves and several functions for families of dependent random probability models. See Chapter 8 for some illustrations. The availability of validated software like `DPpackage` will greatly accelerate the move of nonparametric Bayesian inference into the mainstream statistical literature.

Challenges and future developments

Where are we going, after all of this? A famous statistical prediction is that "the twenty-first century will be Bayesian." This comes from Lindley's preface to the English edition of de Finetti (1974), and has since been repeated with modifications and different degrees of boldness by various observers of and partakers in the principles and practice of statistics; thus the *Statistica Sinica* journal devoted a full issue (2007, no. 2) to this anticipation of the Bayesian century, for example. The present book may be seen as yet another voice in the chorus, promising increased frequency of nonparametric versions of Bayesian methods. Along with implications of certain basic principles, involving the guarantee of uncovering each possible truth with enough data (not only those truths that are associated with parametric models), then, in combination with the increasing versatility and convenience of streamlined software, the century ahead looks decidedly both Bayesian and nonparametric.

There are of course several challenges, associated with problems that have not yet been solved sufficiently well or that perhaps have not yet been investigated at the required level of seriousness. We shall here be bold enough to identify some of these challenges.

Efron (2003) argues that the brightest statistical future may be reserved for *empirical Bayes* methods, as tentatively opposed to the pure Bayes methodology that Lindley and others envisage. This points to the identifiable stream of Bayesian nonparametrics work that is associated with careful setting and fine-tuning of all the algorithmic parameters involved in a given type of construction – the parameters involved in a Dirichlet or beta process, or in an application of quantile pyramids modeling, etc. A subset of such problems may be attacked via empirical Bayes strategies (estimating these hyper parameters via current or previously available data) or by playing the Bayesian card at a yet higher and more complicated level, i.e. via background priors for these hyper parameters.

Another stream of work that may be surfacing is that associated with replacing difficult and slow-converging MCMC type algorithms with quicker, accurate approximations. Running MCMC in high dimensions, as for several methods associated with models treated in this book, is often fraught with difficulties related to convergence diagnostics etc. Inventing methods that somehow sidestep the need for MCMC is therefore a useful endeavour. For good attempts in that direction, for at least some useful and broad classes of models, see Skaug and Fournier (2006) and Rue, Martino and Chopin (2009).

Gelman (2008), along with discussants, considers important objections to the theory and applications of Bayesian analysis; this is also worthwhile reading because the writers in question belong to the Bayesian camp themselves. The themes they discuss, chiefly in the framework of parametric Bayes, are a fortiori valid for nonparametric Bayes as well.

We mentioned above the “two cultures” of modern statistics, associated respectively with the close interpretation of model parameters and the use of automated black boxes. There are yet further schools or cultures, and an apparent growth area is that broadly associated with *causality*. There are difficult aspects of theories of statistical causality, both conceptually and model-wise, but the resulting methods see steadily more application in for example biomedicine, see e.g. Aalen and Frigessi (2007), Aalen, Borgan and Gjessing (2008, Chapter 9) and Pearl (2009). We predict that Bayesian nonparametrics will play a more important role in such directions.

Acknowledgements The authors are grateful to the Isaac Newton Institute for Mathematical Sciences for making it possible for them to organize a broadly scoped program on nonparametric Bayesian methods during August 2007. The efforts and professional skills of the INI were particularly valuable regarding the international workshop that was held within this program, with more than a hundred participants. They also thank Igor Prünster for his many helpful efforts and contributions in connection with the INI program and the tutorial lectures.

The authors also gratefully acknowledge support and research environments conducive to their researches in their home institutions: Department of Mathematics and the Centre for Innovation "Statistics for Innovation" at the University of Oslo, Department of Statistics at Oxford University, Department of Biostatistics at the University of Texas M. D. Anderson Cancer Center, and Institute of Mathematics, Statistics and Actuarial Science at the University of Kent, respectively. They are grateful to Andrew Gelman for constructive suggestions, and finally are indebted to Diana Gillooly at Cambridge University Press for her consistently constructive advice and for displaying the right amount of impatience.

References

Aalen, O. O., Borgan, Ø. and Gjessing, H. (2008). *Survival and Event History Analysis: A Process Point of View.* New York: Springer-Verlag.

Aalen, O. O. and Frigessi, A. (2007). What can statistics contribute to a causal understanding? *Scandinavian Journal of Statistics*, **34**, 155–68.

Baladandayuthapani, V., Mallick, B. K. and Carroll, R. J. (2005). Spatially adaptive Bayesian penalized regression splines (Psplines). *Journal of Computational and Graphical Statistics*, **14**, 378–94.

Bernshteĭn, S. (1917). *Theory of Probability* (in Russian). Moscow: Akademi Nauk.

Breiman, L. (2001). Statistical modeling: The two cultures (with discussion and a rejoinder). *Statistical Science*, **16**, 199–231.

Breiman, L., Friedman, J., Olshen, R. A. and Stone, C. J. (1984). *Classification and Regression Trees.* Monterey, Calif.: Wadsworth Press.

Chipman, H. A., George, E. I. and McCulloch, R. E. (2007). BART: Bayesian additive regression trees. *Technical Report*, Graduate School of Business, University of Chicago.

Claeskens, G. and Hjort, N. L. (2008). *Model Selection and Model Averaging.* Cambridge: Cambridge University Press.

Crainiceanu, C. M., Ruppert, D. and Wand, M. P. (2005). Bayesian analysis for penalized spline regression using WinBUGS. *Journal of Statistical Software*, **14**, 1–24. http://www.jstatsoft.org/v14/i114.

Denison, D. G. T., Mallick, B. K. and Smith, A. F. M. (1998). Automatic Bayesian curve fitting. *Journal of the Royal Statistical Society, Series B*, **60**, 333–50.

Dey, D., Müller, P. and Sinha, D. (1998). *Practical Nonparametric and Semiparametric Bayesian Statisics*. New York: Springer-Verlag.

Diaconis, P. and Freedman, D. A. (1986a). On the consistency of Bayes estimates (with discussion). *Annals of Statistics*, **14**, 1–67.

Diaconis, P. and Freedman, D. A. (1986b). On inconsistent Bayes estimates of location. *Annals of Statistics*, **14**, 68–87.

DiMatteo, I., Genovese, C. R. and Kass, R. F. (2001). Bayesian curve-fitting with free-knot splines. *Biometrika*, **88**, 1055–71.

Doksum, K. A. (1972). Decision theory for some nonparametric models. *Proceedings of the Sixth Berkeley Symposium on Mathematical Statistics*, **1**, 331–44.

Doksum, K. A. (1974). Tailfree and neutral random probabilities and their posterior distribution. *Annals of Probability*, **2**, 183–201.

Dubins, L. E. and Freedman, D. A. (1966). Random distribution functions. *Proceedings of the Fifth Berkeley Symposium on Mathematical Statistics*, **2**, 183–214.

Efron, B. (2003). Robbins, empirical Bayes and microarrays. *Annals of Statistics*, **31**, 366–78.

Fabius, J. (1964). Asymptotic behavior of Bayes estimates. *Annals of Mathematical Statistics*, **35**, 846–56.

Ferguson, T. S. (1973). A Bayesian analysis of some nonparametric problems. *Annals of Statistics*, **1**, 209–30.

Ferguson, T. S. (1974). Prior distributions on spaces of probability measures. *Annals of Statistics*, **2**, 615–29.

de Finetti, B. D. (1974). *Theory of Probability*, Volume 1. Chichester: Wiley.

Freedman, D. A. (1963). On the asymptotic behavior of Bayes estimates in the discrete case. *Annals of Mathematical Statistics*, **34**, 1386–403.

Gelfand, A. E., Guindani, M. and Petrone, S. (2008). Bayesian nonparametric modeling for spatial data analysis using Dirichlet processes (with discussion and a rejoinder). In *Bayesian Statistics 8*, ed. J. Bernardo, J. O. Berger, and A. F. M. Smith, 175–200. Oxford: Oxford University Press.

Gelfand, A. E. and Smith, A. F. M. (1990). Sampling-based approaches to calculating marginal densities. *Journal of the American Statistical Association*, **85**, 398–409.

Gelman, A. (2008). Objections to Bayesian statistics (with discussion and a rejoinder). *Bayesian Analysis 3*, ed. J. Bernado et al., 445–78. Oxford: Oxford University Press.

Ghosh, J. K. and Ramamoorthi, R. V. (2003). *Bayesian Nonparametrics*. New York: Springer-Verlag.

Good, I. J. (1959). 46656 varieties of Bayesians. *American Statistician*, **25**, 62–63. Reprinted in *Good Thinking*, Minneapolis; Minn.: University of Minnesota Press, 1982, pp. 20–21.

Good, I. J. and Gaskins, R. A. (1971). Nonparametric roughness penalties for probability densities. *Biometrika*, **58**, 255–77.

Gramacy, R. B. (2007). tgp: An R package for Bayesian nonstationary, semiparametric nonlinear regression and design by treed Gaussian process models. *Journal of Statistical Software*, **19**.

Gramacy, R. B. and Lee, H. K. H. (2008). Bayesian treed Gaussian process models with an application to computer modeling. *Journal of the American Statistical Association*, **103**, 1119–30.

Green, P. J. and Richardson, S. (2001). Modelling heterogeneity with and without the Dirichlet process. *Scandinavian Journal of Statistics*, **28**, 355–75.

Hjort, N. L. (1990). Nonparametric Bayes estimators based on Beta processes in models for life history data. *Annals of Statistics*, **18**, 1259–94.

Hjort, N. L. (2003). Topics in nonparametric Bayesian statistics (with discussion). In *Highly Structured Stochastic Systems*, ed. P. J. Green, N. L. Hjort, and S. Richardson, 455–87. Oxford: Oxford University Press.

Hjort, N. L. and Walker, S. G. (2009). Quantile pyramids for Bayesian nonparametrics. *Annals of Statistics*, **37**, 105–31.

Holmes, C. C. and Mallick, B. (2000). Bayesian wavelet networks for nonparametric regression. *IEEE Transactions on Neural Networks*, **11**, 27–35.

Ibrahim, J. G., Chen, M.-H. and Sinha, D. (2001). *Bayesian Survival Analysis*. New York: Springer-Verlag.

Jara, A. (2007). Applied Bayesian non- and semi-parametric inference using DPpackage. *Rnews*, **7**, 17–26.

Kraft, C. H. and van Eeden, C. (1964). Bayesian bio-assay. *Annals of Mathematical Statistics*, **35**, 886–90.

Laplace, P. S. (1810). Mémoire sure les formules qui sont fonctions de très grands nombres et sur leurs applications aux probabilités. *Oeuvres de Laplace*, **12**, 301–45.

Le Cam, L. and Yang, G. L. (1990). *Asymptotics in Statistics: Some Basic Concepts*. New York: Springer-Verlag.

Lee, H. K. H. (2004). *Bayesian Nonparametrics via Neural Networks*. Philadephia, Pa.: ASA-SIAM.

Lehmann, E. L. (1975). *Nonparametrics: Statistical Methods Based on Ranks*. San Francisco, Calif.: Holden-Day.

Lindley, D. V. (1972). *Bayesian Statistics: A Review*. Regional Conference Series in Applied Mathematics. Philadelphia, Pa.: SIAM.

von Mises, R. (1931). *Wahrscheinlichkeitsrechnung*. Berlin: Springer.

Müller, P. and Quintana, F. A. (2004). Nonparametric Bayesian data analysis. *Statistical Science*, **19**, 95–110.

Neal, R. (1999). Regression and classification using Gaussian process priors. In *Bayesian Statistics 6*, ed. J. M. Bernardo, J. O. Berger, A. P. Dawid and A. F. M. Smith, 69–95. Oxford: Oxford University Press.

Oakley, J. and O'Hagan, A. (2002). Bayesian inference for the uncertainty distribution of computer model outputs. *Biometrika*, **89**, 769–84.

Pearl, J. (2009). *Causality: Models, Reasoning and Inference*, 2nd edition. Cambridge: Cambridge University Press.

Rasmussen, C. E. and Williams, C. K. I. (2006). *Gaussian Processes for Machine Learning*. Cambridge, Mass.: MIT Press.

R Development Core Team. (2006). *R: A Language and Environment for Statistical Computing*. Vienna: R Foundation for Statistical Computing. `www.R-project.org`.

Rue, H., Martino, S. and Chopin, N. (2009). Approximate Bayesian inference for latent Gaussian models by using integrated nested Laplace approximations (with discussion and a rejoinder). *Journal of the Royal Statistical Society Series, Series B*, **71**, 319–72.

Ruppert, D., Wand, M. P. and Carroll, R. J. (2003). *Semiparametric Regression*. Cambridge: Cambridge University Press.

Schwartz, L. (1965). On Bayes procedures. *Zeitschrift für Wahrscheinlichkeitstheorie und verwandte Gebiete*, **4**, 10–26.

Skaug, H. J. and Fournier, D. A. (2006). Automatic approximation of the marginal likelihood in non-Gaussian hierarchical models. *Computational Statistics and Data Analysis*, **5**, 699–709.

Smith, M. and Kohn, R. (1996). Nonparametric regression using Bayesian variable selection. *Journal of Econometrics*, **75**, 317–43.
Tibshirani, R. J. (1996). Regression shrinkage and selection via the lasso. *Journal of the Royal Statistical Society Series, Series B*, **58**, 267–88.
Walker, S. G., Damien, P., Laud, P. W. and Smith, A. F. M. (1999). Bayesian nonparametric inference for random distributions and related functions (with discussion and a rejoinder). *Journal of the Royal Statistical Society Series, Series B*, **61**, 485–528.
Wasserman, L. (2006). *All of Nonparametric Statistics: A Concise Course in Nonparametric Statistical Inference*. New York: Springer-Verlag.
Wasserman, L. (2008). Comment on article by Gelman. *Bayesian Analysis* **3**, ed. J. Bernado et al., 463–6. Oxford: Oxford University Press.

1

Bayesian nonparametric methods: motivation and ideas

Stephen G. Walker

It is now possible to demonstrate many applications of Bayesian nonparametric methods. It works. It is clear, however, that nonparametric methods are more complicated to understand, use and derive conclusions from, when compared to their parametric counterparts. For this reason it is imperative to provide specific and comprehensive motivation for using nonparametric methods. This chapter aims to do this, and the discussions in this part are restricted to the case of independent and identically distributed (i.i.d.) observations. Although such types of observation are quite specific, the arguments and ideas laid out in this chapter can be extended to cover more complicated types of observation. The usefulness in discussing i.i.d. observations is that the maths is simplified.

1.1 Introduction

Even though there is no physical connection between observations, there is a real and obvious reason for creating a dependence between them from a modeling perspective. The first observation, say X_1, provides information about the unknown density f from which it came, which in turn provides information about the second observation X_2, and so on. How a Bayesian learns is her choice but it is clear that with i.i.d. observations the order of learning should not matter and hence we enter the realms of *exchangeable* learning models. The mathematics is by now well known (de Finetti, 1937; Hewitt and Savage, 1955) and involves the construction of a prior distribution $\Pi(df)$ on a suitable space of density functions. The learning mechanism involves updating $\Pi(df)$ as data arrive, so that after n observations beliefs about f are now encapsulated in the posterior distribution, given by

$$\Pi(df|X_1, \ldots, X_n) = \frac{\prod_{i=1}^n f(X_i)\,\Pi(df)}{\int \prod_{i=1}^n f(X_i)\,\Pi(df)}$$

and this in turn provides information about the future observation X_{n+1} via the predictive density

$$f(X_{n+1}|X_1, \ldots, X_n) = \int f(X_{n+1})\,\Pi(df|X_1, \ldots, X_n).$$

From this it is easy to see that the prior represents what has been learnt about the unknown density function without the presence of any of the observations. Depending on how much is known at this point, that is with no observations, the strength of the prior ranges from very precise with a lot of information, to so-called noninformative or default priors which typically are so disperse that they are even improper (see e.g. Kass and Wasserman, 1996).

This prior distribution is a single object and is a prior distribution on a suitable space of density (or equivalent) functions. Too many Bayesians think of the notion of a likelihood and a prior and this can be a hindrance. The fundamental idea is the construction of random density functions, such as normal shapes, with random means and variances; or the infinite-dimensional exponential family, where probabilities are assigned to the infinite collection of random parameters. It is instructive to think of all Bayesians as constructing priors on spaces of density functions, and it is clear that this is the case. The Bayesian nonparametric statistician is merely constructing random density functions with unrestricted shapes.

This is achieved by modeling random density functions, or related functions such as distribution functions and hazard functions, using stochastic processes; Gaussian processes and independent increment processes are the two most commonly used. The prior is the law governing the stochastic process. The most commonly used is the Dirichlet process (Ferguson, 1973) which has sample paths behaving almost surely as a discrete distribution function. They appear most often as the mixing distribution generating random density functions: the so-called mixture of Dirichlet process model (Lo, 1984), which has many pages dedicated to it within this book. This model became arguably the most important prior for Bayesian nonparametrics with the advent of sampling based approaches to Bayesian inference, which arose in the late 1980s (Escobar, 1988).

The outline of this chapter is as follows. In Section 1.2 we consider the important role that Bayesian nonparametrics plays. Ideas for providing information for nonparametric priors are also discussed. Section 1.3 discusses how many of the practices and low-dimensional activities of Bayesians can be carried out coherently under the umbrella of the nonparametric model. The special case when the nonparametric posterior is taken as the Bayesian bootstrap is considered. Section 1.4 discusses the importance of asymptotic studies. Section 1.5 is a direct consequence of recent consistency studies which put the model assumptions and true sampling assumptions at odds with each other. This section provides an alternative derivation of the Bayesian posterior distribution using loss functions; as such it is no less a rigorous approach to constructing a learning model than is the traditional approach using the Bayes theorem. So Section 1.5 can be thought of as "food for thought." Finally, Section 1.6 concludes with a brief discussion.

1.2 Bayesian choices

Many of the questions posed to the nonparametric methods are of the type "what if this and what if that?" referring to the possibility that the true density is normal or some other low-dimensional density and so using many parameters is going to be highly inefficient. In truth, it is these questions that are more appropriately directed to those who consistently use low-dimensional densities for modeling: "what if the model is not normal?"

However, there was a time, and not so long ago, in fact pre–Markov chain Monte Carlo, when Bayesian methods were largely restricted to a few parametric models, such as the normal, and the use of conjugate prior distributions. Box and Tiao (1973) was as deep as it got. It is therefore not surprising that in this environment, where only simple models were available, the ideas of model selection and model comparison took hold, for the want of something to do and a need to compare log–normal and Weibull distributions. Hence, such model assessments were vital, irrespective of any formal views one may have had about the theory of Bayesian methods (see Bernardo and Smith, 1994, Chapter 2). But it is not difficult to argue that Bayesian model criticism is unsound, and the word that is often used is *incoherent*.

To argue this point, let us keep to the realm of independent and identically distributed observations. In this case, the prior distribution is a probability measure on a space of density functions. This is true for all Bayesians, even those relying on the normal distribution, in which case the Bayesian is putting probability one on the shape of the density function matching those of the normal family.

There is more responsibility on the Bayesian: she gets more out in the form of a posterior distribution on the object of interest. Hence more care needs to be taken in what gets put into the model in the first place. For the posterior to mean anything it must be representing genuine posterior beliefs, solely derived by a combination of the data and prior beliefs via the use of the Bayes theorem. Hence, the prior used must genuinely represent prior beliefs (beliefs without data). If it does not, how can the posterior represent posterior beliefs? So a "prior" that has been selected post data via some check and test from a set of possible "prior" distributions cannot represent genuine prior beliefs. This is obvious, since no one of these "priors" can genuinely represent prior beliefs. The posterior distributions based on such a practice are meaningless.

The prior must encapsulate prior beliefs and be large enough to accommodate all uncertainties. As has been mentioned before, years back prior distributions could not be enlarged to accommodate such problems, and the incoherence of model (prior) selection was adopted for pragmatic reasons, see Box (1980). However, nowadays, it is quite straightforward to build large prior distributions and to undertake prior to posterior analysis. How large a prior should be is a clear matter. It is large enough so that no matter what subsequently occurs, the prior is not checked. Hence, in may cases, it is only going to be a nonparametric model that is going to suffice.

If a Bayesian has a prior distribution and suspects there is additional uncertainty, there are two possible actions. The first is to consider an alternative prior and then select one or the other after the data have been observed. The second action is to enlarge the prior before observing the data to cover the additional uncertainty. It is the latter action which is correct and coherent.

Some Bayesians would argue that it is too hard a choice to enlarge the prior or work with nonparametric priors, particularly in specifying information or putting beliefs into nonparametric priors. If this is the case, though I do not believe it to be true, then it is a matter of further investigation and research to overcome the difficulties rather than to lapse into pseudo-Bayesian and incoherent practices.

To discuss the issue of pinning down a nonparametric prior we can if needed do this in a parametric frame of mind. For the nonparametric model one typically has two functions to specify which relate to $\mu_1(x) = \mathrm{E} f(x)$ and $\mu_2(x) = \mathrm{E} f^2(x)$. If it is possible to specify such functions then a nonparametric prior has typically been pinned down. Two such functions are easy to specify. They can, for example, be obtained from a parametric model, even the normal, in which case one would take

$$\mu_1(x) = \int \mathrm{N}(x|\theta, \sigma^2)\, \pi(d\theta, d\sigma)$$

$$\mu_2(x) = \int \mathrm{N}^2(x|\theta, \sigma^2)\, \pi(d\theta, d\sigma),$$

for some probability measure $\pi(d\theta, d\sigma)$. The big difference now is that a Bayesian using this normal model, i.e.

$$X \sim \mathrm{N}(\theta, \sigma^2) \quad \text{and} \quad (\theta, \sigma) \sim \pi(\theta, \sigma),$$

would be restricted to normal shapes, whereas the nonparametric Bayesian, whose prior beliefs about μ_1 and μ_2, equivalently $\mathrm{E} f(x)$ and $\mathrm{Var} f(x)$, coincide with the parametric Bayesian, has unrestricted shapes to work with.

A common argument is that it is not possible to learn about all the parameters of a nonparametric model. This spectacularly misses the point. Bayesian inference is about being willing and able to specify all uncertainties into a prior distribution. If one does not like the outcome, do not be a Bayesian. Even a parametric model needs a certain amount of data to learn anything reasonable and the nonparametric model, which reflects greater starting uncertainty than a parametric model, needs more data to overcome the additional starting uncertainty. But it is not right to wish away the prior uncertainty or purposefully to underestimate it.

1.3 Decision theory

Many of the Bayesian procedures based on incomplete priors (i.e. priors for which all uncertainty has not been taken into account) can be undertaken coherently (i.e. using a complete prior) using decision theory. Any selection of parametric models can be done under the umbrella of the complete prior. This approach makes extensive use of the utility function for assessing the benefit of actions (such as model selection etc.) when one has presumed a particular value for the correct but unknown density function. Let us consider an example. Which specific density from a family of densities indexed by a parameter $\theta \in \Theta$ is the best approximation to the data?

If the parametric family of densities is $\{f(x;\theta)\}$, then the first task is to choose a utility function which describes the reward in selecting θ, for the parameter space is the action space, when f is the true density. Basing this on a distance between densities seems appropriate here, so we can take

$$u(f,\theta) = -d(f(\cdot;\theta), f(\cdot)).$$

The prior is the nonparametric one, or the complete prior $\Pi(\mathrm{d}f)$, and so making decisions on the basis of the maximization of expected utility, the choice of θ is $\widehat{\theta}$ which maximizes

$$U_n(\theta) = -\int d(f(\cdot;\theta), f(\cdot))\, \Pi(\mathrm{d}f|X_1,\ldots,X_n).$$

An interesting special case arises when we take d to be based on the Kullback–Leibler divergence; that is $d(g,f) = \int g \log(g/f)$ in which case we would choose $\widehat{\theta}$ to maximize

$$\tilde{U}_n(\theta) = \int \log f(x;\theta)\, f_n(\mathrm{d}x)$$

where f_n is the nonparametric predictive density, given by

$$f_n(x) = \int f(x)\, \Pi(\mathrm{d}f|X_1,\ldots,X_n).$$

Furthermore, taking $\Pi(\mathrm{d}f)$ to be the Bayesian bootstrap (Rubin, 1981), so that f_n is the density with point mass $1/n$ at each of the data points, then

$$\tilde{U}_n(\theta) = n^{-1}\sum_{i=1}^{n} \log f(X_i;\theta)$$

and so $\widehat{\theta}$ is the maximum likelihood estimator.

There are many other types of lower dimensional decisions that can be made under the larger prior/posterior; see Gutièrrez-Peña and Walker (2005). As an example, suppose it is required to construct a probability on Θ space when the true posterior is $\Pi(\mathrm{d}f|X_1,\ldots,X_n)$. It is necessary to link up a random f from this posterior with a random θ from Θ space. This can be done by taking θ to maximize $u(f, \theta)$. An interesting special case arises when the posterior is once again taken to be the Bayesian bootstrap in which case we can take

$$f_n(\mathrm{d}x) = \sum_{i=1}^{n} w_i \, \delta_{X_i}(\mathrm{d}x),$$

where the $(w_1, \ldots, w_n)$ are from a Dirichlet distribution with parameters all equal to 1. Therefore, a distribution on Θ space can be obtained by repeated simulation of the weights from the Dirichlet distribution and taking θ to maximize

$$\sum_{i=1}^{n} w_i \, \log f(X_i; \theta).$$

This is precisely the weighted likelihood bootstrap approach to Bayesian inference proposed by Newton and Raftery (1994).

To set up the scene for the next section, let us note that if a Bayesian is making such assessments on utilities, in order to undertake decision theory, then she must be willing to think about the true density function and that this comes from a set of possibilities. How is it possible to make such judgments while having discarded the notion of a true density function?

1.4 Asymptotics

Traditionally, Bayesians have shunned this aspect of statistical inference. The prior and data yield the posterior and the subjectiveness of this strategy does not need the idea of what happens if further data arise. Anyway, there was the theorem of Doob (1949), but like all other Bayesian computations from the past, this theorem involves assuming that the marginal distribution of the observations depends explicitly on and is fully specified by the chosen prior distribution, that is

$$p(X_1, \ldots, X_n) = \int \prod_{i=1}^{n} f(X_i) \, \Pi(\mathrm{d}f).$$

It is unrealistic to undertake asymptotic studies, or indeed any other Bayesian studies, based on this assumption, since it is not true. Doob's theorem relies on this assumption. Even though one knows that this model is mathematically incorrect, it does serve as a useful learning model, as discussed earlier.

On the other hand, it is correct to assume the observations are independent and identically distributed from some true density function f_0 and to undertake the mathematics on this assumption. One is then asking that the posterior distribution accumulates in suitable neighborhoods of this true density function.

This exposes the Bayesian model as being quite different from the correct assumption. There is no conflict here in the discrepancy between the true assumption and the model assumption. The Bayesian model is about learning from observations in a way that the order in which they arrive does not matter (exchangeability). The first observation provides information about the true density function and this in turn provides information about the second observation and so on. The Bayesian writes down how this learning is achieved and specifically how an observation provides information about the true density function. In this approach one obviously needs to start with initial or prior information about the true density function.

In short, the Bayesian believes the data are i.i.d. from some true density function f_0 and then writes down an exchangeable learning model as to how they see the observations providing information about f_0.

So why is consistency important? The important point is that the prior, which fully specifies the learning model, is setting up the learning model. In a way it is doing two tasks. One is representing prior beliefs, learnt about f_0 before or without the presence of data, and the second is fully specifying the learning model. It is this latter task that is often neglected by subjective Bayesians.

Hence, the learning part of the model needs to be understood. With an unlimited amount of data the Bayesian must expect to be able to pin down the density generating her observations exactly. It is perfectly reasonable to expect that as data arrive the learning is going in the right direction and that the process ends up at f_0. If it does not then the learning model (prior) has not been set well, even though the prior might be appropriate as representing prior beliefs.

The basic idea is to ensure that

$$\Pi(d(f, f_0) > \epsilon | X_1, \ldots, X_n) \to 0 \ \text{a.s.}\ F_0^\infty$$

where d is some measure of distance between densities. It is typically taken to be the Hellinger distance since this favors the mathematics. Conditions are assigned to Π to ensure this happens and involve a support condition and a further condition which ensures that the densities which can track the data too closely are given sufficiently low prior mass, see Chapter 2.

However, an alternative "likelihood," given by

$$l_n^{(\alpha)} = \prod_{i=1}^{n} f(X_i)^\alpha$$

for any $0 < \alpha < 1$ yields Hellinger consistency with only a support condition. Can this approach be justified? It possibly can. For consider a cumulative loss function approach to posterior inference, as in the next section.

1.5 General posterior inference

For observables $X_1, \ldots, X_n$, which result in loss $l(a, X_i)$ for each i under action a, the optimal choice of action $\widehat{a}$ minimizes the cumulative loss function

$$L(a; \mathbf{X}) = \sum_{i=1}^{n} l(a, X_i).$$

This is standard theory and widely used in practice. We will not be regarding the sequential decision problem where each observation leads to a decision a_i in which case the cumulative loss function is

$$L(\mathbf{a}; \mathbf{X}) = \sum_{i=1}^{n} l(a_i, X_i),$$

see, for example, Merhav and Feder (1998). Hence, we assume the observations arise as a complete package and one decision or action is required.

We will regard, as we have throughout the chapter, the X_i as independent and identically distributed observations from f_0. Most decision approaches to statistical inference now treat f_0 as the target and construct loss functions, equivalently utility functions, which provide estimators for f_0.

Here we are interested in constructing a "posterior" distribution which is obtained via the minimization of a loss function. If the loss function can be justified then an alternative derivation of the Bayesian approach (i.e. the derivation of the Bayesian posterior) is available which is simple to understand.

The prior distribution $\Pi(\mathrm{d}f)$, a probability on a space of density functions, will solely be used to represent prior beliefs about f_0, but an alternative learning model will be established. So there are $n+1$ pieces of information $(\Pi, X_1, \ldots, X_n)$ and the cumulative loss in choosing $\mu(\mathrm{d}f)$ as the posterior distribution is

$$L(\mu; (\Pi, \mathbf{X})) = \sum_{i=1}^{n} l_X(\mu, X_i) + l(\mu, \Pi),$$

where l_X and l are as yet unspecified loss functions. Hence we treat observables and prior as information together and find a posterior by minimizing a cumulative loss function.

Such a loss function is not unusual if one replaces μ by f, or more typically in a parametric approach by θ, and f is taken as the density $f(\cdot; \theta)$. The prior is then written as $\pi(\theta)$. Then loss functions of the type

$$L(\theta; (\pi, \mathbf{X})) = \sum_{i=1}^{n} l_X(\theta, X_i) + l(\theta, \pi)$$

are commonplace. Perhaps the most important loss function here is the self-information loss function, so that

$$l_X(\theta, X) = -\log f(X;\theta)$$

and

$$l(\theta, \pi) = -\log \pi(\theta).$$

Minimizing $L(\theta; (\pi, \mathbf{X}))$ yields the posterior mode.

Hence, the loss function, replacing θ with μ, is appropriate if interest is in finding a posterior distribution. We will first concentrate on $l(\mu, \Pi)$. To understand this we need to understand what Π is. It represents information, information about the unknown sampling distribution function which is translated into a probability measure Π. Hence, for any suitable set A, the prior belief that f lies in the set A is given by $\Pi(A)$. We need to assess the loss in information in using μ to represent prior beliefs rather than using Π. This loss in information is well known to be evaluated as

$$D(\mu||\Pi) = \int \mu(\mathrm{d}f) \log \{\mu(\mathrm{d}f)/\Pi(\mathrm{d}f)\}$$

and hence we take $l(\mu, \Pi) = D(\mu||\Pi)$.

For the loss function $l_X(\mu, X)$ we have a resource which is first to construct the loss function $l_X(f, X)$ and then to rely on the fact that the expected loss, if μ is chosen as representing beliefs about f, is given by the expectation of $l_X(f, X)$ with respect to $\mu(\mathrm{d}f)$; and so we take

$$l_X(\mu, X) = \int l_X(f, X)\, \mu(\mathrm{d}f).$$

Hence,

$$L(\mu; (\Pi, \mathbf{X})) = -\sum_{i=1}^{n} \int \log f(X_i)\, \mu(\mathrm{d}f) + D(\mu||\Pi)$$

and the solution to this problem is given by

$$\widehat{\mu}(\mathrm{d}f) = \Pi(\mathrm{d}f|X_1, \ldots, X_n),$$

the Bayesian posterior distribution derived via the Bayes theorem.

More generally, we can take a weighting of the two types of loss function so that now

$$L(\mu; (\Pi, \mathbf{X})) = -\alpha_n \sum_{i=1}^{n} \int \log f(X_i)\, \mu(\mathrm{d}f) + D(\mu||\Pi)$$

for $\alpha_n \geq 0$. The solution to this minimization problem is given by

$$\widehat{\mu}(\mathrm{d}f) = \Pi_n(\mathrm{d}f) = \frac{\prod_{i=1}^n f(X_i)^{\alpha_n}\,\Pi(\mathrm{d}f)}{\int \prod_{i=1}^n f(X_i)^{\alpha_n}\,\Pi(\mathrm{d}f)}.$$

Such a pseudo-posterior, with $\alpha_n = \alpha \in (0, 1)$, has previously been considered by Walker and Hjort (2001) for ensuring a strongly consistent sequence of distribution functions, provided f_0 is in the Kullback–Leibler support of Π. That is, for $\alpha_n = \alpha \in (0, 1)$ it is that

$$\Pi_n(A_\epsilon) \to 0$$

with probability one for all $\epsilon > 0$, where

$$A_\epsilon = \{f : d_1(f_0, f) > \epsilon\}$$

and d_1 denotes the L_1 distance between density functions.

There are some special cases that arise.

(a) $\alpha_n = 0$, then $\Pi_n = \Pi$.
(b) $\alpha_n = 1$, then Π_n is the "correct" Bayesian posterior distribution.
(c) $\alpha_n = \alpha \in (0, 1)$, Π_n is the pseudo-posterior of Walker and Hjort (2001).

Indeed, the choice $\alpha_n = \alpha \in (0, 1)$ could well be seen as one such subjective choice for the posterior, guaranteeing strong consistency, which is not guaranteed with $\alpha = 1$. A choice of $\alpha_n = \alpha \in (0, 1)$ reduces the influence of the data, and keeps a Π_n closer to the prior than does the choicc of $\alpha_n = 1$. This suggests that a prudent strategy would be to allow α_n to increase to 1 as $n \to \infty$. But at what rate? We will work out the fastest rate which maintains consistency.

So, now let

$$\Pi_n(A_\epsilon) = \frac{\int_{A_\epsilon} R_n(f)^{\alpha_n}\,\Pi(\mathrm{d}f)}{\int R_n(f)^{\alpha_n}\,\Pi(\mathrm{d}f)},$$

where

$$R_n(f) = \prod_{i=1}^n f(X_i)/f_0(X_i)$$

and define

$$I_n = \int R_n(f)^{\alpha_n}\,\Pi(\mathrm{d}f).$$

There has been a lot of recent work on establishing conditions under which we have, for some fixed $c > 0$,

$$J_n > \exp(-cn\epsilon_n^2)$$

in probability, where $\epsilon_n \to 0$ and $n\epsilon_n^2 \to \infty$ and

$$J_n = \int R_n(f)^{1/2}\, \Pi(\mathrm{d}f),$$

see Chapter 2. Essentially, ϵ_n depends on the concentration of the prior Π around f_0. Although Walker and Hjort establish ϵ_n with

$$K_n = \int R_n(f)\, \Pi(\mathrm{d}f),$$

the same rate for ϵ_n can also be found with J_n for some different constant c. Then, for $\alpha_n > 1/2$,

$$I_n > \left\{\int R_n(f)^{1/2}\, \Pi(\mathrm{d}f)\right\}^{2\alpha_n} = J_n^{2\alpha_n}$$

and so $I_n > \exp(-2cn\epsilon_n^2\alpha_n)$ in probability.

Now let

$$L_n = \int_{A_\epsilon} R_n(f)^{\alpha_n}\, \Pi(\mathrm{d}f)$$

where $A_\epsilon = \{f : d_{\mathrm{H}}(f_0, f) > \epsilon\}$ and $d_{\mathrm{H}}(f_0, f)$ is the Hellinger distance between densities f_0 and f; that is

$$d_{\mathrm{H}}(f_0, f) = \left\{\int \left(\sqrt{f_0} - \sqrt{f}\right)^2\right\}^{1/2}$$

and note that

$$\mathrm{E}\sqrt{f(X_1)/f_0(X_1)} = \int \sqrt{f_0\, f} = 1 - d_{\mathrm{H}}^2(f_0, f)/2.$$

Also note that, for $\alpha > 1/2$, we have

$$\begin{aligned} \mathrm{E}\{(f(X_1)/f_0(X_1)\}^{\alpha} &= \textstyle\int (f/f_0)^{\alpha}\, f_0 \\ &= \textstyle\int (f_0/f)^{1-\alpha}\, f \\ &= \textstyle\int \left(\sqrt{f_0/f}\right)^{2(1-\alpha)}\, f \end{aligned}$$

and so

$$\mathrm{E}\{(f(X_1)/f_0(X_1)\}^{\alpha} \le \left(1 - d_{\mathrm{H}}^2(f_0, f)/2\right)^{2(1-\alpha)}.$$

Then it is easy to see that

$$\mathrm{E}(L_n) < \exp\{-n(1-\alpha_n)\epsilon^2\}.$$

Hence, provided $n(1-\alpha_n) \to \infty$, we have

$$L_n < \exp\{-n(1-\alpha_n)\epsilon^2/2\}$$

in probability. Therefore,

$$\Pi_n(A_\epsilon) < \exp\{-n((1-\alpha_n)\epsilon^2/2 - 2c\epsilon_n^2\alpha_n)\}$$

in probability and so we are looking to choose α_n such that

$$n\left[(1-\alpha_n)\epsilon^2 - 4c\epsilon_n^2\alpha_n\right] \to \infty$$

for all $\epsilon > 0$. We can therefore take

$$\alpha_n = 1 - \psi_n\epsilon_n^2$$

for any $\psi_n \to \infty$ satisfying $\psi_n\epsilon_n^2 \to 0$. For example, if $\epsilon_n^2 = (\log n)/n$ then we can take $\psi_n = \log n$ and so $\alpha_n = 1 - (\log n)^2/n$.

1.6 Discussion

At the heart of this chapter is the idea of thinking about the prior as the probability measure that arises on spaces of density functions, namely $\Pi(\mathrm{d}f)$, and such a prior can be written this way even if one is using normal distributions.

The argument of this chapter is that the Bayesian model is a learning model and not incompatible with the assumption that observations are i.i.d. from some density f_0. An interesting point of view in light of this finding is the general construction of posterior distributions via the use of loss functions. The posterior via Bayes theorem arises naturally, as do alternative learning models, which have the advantage that the learning is consistent, having chosen $\alpha_n = \alpha < 1$, which is not automatically the case for $\alpha = 1$.

Having said this, posterior inference via MCMC, which is wholly necessary, is quite difficult for any case save $\alpha = 1$. For example, try and undertake posterior inference for the Dirichlet mixture model with $\alpha < 1$.

References

Bernardo, J. M. and Smith, A. F. M. (1994). *Bayesian Theory*. Chichester: Wiley.

Box, G. E. P. (1980). Sampling and Bayes inference in scientific modeling and robustness (with discussion). *Journal of the Royal Statistical Society, Series A*, **143**, 383–430.

Box, G. E. P. and Tiao, G. C. (1973). *Bayesian Inference in Statistical Analysis*. Reading, Mass.: Addison-Wesley.

Doob, J. L. (1949). Application of the theory of martingales. In *Le Calcul des Probabilités et ses Applications*, Colloques Internationaux du Centre National de la Recherche Scientifique, 13, 23–37. Paris: CNRS.

Escobar, M. D. (1988). Estimating the means of several normal populations by nonparametric estimation of the distribution of the means. *Unpublished Ph.D. Dissertation*, Department of Statistics, Yale University.

Ferguson, T. S. (1973). A Bayesian analysis of some nonparametric problems. *Annals of Statistics*, **1**, 209–30.

de Finetti, B. (1937). La prevision: ses lois logiques, ses sources subjectives. *Annales de l'Institut Henri Poincaré*, **7**, 1–68.

Gutièrrez-Peña, E. and Walker, S. G. (2005). Statistical decision problems and Bayesian nonparametric methods. *International Statistical Review*, **73**, 309–30.

Hewitt, E. and Savage, L. J. (1955). Symmetric measures on Cartesian products. *Transactions of the American Mathematical Society*, **80**, 470–501.

Kass, R. E. and Wasserman, L. A. (1996). The selection of prior distributions by formal rules. *Journal of the American Statistical Association*, **91**, 1343–70.

Lo, A. Y. (1984). On a class of Bayesian nonparametric estimates: I. Density estimates. *Annals of Statistics*, **12**, 351–57.

Merhav, N. and Feder, M. (1998). Universal prediction. *IEEE Transactions on Information Theory*, **44**, 2124–47.

Newton, M. A. and Raftery, A. E. (1994). Approximate Bayesian inference by the weighted likelihood bootstrap (with discussion). *Journal of the Royal Statistical Society, Series B*, **56**, 3–48.

Rubin, D. B. (1981). The Bayesian bootstrap. *Annals of Statistics*, **9**, 130–34.

Walker, S. G. and Hjort, N. L. (2001). On Bayesian consistency. *Journal of the Royal Statistical Society, Series B*, **63**, 811–21.

2

The Dirichlet process, related priors and posterior asymptotics

Subhashis Ghosal

Here we review the role of the Dirichlet process and related prior distribtions in nonparametric Bayesian inference. We discuss construction and various properties of the Dirichlet process. We then review the asymptotic properties of posterior distributions. Starting with the definition of posterior consistency and examples of inconsistency, we discuss general theorems which lead to consistency. We then describe the method of calculating posterior convergence rates and briefly outline how such rates can be computed in nonparametric examples. We also discuss the issue of posterior rate adaptation, Bayes factor consistency in model selection and Bernshteĭn–von Mises type theorems for nonparametric problems.

2.1 Introduction

Making inferences from observed data requires modeling the data-generating mechanism. Often, owing to a lack of clear knowledge about the data-generating mechanism, we can only make very general assumptions, leaving a large portion of the mechanism unspecified, in the sense that the distribution of the data is not specified by a finite number of parameters. Such nonparametric models guard against possible gross misspecification of the data-generating mechanism, and are quite popular, especially when adequate amounts of data can be collected. In such cases, the parameters can be best described by functions, or some infinite-dimensional objects, which assume the role of parameters. Examples of such infinite-dimensional parameters include the cumulative distribution function (c.d.f.), density function, nonparametric regression function, spectral density of a time series, unknown link function in a generalized linear model, transition density of a Markov chain and so on. The Bayesian approach to nonparametric inference, however, faces challenging issues since construction of prior distribution involves specifying appropriate probability measures on function spaces where the parameters lie. Typically, subjective knowledge about the minute details of the distribution on these infinite-dimensional spaces is not available for nonparametric problems. A prior distribution is generally chosen based on tractability, computational convenience and desirable frequentist behavior, except that some key parameters of the prior may be chosen subjectively. In particular, it is desirable that a chosen prior is spread all over the parameter space, that is, the prior has large topological *support*. Together with additional conditions, large support of a prior helps the corresponding posterior distribution to have good frequentist properties in large samples. To study frequentist properties, it is assumed that there is a true value of the unknown parameter which governs the distribution of the generated data.

We are interested in knowing whether the posterior distribution eventually concentrates in the neighborhood of the true value of the parameter. This property, known as *posterior consistency*, provides the basic frequentist validation of a Bayesian procedure under consideration, in that it ensures that with a sufficiently large amount of data, it is nearly possible to discover the truth accurately. Lack of consistency is extremely undesirable, and one should not use a prior if the corresponding posterior is inconsistent. However, consistency is satisfied by many procedures, so typically more effort is needed to distinguish between consistent procedures. The speed of convergence of the posterior distribution to the true value of the parameter may be measured by looking at the smallest shrinking ball around the true value which contains posterior probability nearly one. It will be desirable to pick up the prior for which the size of such a shrinking ball is the minimum possible. However, in general it is extremely hard to characterize size exactly, so we shall restrict ourselves only to the rate at which a ball around the true value can shrink while retaining almost all of the posterior probability, and call this the *rate of convergence* of the posterior distribution. We shall also discuss adaptation with respect to multiple models, consistency for model selection and *Bernshteĭn–von Mises theorems*.

In the following sections, we describe the role of the *Dirichlet process* and some related prior distributions, and discuss their most important properties. We shall then discuss results on convergence of posterior distributions, and shall often illustrate results using priors related to the Dirichlet process. At the risk of being less than perfectly precise, we shall prefer somewhat informal statements and informal arguments leading to these results. An area which we do not attempt to cover is that of Bayesian survival analysis, where several interesting priors have been constructed and consistency and rate of convergence results have been derived. We refer readers to Ghosh and Ramamoorthi (2003) and Ghosal and van der Vaart (2010) as general references for all topics discussed in this chapter.

2.2 The Dirichlet process

2.2.1 Motivation

We begin with the simplest nonparametric inference problem for an uncountable sample space, namely, that of estimating a probability measure (equivalently, a c.d.f.) on the real line, with independent and identically distributed (i.i.d.) observations from it, where the c.d.f. is completely arbitrary. Obviously, the classical estimator, the empirical distribution function, is well known and is quite satisfactory. A Bayesian solution requires describing a random probability measure and developing methods of computation of the posterior distribution. In order to understand the idea, it is fruitful to look at the closest parametric relative of the problem, namely the multinomial model. Observe that the multinomial model specifies an arbitrary probability distribution on the sample space of finitely many integers, and that a multinomial model can be derived from an arbitrary distribution by grouping the data in finitely many categories. Under the operation of grouping, the data are reduced to counts of these categories. Let $(\pi_1, \dots, \pi_k)$ be the probabilities of the categories with frequencies $n_1, \dots, n_k$. Then the likelihood is proportional to $\pi_1^{n_1} \cdots \pi_k^{n_k}$. The form of the likelihood matches with the form of the finite-dimensional Dirichlet prior, which has density † proportional to $\pi_1^{c_1 - 1} \cdots \pi_k^{c_k - 1}$. Hence the posterior density is proportional to $\pi_1^{n_1 + c_1 - 1} \cdots \pi_k^{n_k + c_k - 1}$, which is again a Dirichlet distribution.

With this nice conjugacy property in mind, Ferguson (1973) introduced the idea of a Dirichlet process – a probability distribution on the space of probability measures which induces finite-dimensional Dirichlet distributions when the data are grouped. Since grouping can be done in many different ways, reduction to a finite-dimensional Dirichlet distribution should hold under any grouping mechanism. In more precise terms, this means that for any finite measurable partition $\{B_1, \dots, B_k\}$ of $\mathbb{R}$, the joint distribution of the probability vector $(P(B_1), \dots, P(B_k))$ is a finite-dimensional Dirichlet distribution. This is a very rigid requirement. For this to be true, the parameters of the finite-dimensional Dirichlet distributions need to be very special. This is because the joint distribution of $(P(B_1), \dots, P(B_k))$ should agree with other specifications such as those derived from the joint distribution of the probability vector $(P(A_1), \dots, P(A_m))$ for another partition $\{A_1, \dots, A_m\}$ finer than $\{B_1, \dots, B_k\}$, since any $P(B_i)$ is a sum of some $P(A_j)$. A basic property of a finite-dimensional Dirichlet distribution is that the sums of probabilities of disjoint chunks again give rise to a joint Dirichlet distribution whose parameters are obtained by adding the parameters of the original Dirichlet distribution. Letting $\alpha(B)$ be the parameter corresponding to $P(B)$ in the specified Dirichlet joint distribution, it thus follows that $\alpha(\cdot)$ must be an additive set function. Thus it is a prudent strategy to let α actually be a measure. Actually, the countable additivity of α will be needed to bring in countable additivity of the random P constructed in this way. The whole idea can be generalized to an abstract Polish space.

† Because of the restriction $\sum_{i=1}^{k} \pi_i = 1$, the density has to be interpreted as that of the first $k - 1$ components.

Definition 2.1 Let α be a finite measure on a given Polish space $\mathfrak{X}$. A random measure P on $\mathfrak{X}$ is called a Dirichlet process if for every finite measurable partition $\{B_1, \ldots, B_k\}$ of $\mathfrak{X}$, the joint distribution of $(P(B_1), \ldots, P(B_k))$ is a k-dimensional Dirichlet distribution with paramaeters $\alpha(B_1), \ldots, \alpha(B_k)$.

We shall call α the base measure of the Dirichlet process, and denote the Dirichlet process measure by $\mathcal{D}_\alpha$.

Even for the case when α is a measure so that joint distributions are consistently specified, it still remains to be shown that the random set function P is a probability measure. Moreover, the primary motivation for the Dirichlet process was to exploit the conjugacy under the grouped data setting. Had the posterior distribution been computed based on conditioning on the counts for the partitioning sets, we would clearly retain the conjugacy property of finite-dimensional Dirichlet distributions. However, as the full data are available under the setup of continuous data, a gap needs to be bridged. We shall see shortly that both issues can be resolved positively.

2.2.2 Construction of the Dirichlet process

Naive construction

At first glance, because joint distributions are consistently specified, viewing P as a function from the Borel σ-field $\mathscr{B}$ to the unit interval, a measure with the specified marginals can be constructed on the uncountable product space $[0, 1]^{\mathscr{B}}$ with the help of Kolmogorov's consistency theorem. Unfortunately, this simple strategy is not very fruitful for two reasons. First, the product σ-field on $[0, 1]^{\mathscr{B}}$ is not rich enough to contain the space of probability measures. This difficulty can be avoided by working with outer measures, provided that we can show that P is a.s. countably additive. For a given sequence of disjoint sets A_n, it is indeed true that $P(\cup_{n=1}^{\infty} A_n) = \sum_{n=1}^{\infty} P(A_n)$ a.s. Unfortunately, the null set involved in the a.s. statement is dependent on the sequence A_n, and since the number of such sequences is uncountable, the naive strategy using the Kolmogorov consistency theorem fails to deliver the final result.

Construction using a countable generator

To save the above construction, we need to work with a countable generating field $\mathscr{F}$ for $\mathscr{B}$ and view each probability measure P as a function from $\mathscr{F}$ to $[0, 1]$. The previously encountered measure theoretic difficulties do not arise on the countable product $[0, 1]^{\mathscr{F}}$.

Construction by normalization

There is another construction of the Dirichlet process which involves normalizing a *gamma process* with intensity measure α. A gamma process is an independent increment process whose existence is known from the general theory of *Lévy processes*. The gamma process representation of the Dirichlet process is particularly useful for finding the distribution of the mean functional of P and estimating of the tails of P when P follows a Dirichlet process on $\mathbb{R}$.

2.2.3 Properties

Once the Dirichlet process is constructed, some of its properties are immediately obtained.

Moments and marginal distribution

Considering the partition $\{A, A^c\}$, it follows that $P(A)$ is distributed as $\text{Beta}(\alpha(A), \alpha(A^c))$. Thus in particular, $\text{E}(P(A)) = \alpha(A)/(\alpha(A) + \alpha(A^c)) = G(A)$, where $G(A) = \alpha(A)/M$, a probability measure and $M = \alpha(\mathbb{R})$, the total mass of α. This means that if $X|P \sim P$ and P is given the measure $\mathcal{D}_\alpha$, then the marginal distribution of X is G. We shall call G the *center measure*. Also, observe that $\text{Var}(P(A)) = G(A)G(A^c)/(M+1)$, so that the prior is more tightly concentrated around its mean when M is larger, that is, the prior is more precise. Hence the parameter M can be regarded as the *precision parameter*. When P is distributed as the Dirichlet process with base measure $\alpha = MG$, we shall often write $P \sim \text{DP}(M, G)$.

Linear functionals

If ψ is a G-integrable function, then $\text{E}(\int \psi \mathrm{d}P) = \int \psi \mathrm{d}G$. This holds for indicators from the relation $\text{E}(P(A)) = G(A)$, and then standard measure theoretic arguments extend this sequentially to simple measurable functions, nonnegative measurable functions and finally to all integrable functions. The distribution of $\int \psi \mathrm{d}P$ can also be obtained analytically, but this distribution is substantially more complicated than beta distribution followed by $P(A)$. The derivation involves the use of a lot of sophisticated machinery. Interested readers are referred to Regazzini, Guglielmi and Di Nunno (2002), Hjort and Ongaro (2005), and references therein.

Conjugacy

Just as the finite-dimensional Dirichlet distribution is conjugate to the multinomial likelihood, the Dirichlet process prior is also conjugate for estimating a completely unknown distribution from i.i.d. data. More precisely, if $X_1, \ldots, X_n$ are i.i.d. with distribution P and P is given the prior $\mathcal{D}_\alpha$, then the posterior distribution of P given $X_1, \ldots, X_n$ is $\mathcal{D}_{\alpha+\sum_{i=1}^n \delta_{X_i}}$.† To see this, we need to show that for any measurable finite partition $\{A_1, \ldots, A_k\}$, the posterior distribution of $(P(A_1), \ldots, P(A_k))$

† Of course, there are other versions of the posterior distribution which can differ on a null set for the joint distribution.

given $X_1, \dots, X_n$ is k-dimensional Dirichlet with parameters $\alpha(A_j) + N_j$, where $N_j = \sum_{i=1}^n 1\!\!1\{X_i \in A_j\}$, the count for A_j, $j = 1, \dots, k$. This certainly holds by the conjugacy of the finite-dimensional Dirichlet prior with respect to the multinomial likelihood had the data been coarsened to only the counts $N_1, \dots, N_k$. Therefore, the result will follow if we can show that the additional information contained in the original data $X_1, \dots, X_n$ is irrelevant as far as the posterior distribution of $(P(A_1), \dots, P(A_k))$ is concerned. One can show this by first considering a partition $\{B_1, \dots, B_m\}$ finer than $\{A_1, \dots, A_k\}$, computing the posterior distribution of $(P(B_1), \dots, P(B_m))$ given the counts of $\{B_1, \dots, B_m\}$, and marginalizing to the posterior distribution of $(P(A_1), \dots, P(A_k))$ given the counts of $\{B_1, \dots, B_m\}$. By the properties of finite-dimensional Dirichlet, this coincides with the posterior distribution of $(P(A_1), \dots, P(A_k))$ given the counts of $\{A_1, \dots, A_k\}$. Now making the partitions infinitely finer and applying the martingale convergence theorem, the final result is obtained.

Posterior mean

The above expression for the posterior distribution combined with the formula for the mean of a Dirichlet process imply that the posterior mean of P given $X_1, \dots, X_n$ can be expressed as

$$\tilde{\mathbb{P}}_n = \mathrm{E}(P|X_1, \dots, X_n) = \frac{M}{M+n} G + \frac{n}{M+n} \mathbb{P}_n, \tag{2.1}$$

a convex combination of the prior mean and the empirical distribution. Thus the posterior mean essentially shrinks the empirical distribution towards the prior mean. The relative weight attached to the prior is proportional to the total mass M, giving one more reason to call M the precision parameter, while the weight attached to the empirical distribution is proportional to the number of observations it is based on.

Limits of the posterior

When n is kept fixed, letting $M \to 0$ may be regarded as making the prior imprecise or noninformative. The limiting posterior, namely $\mathcal{D}_{\sum_{i=1}^n \delta_{X_i}}$, is known as the Bayesian bootstrap. Samples from the Bayesian bootstrap are discrete distributions supported at only the observation points whose weights are distributed according to the Dirichlet distribution, and hence the *Bayesian bootstrap* can be regarded as a resampling scheme which is smoother than Efron's bootstrap. On the other hand, when M is kept fixed and n varies, the asymptotic behavior of the posterior mean is entirely controlled by that of the empirical distribution. In particular, the c.d.f. of $\tilde{\mathbb{P}}_n$ converges uniformly to the c.d.f. of the true distribution P_0 and $\sqrt{n}(\tilde{\mathbb{P}}_n - P_0)$ converges weakly to a Brownian bridge process. Further, for any set A, the posterior variance of $P(A)$ is easily seen to be $O(n^{-1})$ as $n \to \infty$. Hence Chebyshev's inequality implies that the posterior distribution of $P(A)$ approaches the degenerate distribution at $P_0(A)$, that is, the posterior distribution of $P(A)$ is consistent at P_0, and the rate of this convergence is $n^{-1/2}$. Shortly, we shall see that the entire posterior of P is also consistent at P_0.

Lack of smoothness

The presence of the point masses δ_{X_i} in the base measure of the posterior Dirichlet process gives rise to some peculiar behavior. One such property is the total disregard of the topology of the sample space. For instance, if A is a set such that many observations fall close to it but A itself does not contain any observed point, then the posterior mean of $P(A)$ is smaller than its prior mean. Thus the presence of observations in the vicinity does not enhance the assessment of the probability of a set unless the observations are actually contained there. Hence it is clear that the Dirichlet process is somewhat primitive in that it does not offer any smoothing, quite unlike the characteristic of a Bayes estimator.

Negative correlation

Another peculiar property of the Dirichlet process is negative correlation between probabilities of any two disjoint sets. For a random probability distribution, one may expect that the masses assigned to nearby places increase or decrease together, so the blanket negative correlation attached by the Dirichlet process may be disappointing. This again demonstrates that the topology of the underlying space is not considered by the Dirichlet process in its mass assignment.

Discreteness

A very intriguing property of the Dirichlet process is the discreteness of the distributions sampled from it, even when G is purely nonatomic. This property also has its roots in the expression for the posterior of a Dirichlet process. To see why this is so, observe that a distribution P is discrete if and only if $P(x : P\{x\} > 0) = 1$. Now, considering the model $X|P \sim P$ and P given $\mathcal{D}_\alpha$ measure, the property holds if

$$(\mathcal{D}_\alpha \times P)\{(P, x) : P\{x\} > 0\} = 1. \tag{2.2}$$

The assertion is equivalent to

$$(G \times \mathcal{D}_{\alpha+\delta_x})\{(x, P) : P\{x\} > 0\} = 1 \tag{2.3}$$

as G is the marginal of X and the conditional distribution of $P|X$ is $\mathcal{D}_{\alpha+\delta_X}$. The last relation holds, since the presence of the atom at x in the base measure of the posterior Dirichlet process ensures that almost all random P sampled from the posterior process assigns positive mass to the point x. Thus the discreteness property is the consequence of the presence of an atom at the observation in the base measure of the posterior Dirichlet process.

The discreteness property of the Dirichlet process may be disappointing if one's perception of the true distribution is nonatomic, such as when it has a density. However, discreteness itself may not be an obstacle to good convergence properties of estimators, considering the fact that the empirical distribution is also discrete but converges uniformly to any true distribution.

Support

Even though only discrete distributions can actually be sampled from a Dirichlet process, the topological support of the Dirichlet measure $\mathcal{D}_\alpha$, which is technically the smallest closed set of probability one, could be quite big. The support is actually characterized as all probability measures P^* whose supports are contained in that of G, that is,

$$\text{supp}(\mathcal{D}_\alpha) = \{P^* : \text{supp}(P^*) \subset \text{supp}(G)\}. \tag{2.4}$$

In particular, if G is fully supported, like the normal distribution on the line, then trivially every probability measure is in the support of $\mathcal{D}_\alpha$. To see why (2.4) is true, first observe that any supported P^* must have $P^*(A) = 0$ if A is disjoint from the support of G, which implies that $G(A) = 0$ and so $P(A) = 0$ a.s. $[\mathcal{D}_\alpha]$. For the opposite direction, we use the fact that weak approximation will hold if probabilities of a fine partition are approximated well, and this property can be ensured by the nonsingularity of the Dirichlet distribution with positive parameters.

Self-similarity

Another property of the Dirichlet process which distinguishes it from other processes is the self-similarity property described as follows. Let A be any set with $0 < G(A) < 1$, which ensures that $0 < P(A) < 1$ for almost all Dirichlet process samples. Let $P|_A$ be the restriction of P to A, that is, the probability distribution defined by $P|_A(B) = P(A \cap B)/P(A)$, and similarly $P|_{A^c}$ is defined. Then the processes $\{P(A), P(A^c)\}$, $P|_A$ and $P|_{A^c}$ are mutually independent, and moreover $P|_A$ follows $\text{DP}(MG(A), G|_A)$. Thus the assertion says that at any given locality A, how mass is distributed within A is independent of how mass is distributed within A^c, and both mass distribution processes are independent of how much total mass is assigned to the locality A. Further, the distribution process within A again follows a Dirichlet process with an appropriate scale. The property has its roots in the connection between independent gamma variables and the Dirichlet distributed variable formed by their ratios: if $X_1, \dots, X_k$ are independent gamma variables, then $X = \sum_{i=1}^k X_i$ and $(X_1/X, \dots, X_k/X)$ are independent. The self-similarity property has many interesting consequences, an important one being that a Dirichlet process may be generated by sequentially distributing mass independently to various subregions following a tree structure. The independence at various levels of allocation, known as the *tail-freeness* property, is instrumental in obtaining large weak support of the prior and weak consistency of posterior. In fact, the Dirichlet process is the only *tail-free process* where the choice of the partition does not play a role.

Limit types

When we consider a sequence of Dirichlet processes such that the center measures converge to a limit G, then there can be three types of limits:

(i) if the total mass goes to infinity, the sequence converges to the prior degenerate at G;
(ii) if the total mass goes to a finite nonzero number M, then the limit is $\mathrm{DP}(M, G)$;
(iii) if the total mass goes to 0, the limiting process chooses a random point from G and puts the whole mass 1 at that sampled point.

To show the result, one first observes that tightness is automatic here because of the convergence of the center measures, while finite dimensionals are Dirichlet distributions, which converge to the appropriate limit by convergence of all mixed moments. The property has implications in two different scenarios: the Dirichlet posterior converges weakly to the Bayesian bootstrap when the precision parameter goes to zero, and converges to the degenerate measure at P_0 as the sample size n tends to infinity, where P_0 is the true distribution. Thus the entire posterior of P is weakly consistent at P_0, and the convergence automatically strengthens to convergence in the Kolmogorov–Smirnov distance, much in the tone with the Glivenko–Cantelli theorem for the empirical distribution. The result is extremely intriguing in that no condition on the base measure of the prior is required; consistency holds regardless of the choice of the prior, even when the true distribution is not in the support of the prior. This is very peculiar in the Bayesian context, where having the true distribution in the support of the prior is viewed as the minimum condition required to make the posterior distribution consistent. The rough argument is that when the prior excludes a region, the posterior, obtained by multiplying the prior with the likelihood and normalizing, ought to exclude that region. In the present context, the family is undominated and the posterior is not obtained by applying the Bayes theorem, so the paradox is resolved.

Dirichlet samples and ties

As mentioned earlier, the Dirichlet process samples only discrete distributions. The discreteness property, on the other hand, is able to generate ties in the observations and is extremely useful in clustering applications. More specifically, the marginal joint distribution of n observations $(X_1, \ldots, X_n)$ from P which is sampled from DP(M, G) may be described sequentially as follows. Clearly, $X_1 \sim G$ marginally. Now

$$X_2|P, X_1 \sim P \quad \text{and} \quad P|X_1 \sim \mathrm{DP}\left(M+1, \frac{M}{M+1}G + \frac{1}{M+1}\delta_{X_1}\right), \tag{2.5}$$

which implies, after eliminating P, that $X_2|X_1 \sim \frac{M}{M+1}G + \frac{1}{M+1}\delta_{X_1}$, that is, the distribution of X_2 given X_1 can be described as duplicating X_1 with probability $1/(M+1)$ and getting a fresh draw from G with probability $M/(M+1)$. Continuing this argument to X_n given $X_1, \ldots, X_{n-1}$, it is clear that X_n will duplicate any previous X_i with probability $1/(M+n-1)$ and will obtain a fresh draw from G with probability $M/(M+n-1)$. Of course, many of the previous X_i are equal among themselves, so the conditional draw can be characterized as setting to θ_j with probability $n_j/(M+n-1)$, where the θ_j are distinct values of $\{X_1, \ldots, X_{n-1}\}$ with frequencies n_j respectively, $j = 1, \ldots, k$, and as before, a fresh draw from G with probability $M/(M+n-1)$:

$$X_n|X_1, \ldots, X_{n-1} \sim \begin{cases} \delta_{\theta_j} & \text{with probability } \frac{n_j}{M+n-1} \quad j = 1, \ldots, k \\ G & \text{with probability } \frac{M}{M+n-1}, \end{cases}$$

where k is the number of distinct observations in $X_1, \ldots, X_{n-1}$ and $\theta_1, \ldots, \theta_k$ are those distinct values. Also observe that, since $(X_1, \ldots, X_n)$ are exchangeable, the same description applies to any X_i given X_j, $j = 1, \ldots, i-1, i+1, \ldots, n$. This procedure, studied in Blackwell and MacQueen (1973), is known as the generalized Pólya urn scheme. This will turn out to have a key role in the development of Markov chain Monte Carlo (MCMC) procedures for latent variables sampled from a Dirichlet process, as in Dirichlet mixtures discussed shortly.

Because of ties in the above description, the number of distinct observations, the total number of fresh draws from G including the first, is generally much smaller than n. The probabilities of drawing a fresh observation at steps $1, 2, \ldots, n$ are $1, M/(M+1), \ldots, M/(M+n-1)$ respectively, and so the expected number of distinct values K_n is

$$\mathrm{E}(K_n) = \sum_{i=1}^{n} \frac{M}{M+i-1} \sim M \log \frac{n}{M} \quad \text{as } n \to \infty. \tag{2.6}$$

Moreover, one can obtain the exact distribution of K_n, and its normal and Poisson approximation, quite easily. The logarithmic growth of K_n induces sparsity that is often used in machine learning applications.

Sethuraman stick-breaking representation

The Dirichlet process DP(M, G) also has a remarkable representation known as the Sethuraman (1994) representation:

$$P = \sum_{i=1}^{\infty} V_i \delta_{\theta_i}, \quad \theta_i \overset{\text{i.i.d.}}{\sim} G, \quad V_i = \left[\prod_{j=1}^{i-1}(1 - Y_j)\right] Y_i, \quad Y_i \overset{\text{i.i.d.}}{\sim} \text{Beta}(1, M). \tag{2.7}$$

Thus $P = Y_1\delta_{\theta_1} + (1 - Y_1)\sum_{i=2}^{\infty} V_i'\delta_{\theta_{i+1}}$, where $V_i' = [\prod_{j=2}^{i}(1 - Y_j)]Y_{i+1}$, so that

$$P =_d Y_1\delta_{\theta_1} + (1 - Y)P. \tag{2.8}$$

This distributional equation is equivalent to the representation (2.7), and can be used to derive various properties of the random measure defined by (2.7) and to generate such a process by MCMC sampling. The weights V_i attached to the points $\theta_1, \theta_2, \ldots$ respectively may be viewed as the result of breaking a stick of unit length randomly in infinite fragments as follows. First break the stick at a location $Y_1 \sim$ Beta$(1, M)$ and assign the mass Y_1 to a random point $\theta_1 \sim G$. The remaining mass $(1 - Y_1)$ is the split in the proportion $Y_2 \sim$ Beta$(1, M)$ and the net mass $(1 - Y_1)Y_2$ is assigned to a random point $\theta_2 \sim G$. This process continues infinitely many times to complete the assignment of the whole mass to countably many points. What is intriguing is that the resulting process is actually DP(M, G). To get a rough idea why this is so, recall that for any random distribution P and $\theta \sim P$, the prior for P is equal to the mixture of the posterior distribution $P|\theta$ where θ follows its marginal distribution. In the context of the Dirichlet process, this means that $\mathcal{D}_\alpha = \int \mathcal{D}_{\alpha+\delta_\theta} dG(\theta)$. Now if P is sampled from $\mathcal{D}_{\alpha+\delta_\theta}$, then $P\{\theta\} \sim$ Beta$(1, M)$ assuming that α is nonatomic. Thus the random P has a point mass at θ of random magnitude distributed as $Y \sim$ Beta$(1, M)$. With the remaining probability, P is spread over $\{\theta\}^c$, and $P|_{\{\theta\}^c} \sim$ DP(M, G) independently of $P\{\theta\}$ by the self-similarity property of the Dirichlet process, that is $P|_{\{\theta\}^c} =_d P$. This implies that the DP(M, G) satisfies the distributional equation (2.8), where $Y \sim$ Beta$(1, M), \theta \sim G$ and are mutually independent of P. The solution of the equation can be shown to be unique, so the process constructed through the stick-breaking procedure described above must be DP(M, G).

Sethuraman's representation of the Dirichlet process has far reaching significance. First, along with an appropriate finite stage truncation, it allows us to generate a Dirichlet process approximately. This is indispensable in various complicated applications involving Dirichlet processes, where analytic expressions are not available, so that posterior quantities can be calculated only by simulating them from their posterior distribution. Once a finite stage truncation is imposed, for computational purposes, the problem can be treated essentially as a parametric problem for which general MCMC techniques such as Metropolis–Hastings algorithms and reversible jump MCMC methods can be applied. Another advantage of the sum representation is that new random measures can be constructed by changing the stick-breaking distribution from Beta$(1, M)$ to other possibilities. One example is the *two-parameter Poisson–Dirichlet process* where actually the stick-breaking distribution varies with the stage. Even more significantly, for more complicated applications involving covariates, dependence can be introduced among several random measures which are marginally Dirichlet by allowing dependence in their support points θ, or their weights V or both.

Mutual singularity

There are many more interesting properties of the Dirichlet process, for example any two Dirichlet processes are mutually singular unless their base measures share same atoms; see Korwar and Hollander (1973). In particular, the prior and the posterior Dirichlet processes are mutually singular if the prior base measure is nonatomic. This is somewhat peculiar because the Bayes theorem, whenever applicable, implies that the posterior is absolutely continuous with respect to the prior distribution. Of course, the family under consideration is undominated, so the Bayes theorem does not apply in the present context.

Tail of a Dirichlet process

We end this section by mentioning the behavior of the tail of a Dirichlet process. Since $\mathrm{E}(P) = G$, one may think that the tails of G and the random P are equal on average. However, this is false as the tails of P are much thinner almost surely. Mathematically, this is quite possible as the thickness of the tail is an asymptotic property. The exact description of the tail involves long expressions, so we do not present it here; see Doss and Sellke (1982). However, it may be mentioned that if G is standard normal, the tail of $P(X > x)$ is thinner than $\exp[-\mathrm{e}^{x^2/2}]$ for all sufficiently large x a.s., much thinner than the original Gaussian tail. In a similar manner, if G is standard Cauchy, the corresponding random P has finite moment generating functions, even though the Cauchy distribution does not even have a mean.

2.3 Priors related to the Dirichlet process

Many processes constructed using the Dirichlet process are useful as prior distributions under a variety of situations. Below we discuss some of these processes.

2.3.1 Mixtures of Dirichlet processes

In order to elicit the parameters of a Dirichlet process DP(M, G), as the center measure G is also the prior expectation of P, it is considered as the prior guess about the parameter P. However, in practice, it is difficult specify a distribution like Nor(0, 1) as the prior guess; it is more natural to propose a parametric family with unknown parameters as one's guess about the data generating mechanism. In other words, the center measure contains additional hyperparameters which are then given somewhat flat prior distributions. The resulting process is thus a *mixture of Dirichlet process* (MDP) studied by Antoniak (1974).

Some of the properties of the MDP are quite similar to the Dirichlet. For instance, samples from an MDP are a.s. discrete. Exact expressions for prior mean and variance may be obtained by conditioning on the hyperparameter and finally integrating it out. However, the self-similarity and tail-freeness properties no longer hold for the MDP.

The posterior distribution based on an MDP prior is again MDP. To see this, observe that conditionally on the hyperparameter θ, the structure of a Dirichlet process, and hence its conjugacy property, is preserved. Thus the posterior is a mixture of these Dirichlet processes, although the posterior distribution of θ changes from its prior density $\pi(\theta)$ to the posterior density $\pi(\theta|\text{data})$. In many applications, the precision parameter in the MDP set-up is kept unchanged and the base measure G_θ admits a density g_θ. In this case, the posterior distribution can be found relatively easily and is given by

$$\pi(\theta|\text{data}) \propto \pi(\theta) \prod_{i=1}^{n} g_\theta(X_i), \tag{2.9}$$

provided that the data are actually sampled from a continuous distribution so that there are no ties among the observations. This is a consequence of the Blackwell–MacQueen urn scheme describing the joint density of $(X_1, \dots, X_n)$ as $\prod_{i=1}^{n} g_\theta(X_i)$, assuming all the X are distinct, and the Bayes theorem.

2.3.2 Dirichlet process mixtures

While the MDP is a parametric mixture of "nonparametric priors," a very different scenario occurs when one mixes parametric families nonparametrically. Assume that given a latent variable θ_i, the observations X_i follow a parametric density $\psi(\cdot, \theta_i)$, $i = 1, \dots, n$, respectively. The unknown quantities, unlike in the parametric inference, are not assumed to be equal, but appear, like random effects, from a distribution P. The resulting marginal density for any X_i is thus $f_P(x) = \int \psi(x; \theta) \mathrm{d}P(\theta)$ and $X_1, \dots, X_n$ are independent. Since P is not known and is completely unrestricted, a Dirichlet process prior may be considered as an appropriate prior for P. This induces a prior for the density f_P known as the *Dirichlet process mixture* (DPM), and serves as an extremely useful Bayesian model for density estimation; see Ferguson (1983) and Lo (1984). The model is very rich under a variety of situations, for instance, if the kernel is a normal density with mean θ and scale σ converging to 0, since for any density $\int \frac{1}{\sigma\sqrt{2\pi}} \mathrm{e}^{-\frac{1}{2}(x-\theta)^2/\sigma^2} f_0(\theta)\mathrm{d}\theta \to f_0(x)$ in L_1-distance.

It is possible to write down the expressions for the posterior mean and the posterior variance of the density $f_P(x)$, but the formulae contain an enormously large number of terms prohibiting any real use of them. Fortunately, computable expressions can be obtained by MCMC methods by simulating the latent variables $(\theta_1, \ldots, \theta_n)$ from their posterior distribution by a scheme very similar to the Blackwell–MacQueen urn scheme, as studied by Escobar and West (1995) and many others. As before, we can describe the distribution of any θ_i given the other θ_j and all X_i. The scheme is structurally quite similar to the original Blackwell–MacQueen scheme. However, the presence of the extra X_i in the conditioning changes the relative weights and the distribution from where a fresh sample is drawn. More precisely, given θ_j, $j \neq i$, only conditioning by X_i matters, which weighs the selection probability of an old θ_j by $\psi(X_i; \theta_j)$, and the fresh draw by $\int \psi(X_i; \theta)\mathrm{d}G(\theta)$, and a fresh draw, whenever obtained, is taken from the "baseline posterior" defined by $\mathrm{d}G_b(\theta) \propto \psi(X_i; \theta)\mathrm{d}G(\theta)$.

The kernel used in forming the DPM can be chosen in different ways depending on purpose. If density estimation on the line is the objective, one may use a location-scale kernel such as the normal density. On the half-line, gamma, log-normal and Weibull mixtures seem to be more appropriate. On the unit interval, mixtures of beta densities can be considered. Sometimes, special shapes can be produced by special types of mixtures. For instance, a decreasing density on the half-line is a scale mixture of uniform densities, so a prior on a decreasing density can be induced by the mixture model technique. A prior on symmetric strongly unimodal densities can be induced using normal scale mixtures.

2.3.3 Hierarchical Dirichlet processes

A curious process is obtained when one models observations X_{ij} coming from totally unknown distributions F_i, and the distributions $F_1, \ldots, F_k$ themselves, treated as unknown parameters, are sampled i.i.d. from a Dirichlet process whose center measure G is itself randomly sampled from a Dirichlet process. Since Dirichlet samples are discrete, the discreteness of G forces $F_1, \ldots, F_k$ to share their atoms, and hence ties will be observed in the values of X even across different groups. This feature is often desirable in some genomic and machine learning applications. Because of the presence of two levels of Dirichlet process, the prior on $F_1, \ldots, F_k$ is known as the *hierarchical Dirichlet process*; see Teh, Jordan, Beal and Blei (2006).

2.3.4 Invariant and conditioned Dirichlet processes

In some applications, an unknown distribution needs to be moulded to satisfy certain invariance requirements, such as symmetry. Since the Dirichlet process supports all types of distributions, one needs to symmetrize a random distribution obtained from the Dirichlet process prior as in Dalal (1979). This technique can be used, for instance, in proposing a prior for the distribution of error, so that a location or regression parameter can be made identifiable. Another alternative is to constrain the distribution to have median zero. A prior for this was obtained in Doss (1985a) by conditioning the Dirichlet process to assign probability $\frac{1}{2}$ to $[0, \infty)$.

2.4 Posterior consistency

2.4.1 Motivation and implications

Now we turn our attention to asymptotic properties of the posterior distributions, and begin with a discussion on posterior consistency. Consider a sequence of statistical experiments parameterized by a possibly infinite-dimensional parameter θ taking values in a separable metric space. Let Π stand for the prior distribution on θ and $\Pi(\cdot|\text{data})$ stand for (a version of) the posterior distribution.

Definition 2.2 The posterior distribution is said to be *consistent* at a given θ_0, or (θ_0, Π) is a consistent pair, if for any neighborhood V of θ_0, $\Pi(\theta \notin V|\text{data}) \to 0$ (in probability or a.s.) as the size of the data tends to infinity when θ_0 is the true value of the parameter.

It may be noted that consistency depends on the choice of a version, but in dominated cases, there is essentially only one version that matters.

The importance of consistency stems from the desire to be able to identify correctly the data-generating mechanism when an unlimited supply of data is available. Even though this is purely a large sample property, an inconsistent posterior is often an indication of seriously incorrect inference, even for moderate sample sizes. Moreover, consistency can be shown to be equivalent with agreement among Bayesians with different sets of priors; see Diaconis and Freedman (1986).

Consistency has several immediate connections with other properties. First, it may be observed that it is not necessary to check convergence for every possible neighborhood; it is enough to consider a class which forms a local sub-base for the topology, that is, a class whose finite intersections form a base for the topology at the true value. Consistency implies existence of a rate of convergence also, that is, a shrinking ball around the true value whose posterior probability tends to one. Consistency also implies existence of an estimator which is consistent in the frequentist sense. To construct such an estimator, one may look at the point such that a small ball around it has maximum posterior probability among all balls of equal radius. Then, since balls around the true parameter have posterior probability tending to one, and the chosen point, by definition, cannot be beaten in this game, we obtain two small balls both of which have posterior probability close to one. Such balls must intersect, since otherwise the total posterior probability would exceed one. Therefore the two center points, the true parameter value and the estimator, must be infinitesimally close. In other words, the estimator constructed by maximizing the posterior probability of a ball of small radius around it is consistent. If posterior consistency holds, then for convex parameter spaces such as the space of densities with the L_1, Hellinger or some bounded metric on it which induces convex neighborhoods, the posterior mean gives another consistent estimator.

2.4.2 Doob's theorem

There is an extremely general theorem by Doob (1948), which essentially ensures consistency under any model where consistent estimators exist (such as when i.i.d. observations are available and the model is identifiable), provided we are happy to live with possible inconsistency on a null set with respect to the prior. While, at the first glance, this may be a cheering general fact, the interpretation of nullity in view of the chosen prior is not satisfactory. It is easy for a null set to be topologically huge. An extreme possibility is exhibited by the prior degenerate at a point θ^*. In this case, consistency fails everywhere except for $\theta_0 = \theta^*$, yet this huge exceptional set is a null set in view of the prior considered. Thus, to study whether consistency holds in a given situation, it is important to give sufficient conditions on the true value of the parameter and the prior which ensure consistency. It may be noted that if the parameter space is countable, Doob's result actually implies consistency everywhere provided all points receive positive prior mass.

2.4.3 Instances of inconsistency

For finite-dimensional parameter spaces, consistency is almost guaranteed, at least for well-behaved parametric families, if the prior density is positive in the neighborhoods of the true parameter. Surprisingly, consistency can fail in infinite-dimensional spaces for quite well-behaved models even for seemingly natural priors. In particular, the condition of assigning positive prior probabilities in usual neighborhoods of the true parameter is not at all sufficient to ensure consistency.

An interesting counterexample was constructed by Freedman (1963) in the context of estimating a discrete distribution on natural numbers, which is the simplest nonparametric estimation problem. Let the true distribution of observations be geometric with parameter $\frac{1}{4}$. Freedman constructed a prior which assigns positive probability to every weak neighborhood of the true distribution, but the posterior concentrates near a wrong value, the geometric distribution with parameter $\frac{3}{4}$. A more striking and counter-intuitive example was constructed more recently in Kim and Lee (2001) in the context of Bayesian survival analysis, where it was shown that among two priors for cumulative hazard function, both with mean equal to the true cumulative hazard, the one with larger prior spread achieves posterior consistency but the one which is more tightly spread leads to inconsistency.

Freedman's example is actually generic in the topological sense. If we look at all possible pairs of true parameter values and priors which lead to consistency, then the collection is extremely narrow when the size is measured topologically. A set F is called *meager* and considered to be topologically small if F can be expressed as a countable union of sets $C_i, i \geq 1$, whose closures $\bar{C}_i$ have empty interior. Freedman (1963) showed that the collection of "good pairs" is meager in the product space.

Should this result scare a Bayesian into abandoning his approach? No. The reason is that we are only concerned about a relatively small collection of priors. Therefore, what happens to most priors does not bother us, as long as we can find a prior incorporating any available subjective features and the corresponding posterior distribution good frequentist properties of the posterior. However, the counterexample and result above warn against careless use of a prior and emphasize the need to prove theorems assuring posterior consistency under mild conditions on the true parameter and the prior.

2.4.4 Approaches to consistency

If the posterior has an explicit expression, it may be possible to prove consistency or rate of convergence by simple Chebyshev-type inequalities. This is often the case in Bayesian survival analysis, where posterior conjugacy holds, for instance, for priors described by a Lévy process. We have also seen that convergence properties of the Dirichlet process give rise to posterior consistency. However, these situations are very special and are not to be expected in all applications.

In the context of estimating a c.d.f., a reasonably general class of priors for which posterior consistency holds is given by the class of tail-free priors considered in Freedman (1963). For the weak topology, convergence can be assessed through convergence of probabilities of the sets in a sufficiently fine partition. Thus one can restrict attention to the finite-dimensional object given by the probability vector corresponding to this fine partition. Interestingly, the posterior distribution of this vector depends only on the counts of the corresponding cells by the tail-freeness property, so the problem reduces to that of estimating parameters in a multinomial distribution, for which consistency holds under the general conditions that the weak support of the tail-free process contains the true distribution. A particularly important tail-free class is given by the Pólya tree process; see Lavine (1992). In this case, the space is split binarily and each time mass is distributed to the left and the right parts according to an independent random variable following a beta distribution, whose parameters can vary freely while the remaining mass is assigned to the corresponding right portion. Such priors have generally large weak support, ensuring consistency.

Although the above result is very interesting, it is also somewhat restrictive in that it is applicable only to the problem of estimating a c.d.f., and only if a tail-free prior is used. The tail-freeness property is very delicate and can be easily destroyed by common operations like symmetrization or mixing. Indeed, inconsistency may occur in this way as Diaconis and Freedman (1986) showed.

2.4.5 Schwartz's theory

A more useful approach, due to Schwartz (1965), is obtained by putting appropriate size restrictions on the model and conditions on the support of the prior in the sense of Kullback–Leibler divergence. Below, we describe Schwartz's theory and its extensions along with some applications, especially to the density estimation problem.

For the general theory, we assume that the family is dominated. Let $p_{\theta,n}(X_1,\ldots,X_n)$ stand for the joint density of observations and Π for the prior distribution. It is possible to let Π depend on n, but to keep ideas simple, we assume that the prior is fixed. Let θ_0 stand for the true value of the parameter. Then the posterior probability of any set B can be written as

$$\Pi(\theta \in B|X_1,\ldots,X_n) = \frac{\int_B \frac{p_{\theta,n}(X_1,\ldots,X_n)}{p_{\theta_0,n}(X_1,\ldots,X_n)} \mathrm{d}\Pi(\theta)}{\int \frac{p_{\theta,n}(X_1,\ldots,X_n)}{p_{\theta_0,n}(X_1,\ldots,X_n)} \mathrm{d}\Pi(\theta)}. \tag{2.10}$$

To establish consistency, we let B be the complement of a neighborhood U of θ_0 and show that the above expression with $B = U^{\mathrm{c}}$ goes to 0 as $n \to \infty$, either in P_{θ_0}-probability or P_{θ_0} a.s. A strategy that often works, especially when the observations are i.i.d., is to show that the numerator in (2.10) converges to zero exponentially fast like $\mathrm{e}^{-\beta n}$ for some $\beta > 0$, while the denominator multiplied by $\mathrm{e}^{\beta n}$ converges to infinity for all $\beta > 0$. Thus we need to give sufficient conditions to ensure these two separate assertions.

Below we assume that the observations are i.i.d. following a density p_θ with respect to a σ-finite measure ν; we shall indicate later how to extend the result to independent non-identically distributed or even dependent observations.

Kullback–Leibler property

Note that the integrand in the denominator of (2.10) can be written as $e^{-n\Lambda_n(\theta,\theta_0)}$, and for large n, $\Lambda_n(\theta,\theta_0) := n^{-1}\sum_{i=1}^n \log(p_{\theta_0}(X_i)/p_\theta(X_i))$ behaves like the *Kullback–Leibler divergence* number given by $K(p_{\theta_0}; p_\theta) = \int p_{\theta_0}\log(p_{\theta_0}/p_\theta)\mathrm{d}\nu$. Thus the integrand is at least as big as $e^{-2n\epsilon}$ for all sufficiently large n if $K(p_{\theta_0}; p_\theta) < \epsilon$, so the contribution of the part $A := \{\theta : K(p_{\theta_0}; p_\theta) < \epsilon\}$ alone to the integral in the denominator of (2.10) is at least $e^{-2n\epsilon}\Pi(\theta : K(p_{\theta_0}; p_\theta) < \epsilon)$. Since $\epsilon > 0$ can be chosen arbitrarily small, it is clear that the term in the denominator multiplied by $e^{n\beta}$ is exponentially big, provided that $\Pi(\theta : K(p_{\theta_0}; p_\theta) < \epsilon) > 0$ for all $\epsilon > 0$. The whole argument can easily be made rigorous by an application of Fatou's lemma. Thus the last condition emerges as a key condition in the study of consistency, and will be referred to as Schwartz's prior positivity condition or the *Kullback–Leibler property* of the prior, or the true parameter is said to be in the *Kullback–Leibler support* of the prior. Note that the condition essentially means that the prior should assign positive probability to any neighborhood of the true parameter, much in the spirit of the "obvious requirement" for consistency. However, the important matter here is that the neighborhood is defined by nearness in terms of the Kullback–Leibler divergence, not in terms of the topology of original interest. In many parametric cases, regularity conditions on the family ensure that a Euclidean neighborhood is contained inside such a Kullback–Leibler neighborhood, so the usual support condition suffices. For infinite-dimensional families, the Kullback–Leibler divergence is usually stronger than the metric locally around θ_0.

Bounding the numerator

To show that the numerator in (2.10) is exponentially small, a naive but straightforward approach would be to bound $\Lambda_n(\theta,\theta_0)$ uniformly over θ lying outside the given neighborhood U. For individual θ, the above quantity stays below a negative number by the law of large numbers and the fact that the Kullback–Leibler divergence is strictly positive. However, to control the integral, one needs to control $\Lambda_n(\theta,\theta_0)$ uniformly over U^c, which poses a tough challenge. If the parameter space is compact, a classical approach used by Wald, which bounds the log-likelihood ratio by a maximum of finitely many terms each of which is a sum of integrable i.i.d. variables with negative expectation, is useful. More modern approaches to bounding the log-likelihood ratio outside a neighborhood involve bounding bracketing entropy integrals, which is also a strong condition.

Uniformly consistent tests

Clearly, bounding an average by the maximum is not the best strategy. Schwartz's ingenious idea is to link the numerator in (2.10) with the power of uniformly exponentially consistent tests for the hypothesis $\theta = \theta_0$ against $\theta \in U^c$. Under the existence of such a test, Schwartz (1965) showed that the ratio of the marginal density of the observation with θ conditioned to lie outside U to the true joint density is exponentially small except on a set with exponentially small sampling probability. This is enough to control the numerator in (2.10) as required.

Uniformly exponentially consistent tests, that is, tests for which both the type I and type II error probabilities go to zero exponentially fast, have been well studied in the literature. If two convex sets of densities C_1 and C_2 are separated by a positive distance at least ϵ in terms of the Hellinger distance, then one can construct a test for the pair of hypotheses $p_\theta \in C_1$ against $p_\theta \in C_2$ whose error probabilities decay like $e^{-cn\epsilon^2}$ for some universal constant $c > 0$. In this result, the sizes of C_1 and C_2 are immaterial; only convexity and their distance matter. Generally, U^c is not convex, so the result does not directly give an exponentially consistent test for testing $\theta = \theta_0$ against the alternative U^c. However, we observe that if U^c can be covered by finitely many convex bodies $C_1, \dots, C_k$, each of which maintains a positive distance from θ_0, then a uniformly exponentially consistent test ϕ_j is obtained for testing $\theta = \theta_0$ against C_j with both error probabilities bounded by $e^{-c'n}$. Then define a test $\phi = \max_j \phi_j$. Clearly, the power of ϕ at any point is better than the power of any of the ϕ_j. In particular, for any $j = 1, \dots, k$, if $p_\theta \in C_j$, then $E_\theta(1-\phi) \le E_\theta(1-\phi_j)$. By the given construction, the latter term is already exponentially small uniformly over C_j. Thus the type II error probability is easily bounded. For the type I error probability, we can bound $E_{\theta_0}\phi \le \sum_{j=1}^k E_{\theta_0}\phi_j \le k e^{-c'n}$, establishing the required exponential bound.

The strategy works nicely in many parametric models where a uniformly exponentially consistent test for the complement of a very large ball can often be obtained directly, so that the remaining compact portion may be covered with a finite number of balls. In infinite-dimensional spaces, this is much harder. When the topology under consideration is the weak topology, U^c can be covered by finitely many convex sets. To see this, observe that a basic open neighborhood U of a true density p_0 is described by conditions on finitely many integrals $\{p : |\int \psi_j p - \int \psi_j p_0| < \epsilon_j, j = 1, \dots, k\}$, which can be written as $\cap_{j=1}^k \{p : \int \psi_j p < \int \psi_j p_0 + \epsilon_j\} \cap \cap_{j=1}^k \{p : \int \psi_j p_0 < \int \psi_j p + \epsilon_j\}$. Thus U^c is a finite union of sets of the form $\{p : \int \psi p \ge \int \psi p_0 + \epsilon\}$ or $\{p : \int \psi p_0 \ge \int \psi p + \epsilon\}$, both of which are convex and separated from p_0.

Entropy and sieves

Unfortunately, the procedure runs into difficulty in infinite-dimensional spaces with stronger topologies, such as for density estimation with the Hellinger or L_1-distance, unless the space of densities under consideration is assumed to be compact. For the space of density functions, the complement of a neighborhood cannot be covered by finitely many balls or convex sets, each maintaining a positive distance from the true one. However, not everything is lost, and much of the idea can be carried out with the help of a technique of truncating the parameter space, depending on the sample size. Observe that in the argument, the type II error probability will not be problematic whenever the final test is greater than individual tests, so it is only the type I error probability which needs to be properly taken care of. From the bound $ke^{-c'n}$, it is clear that one may allow k to depend on n, provided that its growth is slower than $e^{c'n}$, to spare an exponentially small factor.

To formalize the idea, let p denote the density function which itself is treated as the parameter. Let $\mathcal{P}$ be a class of density functions where the possible values of p lie and $p_0 \in \mathcal{P}$ stands for the true density function. For definiteness, we work with the Hellinger distance on $\mathcal{P}$, although the L_1-distance may also be used. In fact, the two metrics define the same notion of convergence. Let $U = \{p : d(p, p_0) < \epsilon\}$ for some given $\epsilon > 0$. Let $\mathcal{P}_n$ be a sequence of subsets of $\mathcal{P}$, also called a *sieve* (possibly depending on ϵ), gradually increasing to $\mathcal{P}$. Let N_n be the number of balls of size $\epsilon/2$ with center in $\mathcal{P}_n$, needed to cover $\mathcal{P}_n$. Any such ball which intersects U^c clearly maintains a distance at least $\epsilon/2$ from p_0. Thus the type I and type II error probability for testing $p = p_0$ against any ball is bounded by $e^{-n\delta}$, where $\delta > 0$ depends on ϵ, and can be made explicit if desired. Then if $\log N_n < n\delta'$ for all n and $\delta' < \delta$, then by the discussion in the preceding paragraph, it follows that the numerator in (2.10) is exponentially small, and hence $\Pi(p \in U^c \cap \mathcal{P}_n | X_1, \ldots, X_n) \to 0$. The number N_n is called the *$\epsilon/2$-covering number* of $\mathcal{P}_n$ with respect to the metric d, which is denoted by $N(\epsilon/2, \mathcal{P}_n, d)$, and its logarithm is known as the *metric entropy*. Thus a bound for the metric entropy of the sieve limited by a suitably small multiple of n implies that the posterior probability of U^c fades out unless it goes to the complement of the sieve $\mathcal{P}_n$. In order to complete the proof of posterior consistency, one must show that $\Pi(p \in \mathcal{P}_n^c | X_1, \ldots, X_n) \to 0$ by other methods. This is sometimes possible by direct calculations. Note that $\mathcal{P}_n^c$ has small prior probability, so we should expect it to have small posterior probability in some sense if the likelihood is bounded appropriately. Unfortunately, small prior probability of $\mathcal{P}_n^c$ does not imply small posterior probability, since the likelihood may increase exponentially, enhancing the posterior probability. However, if the prior probability is exponentially small, then so is the posterior probability, under the Kullback–Leibler positivity condition. This follows quite easily by a simple application of Fubini's theorem applied to the numerator of (2.10) with $B = \mathcal{P}_n^c$. Thus, to establish consistency, one just needs to construct a sieve $\mathcal{P}_n$ such that $\log N(\epsilon/2, \mathcal{P}_n, d) < n\delta' < n\delta$, where δ is defined before, and $\Pi(\mathcal{P}_n^c)$ is exponentially small. In particular, the entropy condition will hold for a sieve $\mathcal{P}_n$ with $\log N(\epsilon/2, \mathcal{P}_n, d) = o(n)$. The main ideas behind the result were developed through the works Schwartz (1965), Barron, Schervish and Wasserman (1999) and Ghosal, Ghosh and Ramamoorthi (1999a). It is also interesting to note that the approach through testing is "optimal." This is because a result of Barron (see Theorem 4.4.3 of Ghosh and Ramamoorthi (2003)) shows that consistency with exponenial speed holds if and only if one can find a sieve whose complement has exponentially small prior probability and a test which has exponentially small error probabilities on the sieve.

2.4.6 Density estimation

The above consistency theorem can be applied to derive posterior consistency in Bayesian density estimation using the commonly used priors such as the Dirichlet mixture or Gaussian processes.

Dirichlet mixtures

To establish the Kullback–Leibler property of a Dirichlet mixture of normal prior at a true density p_0, one approximates p_0 by p_m defined as the convolution of p_0 truncated to $[-m, m]$ for some large m and the normal kernel with a small bandwidth. This convolution, which is itself a normal mixture, approximates p_0 pointwise as well as in the Kullback–Leibler sense under mild conditions on p_0. Now a Kullback–Leibler neighborhood around p_m includes a set which can be described in terms of a weak neighborhood around p_0 truncated to $[-m, m]$. Since the Dirichlet process has large weak support, the resulting weak neighborhood will have positive probability, proving the Kullback–Leibler property. Indeed, the argument applies to many other kernels. To construct appropriate sieves, consider all mixture densities arising from all bandwidths $h > h_n$ and mixing distribution F with $F[-a_n, a_n] > 1 - \delta$, where $a_n/h_n < cn$ for some suitably small $c > 0$. Then the condition for exponentially small prior probability of the complement of the sieve holds if the prior on the bandwidth is such that $\Pi(h < h_n)$ is exponentially small, and the base measure α of the Dirichlet process assigns exponentially small mass to $[-a_n, a_n]^c$. These conditions can be met, for instance, if the prior for h^2 is inverse gamma and the base measure of the Dirichlet process of the mixing distribution is normal. Results of this kind were obtained in Ghosal, Ghosh and Ramamoorthi (1999a), Lijoi, Prünster and Walker (2005), Tokdar (2006) and Wu and Ghosal (2008).

Gaussian processes

A prior for density estimation on a compact interval I can also be constructed from a Gaussian process $\xi(t)$ by exponentiating and normalizing to $e^{\xi(t)}/\int_I e^{\xi(u)}du$.

Assuming that the true density p_0 is positive throughout, it is easy to see that a Kullback–Leibler neighborhood of p_0 contains a set which is described by the uniform neighborhood in terms of $\xi(t)$ about a function $\xi_0(t)$ satisfying $p_0(t) = e^{\xi_0(t)} / \int_I e^{\xi_0(u)} du$. Now, for a Gaussian process, a continuous function $\xi_0(t)$ is in the support if and only if it belongs to the closure of the *reproducing kernel Hilbert space* (RKHS) of the Gaussian process. The Brownian motion and its integrals are Gaussian processes with large RKHS. Another possibility is to consider a Gaussian process with kernel containing a scale which is allowed to assume arbitrarily large positive values (so that the prior is actually a mixture of Gaussian processes). Then under extremely mild conditions on the kernel, the overall support of the process includes all continuous functions. Further, sieves can easily be constructed using Borel's inequality or smoothness properties of Gaussian paths and maximal inequalities for Gaussian processes; see Ghosal and Roy (2006), Tokdar and Ghosh (2007), and van der Vaart and van Zanten (2007, 2008).

Pólya tree processes

To estimate the density of the observations using a Pólya tree prior, consider for simplicity binary partitions used in the mass distribution in the tree structure obtained sequentially by the median, quartiles, octiles etc. of a density λ. Further assume that the parameters of the beta distributions used to split mass randomly are all equal within the same level of the tree, that is, say the parameters are all equal to a_m at level m. Then by a theorem of Kraft (1964), it follows that the random distributions generated by a Pólya tree admit densities a.s. if $\sum_{m=1}^{\infty} a_m^{-1} < \infty$. To establish the Kullback–Leibler property, one needs to strengthen the condition to $\sum_{m=1}^{\infty} a_m^{-1/2} < \infty$ and assume that the density λ has finite entropy in the sense $\int p_0(x) \log \lambda(x) dx < \infty$; see Ghosal, Ghosh and Ramamoorthi (1999b) for a proof. The Kullback–Leibler property implies posterior consistency with respect to the weak topology. However, since the sample paths of a Pólya tree lack appropriate smoothness, it is difficult to control the size of the space where the prior is essentially supported. Under quite strong growth conditions $a_m \sim 8^m$ on the parameters, appropriate sieves were constructed in Barron, Schervish and Wasserman (1999), giving consistency with respect to the Hellinger distance.

2.4.7 Semiparametric applications

Schwartz's consistency theory and its extensions lead to very useful consistency results in Bayesian semiparametric inference. Diaconis and Freedman (1986) and Doss (1985b) gave examples showing that inconsistency may occur when one estimates the location parameter θ in the location model $X = \theta + e$, $e \sim F$, using a symmetrized Dirichlet (respectively, Dirichlet conditioned to have mean zero) prior for F. To understand why this is happening, ignore the issue of symmetrization and represent the prior as a mixture of Dirichlet process with θ acting as a location parameter for the base measure G with p.d.f. g. Then it follows from (2.9) that the likelihood for θ given $X_1, \ldots, X_n$ is $\prod_{i=1}^{n} g(X_i - \theta)$, so the Bayes estimator is similar to the maximum likelihood estimator (MLE) based on the above incorrect "parametric" likelihood, which may or may not give the correct result. Also observe that discreteness of Dirichlet samples prohibits the prior putting any mass in the Kullback–Leibler neighborhoods, so Schwartz's theory does not apply there. However, positive results were obtained in Ghosal, Ghosh and Ramamoorthi (1999b) for a prior which leads to densities. Indeed, if the true error density f_0 is in the Kullback–Leibler support of the prior Π for the density f of F, then the true density of observations $f_0(\cdot - \theta_0)$ is in the support of the prior for the density of observations. Thus the Kullback–Leibler property is not destroyed by location shifts, unlike the fragile tail-freeness property of the Dirichlet process which is lost by symmetrization and location change. For instance, using an appropriate Pólya tree, Dirichlet mixture or Gaussian process prior for f, we can ensure that the distribution of X is consistently estimated in the weak topology. Now within a class of densities with fixed median, the map $(\theta, f) \mapsto f(\cdot - \theta)$ is both-way continuous with respect to the weak topology. Thus consistency for θ follows from the weak consistency for the density of the observations, which is obtained without directly constructing any test or sieves. This clearly shows the power of Schwartz's theory, especially for semiparametric problems, where the infinite-dimensional part is usually not of direct interest.

2.4.8 Non-i.i.d. observations

The theory of posterior consistency can be extended to independent, non-identically distributed variables as well as to some dependent situations. First observe that the denominator in (2.10) can be tackled essentially in the same way when a law of large numbers is applicable to the summands appearing in the log-likelihood ratio. For independent, non-identically distributed random variables, this is possible by Kolmogorov's strong law when variances of the log-likelihood ratio based on each observation can be controlled appropriately. For ergodic Markov processes, a law of large numbers is available too.

To control the numerator, one needs to construct appropriate tests against complements of neighborhoods for a given topology. Such tests have been constructed in the literature for applications such as linear regression with nonparametric error (Amewou-Atisso, Ghosal, Ghosh and Ramamoorthi 2003), binary regression with Gaussian process prior (Ghosal and Roy, 2006), normal regression with Gaussian process prior (Choi and Schervish, 2007), estimation of the spectral density of a time series using Whittle likelihood (Choudhuri, Ghosal and Roy, 2004) and estimating the transition density of a Markov process using Dirichlet mixtures (Tang and Ghosal, 2007). Other important work on consistency includes Diaconis and Freedman (1993) and Coram and Lalley (2006) showing fine balance between consistency and inconsistency in binary regression with a prior supporting only piecewise constant functions. The last two works use direct computation, rather than Schwartz's theory, to prove consistency.

2.4.9 Sieve-free approaches

We end this section by mentioning alternative approaches to consistency which do not require the construction of sieves and uniformly exponentially consistent tests on them.

Martingale method

Consider the Hellinger distance on the space of densities of i.i.d. observations and assume that the Kullback–Leibler property holds. Then, by utilizing a martingale property of marginal densities, Walker (2004) showed that the posterior probability of a set A goes to 0 if the posterior mean of p, when the prior is restricted to A, is asymptotically a positive distance away from the true density p_0. Now, when $A = U^c$, the result is not directly applicable as the posterior mean for the prior restricted to U^c may come close to p_0. To obtain the required result, Walker (2004) covered U^c with countably many balls, and controlled both the size and prior probability of each ball. More precisely, Walker (2004) showed that if for any given $\epsilon > 0$, there is $0 < \delta < \epsilon$ such that each ball has diameter at most δ and the sum of the square root of the prior probabilities of these balls is finite, then the posterior is consistent. The argument also extends to Markov processes as shown by Ghosal and Tang (2006). A lucid discussion on the basis of consistency or inconsistency without referring to sieves is given by Walker, Lijoi and Prünster (2005).

Although the proof of consistency based on the martingale property is very interesting and one does not need to construct sieves, the approach does not lead to any new consistency theorem. This is because the condition on the diameter of each ball and the summability of square roots of prior probabilities imply existence of a sieve whose complement has exponentially small prior probability and which has $\epsilon/2$-Hellinger metric entropy bounded by a small multiple of n, that is, the conditions of the consistency theorem obtained from the extension of Schwartz's theory discussed before, hold.

Power-posterior distribution

A remarkable finding of Walker and Hjort (2001) is that the posterior distribution is consistent only under the Kullback–Leibler property if the likelihood function is raised to a power $\alpha < 1$ before computing the "posterior distribution" using the Bayes formula. The resulting random measure may be called the *power-posterior distribution*. The above assertion follows quite easily by bounding the numerator in the Bayes theorem by Markov's inequality and the denominator by Schwartz's method using the Kullback–Leibler property as described before. The greatest advantage with this approach is that there is no need for any additional size constraint in the form of tests or entropy bounds, and hence there is no need to construct any sieves, provided that one is willing to alter the posterior distribution to make inference. However, usual MCMC methods may be difficult to adapt, especially for density estimation using Dirichlet mixture prior. The Kullback–Leibler property alone can lead to other desirable conclusions such as convergence of sequential predictive densities in relative entropy risk as shown by Barron (1999).

2.5 Convergence rates of posterior distributions

2.5.1 *Motivation, description and consequences*

As mentioned in the introduction, in naive terms, the convergence rate is the size ϵ_n of the smallest ball centered about the true parameter θ_0 such that the posterior probability converges to one. In practice, we often just find one sequence ϵ_n such that the posterior probability of the ball of radius ϵ_n around θ_0 converges to one, so it may be more appropriate to term this "a rate of convergence."

Definition 2.3 Let $X^{(n)}$ be data generated by a model $P_\theta^{(n)}$. We say that a sequence $\epsilon_n \to 0$ is the *convergence rate of the posterior distribution* $\Pi_n(\cdot|X^{(n)})$ at the true parameter θ_0 with respect to a pseudo-metric d if for any $M_n \to \infty$, we have that $\Pi_n(\theta : d(\theta, \theta_0) \geq M_n\epsilon_n) \to 0$ in $P_{\theta_0}^{(n)}$ probability.

Thus, by rate of convergence, we mean only up to a multiplicative constant, thus disregarding the constants appearing in the bound. At present, the available techniques do not guide us to the best possible constants. It is well known that the convergence rate in regular parametric families is $n^{-1/2}$, agreeing with the *minimax rate*, that is the best convergence rate for estimators. For infinite-dimensional models, the rate of convergence may be $n^{-1/2}$ or slower. In many cases, the posterior convergence rates corresponding to well-known priors agree with the corresponding minimax rate, possibly up to logarithmic factors.

There are some immediate consequences of the posterior convergence rate. If the posterior converges at the rate ϵ_n, then as in the previous section, the estimator defined as the center of the ball of radius maximizing the posterior probability converges to the true parameter at rate ϵ_n in the frequentist sense. Since the convergence rate of an estimator cannot be faster than the minimax rate for the problem, it also follows that the posterior convergence rate cannot be better than the minimax rate. Thus achieving the minimax rate can be regarded as the ideal goal. For the special case of density estimation with the L_1 or the Hellinger metric defining the convergence rate, the posterior mean also converges at the same rate at which the posterior converges.

When the parameter space is equipped with the L_2-norm and expressions for the posterior mean $\hat{\theta}$ and variance are explicitly available, it may be possible to derive the convergence rate from Chebyshev's inequality. When θ stands for the mean of an infinite-dimensional normal distribution and an appropriate conjugate normal prior is used, the convergence rate can be calculated easily by explicit calculations. In general, this seems to be difficult, so we need to develop a general theory along the lines of Schwartz's theory for posterior consistency for dominated families.

2.5.2 General theory

Let us first consider the i.i.d. case where observations $X_1, X_2, \ldots \sim p$, and p is given possibly a sequence of prior Π. Let $\epsilon_n \to 0$ be the targeted rate. In density estimation problems, this is slower than $n^{-1/2}$, so we assume that $n\epsilon_n^2 \to \infty$. The basic ideas we use here were developed by Ghosal, Ghosh and van der Vaart (2000), and are similar to those used to study consistency. See also Shen and Wasserman (2001) and Walker, Lijoi and Prünster (2007) for alternative approaches involving somewhat stronger conditions. As in (2.10), we express the posterior probability of $B = \{p : d(p, p_0) \geq \epsilon_n\}$ as a ratio, and show that, under appropriate conditions, (i) the numerator is bounded by $e^{-cn\epsilon_n^2}$, where $c > 0$ can be chosen sufficiently large, while (ii) the denominator is at least $e^{-bn\epsilon_n^2}$.

The above two assertions hold under conditions which can be thought of as the quantitative analog of the conditions ensuring consistency. This is quite expected as a rate statement is a quantitative refinement of consistency.

Prior concentration rate

To take care of the second assertion, we replace the Kullback–Leibler positivity condition by

$$\Pi(B(p_0, \epsilon_n)) := \Pi\{p : K(p_0; p) \leq \epsilon_n^2,\ V(p_0; p) \leq \epsilon_n^2\} \geq e^{-b_1 n\epsilon_n^2}, \qquad (2.11)$$

where $V(p_0; p) = \int p_0(\log(p_0/p))^2$. Thus apart from the fact that the description of the neighborhood also involves the second moment of the log-likelihood ratio, the condition differs from Schwartz's condition in requiring a minimum level of concentration of prior probability around the true density p_0. Intuitively, the likelihood function at p with $K(p_0; p) \leq \epsilon_n^2$, apart from random fluctuations, is at least $e^{-n\epsilon_n^2}$. When the variance $V(p_0; p)$ is also smaller than ϵ_n^2, it can be seen that the random fluctuations can change the lower bound only slightly, to $e^{-b_2 n\epsilon_n^2}$, except on a set of small probability. The prior probability of the set of such p is at least $e^{-b_1 n\epsilon_n^2}$ by (2.11), leading to assertion (ii).

Entropy and tests

To take care of assertion (i), we follow the testing approach of Schwartz. Note that, as the alternative $\{p : d(p, p_0) \geq \epsilon_n\}$ is getting close to the null $p = p_0$, it is not possible to test the pair with exponentially small type I and type II error probabilities uniformly. However, since we now have a better lower bound for the denominator, our purpose will be served if we can test with both type I and type II error probabilities bounded by $e^{-cn\epsilon_n^2}$, where c is larger than b appearing in assertion (ii). By the discussion in the previous section, such error probabilities are possible for convex alternatives which are separated from p_0 by at least ϵ_n in terms of the Hellinger distance.† To construct the final test, one needs to cover the alternative with balls of size $\epsilon_n/2$ and control their number to no more than $e^{c_1 n\epsilon_n^2}$, that is, satisfy the metric entropy condition $\log N(\epsilon_n/2, \mathcal{P}, d) \leq c_1 n\epsilon_n^2$. The smallest ϵ satisfying the inequality $\log N(\epsilon/2, \mathcal{P}, d) \leq n\epsilon^2$ appears in the classical theory of minimax rates in that the resulting rate ϵ_n determines the best possible rate achievable by an estimator. Therefore, if the prior concentration rate can be matched with this ϵ_n, the posterior will converge at the rate at par with the minimax rate.

Sieves

Of course, satisfying the entropy condition is not generally possible unless $\mathcal{P}$ is compact, so we need to resort to the technique of sieves. As before, if there exists a sieve $\mathcal{P}_n$ such that

$$\log N(\epsilon_n/2, \mathcal{P}_n, d) \leq c_1 n\epsilon_n^2, \tag{2.12}$$

then, by replacing ϵ_n by a sufficiently large multiple $M\epsilon_n$, it follows that $\Pi(p \in \mathcal{P}_n : d(p, p_0) \geq \epsilon_n)$ converges to zero. To take care of the remaining part $\mathcal{P}_n^c$, the condition $\Pi(\mathcal{P}_n^c) \leq e^{-c_2 n\epsilon_n^2}$ suffices, completing the proof that the rate of convergence is ϵ_n.

† Alternatively, the L_1-distance can be used, and also the L_2-distance if densities are uniformly bounded.

2.5.3 Applications

The rate theorem obtained above can be applied to various combinations of model and prior.

Optimal rates using brackets

First we observe that optimal rates can often be obtained by the following technique of bracketing applicable for compact families. By an *ϵ-bracketing* of $\mathcal{P}$, we mean finitely many pairs of functions (l_j, u_j), $l_j(\cdot) \le u_j(\cdot)$, $d(u_j, l_j) < \epsilon$, known as *ϵ-brackets*, such that any $p \in \mathcal{P}$ is contained in one of these brackets. The smallest number of ϵ-brackets covering $\mathcal{P}$ is called the *ϵ-bracketing number* of $\mathcal{P}$, denoted by $N_{[\,]}(\epsilon, \mathcal{P}, d)$. Let ϵ_n be the smallest number satisfying $\log N_{[\,]}(\epsilon, \mathcal{P}, d) \le n\epsilon^2$. For all j, find an ϵ_j-bracketing and normalize its upper brackets to p.d.f.s. Now put the discrete uniform distribution on these p.d.f.s and mix these discrete uniform distributions according to a thick-tailed distribution λ_j on the natural numbers. Then the resulting prior automatically satisfies the metric entropy condition and prior concentration condition for the sequence $c\epsilon_n$ for some $c > 0$. Although the *bracketing entropy* $\log N_{[\,]}(\epsilon, \mathcal{P}, d)$ can be larger than the ordinary metric entropy $\log N(\epsilon, \mathcal{P}, d)$, often they are of equal order. In such cases, the construction leads to the optimal rate of convergence of the posterior distribution in the sense that the frequentist minimax rate is achieved. Since the minimax rate is unbeatable, this recipe of prior construction leads to the best possible posterior convergence rate. The construction can be extended to noncompact parameter spaces with the help of sieves.

As for specific applications of the bracketing techniques, consider the Hölder class of densities with smoothness α, which is roughly defined as the class of densities on a compact interval with α-many continuous derivatives. The bracketing entropy grows as $\epsilon^{-1/\alpha}$ in this case. This leads to the rate equation $n\epsilon^2 = \epsilon^{-1/\alpha}$, leading to the rate $n^{-\alpha/(2\alpha+1)}$, agreeing with the corresponding minimax rate. Another example is provided by the class of monotone densities, whose bracketing entropy grows as ϵ^{-1}. The corresponding rate equation is $n\epsilon^2 = \epsilon^{-1}$, again leading to the minimax rate $n^{-1/3}$ for this problem.

Finite-dimensional models

The conditions assumed in the rate theorem are suboptimal in the sense that for parametric applications, or some other situation for which the calculation involves Euclidean spaces, the rate equations lead to the best rate only up to a logarithmic factor. It is possible to remove this extra undesirable factor by refining both the entropy condition and the condition on the concentration of prior probability. The entropy condition can be modified by considering the *local entropy* $\log N(\epsilon/2, \{p \in \mathcal{P}_n : \epsilon \le d(p, p_0) \le 2\epsilon\}, d)$, which is smaller in finite-dimensional models but is as large as the ordinary metric entropy in many nonparametric models. The condition on the prior is modified to

$$\frac{\Pi\{p : j\epsilon_n < d(p, p_0) \leq 2j\epsilon_n\}}{\Pi(B(p_0, \epsilon_n))} \leq \mathrm{e}^{Knj^2\epsilon_n^2} \text{ for all } j. \tag{2.13}$$

With this modification, the posterior convergence rate in parametric families turns out to be $n^{-1/2}$.

Log-spline priors

The improved posterior convergence theorem based on the local entropy condition and (2.13) has a very important application in density estimation with log-spline priors. We form an exponential family of densities by a B-spline basis for α-smooth functions. A prior is induced on the densities through independent uniform priors on the coefficients. The exponential family based on B-splines approximates any α-smooth density within $J^{-\alpha}$, where J is the number of basis elements, so we need to keep increasing J with n appropriately. For a given J, the whole calculation can be done in $\mathbb{R}^J$. The rate equation is then essentially given by $J \sim n\epsilon_n^2$. However, since the convergence rate ϵ_n cannot be better than the rate of approximation $J^{-\alpha}$, the best trade-off is obtained by $J = J_n \sim n^{1/(1+2\alpha)}$ and $\epsilon_n \sim n^{-\alpha/(1+2\alpha)}$.

Applications of rate theorems to Dirichlet mixtures and Gaussian process priors are more involved.

Dirichlet mixtures

For the Dirichlet mixture of normal kernel, we consider two separate cases:

(a) the *supersmooth* case when the true density itself is a mixture of normal with standard deviation lying between two positive numbers,
(b) the *ordinary smooth* case when the true density is twice-continuously differentiable but need not itself be a mixture of normal.

The Dirichlet mixture prior used in the first case restricts the variation of the standard deviation of the normal kernel in between the two known bounds for it. Assume further that both the mixing distribution and the base measure of the Dirichlet process are compactly supported. Then one can approximate a normal mixture within ϵ by a finite mixture of normal with only $\mathrm{O}(\log \frac{1}{\epsilon})$ support points. Then the calculation essentially reduces to that in a simplex of dimension $N = \mathrm{O}(\log \frac{1}{\epsilon})$. Entropy of the N-simplex grows as ϵ^{-N} while the concentration rate of a Dirichlet distribution is $\mathrm{e}^{-cN \log \frac{1}{\epsilon}}$. This shows that the Hellinger metric entropy grows as $\log^2 \frac{1}{\epsilon}$ and the concentration rate of the Dirichlet mixture prior in Kullback–Leibler neighborhoods of size ϵ is $\mathrm{e}^{-c\log^2 \frac{1}{\epsilon}}$. Equating $n\epsilon^2 = \log^2 \frac{1}{\epsilon}$, the best rate of convergence $n^{-1/2} \log n$ is obtained. The compactness conditions assumed above are easy to relax with the consequence of a slight increase in the power of the logarithm. Interestingly, the convergence rate is nearly equal to the parametric convergence rate. Details are given in Ghosal and van der Vaart (2001).

For the ordinary smooth case, one needs to make the scale parameter close to zero with sufficiently high probability, so a sequence of priors for it can be constructed by scaling a fixed prior by some sequence σ_n. A twice-continuously differentiable density can be approximated by such a normal mixture up to σ_n^2. In this case, the estimate of entropy is $\sigma_n^{-1} \log^2 \frac{1}{\epsilon}$ and the prior concentration rate is $\mathrm{e}^{-c\sigma_n^{-1} \log^2 \frac{1}{\epsilon}}$. Equating $\sigma_n^{-1} \log^2 \frac{1}{\epsilon_n}$ with $n\epsilon_n^2$ subject to $\epsilon_n \geq \sigma_n$ gives the optimal frequentist rate $n^{-2/5}$ up to some logarithmic factors. Details are given in Ghosal and van der Vaart (2007b).

Gaussian processes

The rate of convergence for density estimation using a Gaussian process prior was calculated in van der Vaart and van Zanten (2008). In this case, sieves are constructed from Borell's inequality and the prior concentration rate from small ball probability for Gaussian processes. When the true density function is α-smooth, van der Vaart and van Zanten (2008) showed that by using an integrated Brownian motion (or some other similar processes) whose sample paths are also α-smooth, the minimax rate $n^{-\alpha/(2\alpha+1)}$ is achieved. As mentioned in the last section, another way to construct Gaussian processes with large support is to rescale the covariance kernel of the process by a sequence c_n. For large c_n, the procedure can "pack up" the variation of the Gaussian process on a long interval into a Gaussian process on the unit interval, resulting in rougher sample paths. Thus, starting with an infinitely smooth kernel, by the rescaling technique, one can approximate any continuous function. Indeed, it was shown in van der Vaart and van Zanten (2007) that this approximation holds in the right order, while entropies and prior concentration change in tune with the rescaling, resulting in the usual rate $n^{-\alpha/(2\alpha+1)}$ for α-smooth densities.

2.5.4 Misspecified models

The general theory discussed so far assumes that the true density belongs to the model, at least in a limiting sense. If the true density maintains a positive distance from the model, the model is called *misspecified*. Experience with parametric cases suggests that the posterior concentrates around the *Kullback–Leibler projection* of the true density, that is, the density within the model minimizing the Kullback–Leibler divergence from the true density. For general infinite-dimensional cases, a theory of posterior convergence rate for such misspecified models was developed in Kleijn and van der Vaart (2006). In analogy with the well-specified case discussed earlier, one needs to measure the prior concentration rate near the Kullback–Leibler projection and control the size of the sieve. It turns out that some different notion of covering, instead of ordinary metric entropy, is the appropriate concept of size in the misspecified case. The theory can be applied to several examples. An important conclusion from their work is that in the semiparametric linear regression model with unknown error density, the convergence rate of the posterior at the true value of the regression parameter does not suffer any loss if the error density is misspecified, such as a normal density instead of the correct double exponential density.

2.5.5 Non-i.i.d. extensions

Like the theory of consistency, the theory of the convergence rate can also be extended beyond the i.i.d. setup. In Ghosal and van der Vaart (2007a), rate theorems are derived for any general dependence and for any given (sequence of) metric d_n, where one can test the null hypothesis $\theta = \theta_0$ against balls of the type $\{\theta : d_n(\theta, \theta_1) < \xi\epsilon\}$, $d_n(\theta_1, \theta_0) > \epsilon$ and ξ is a universal constant, with both type I and type II error probabilities bounded by $e^{-cn\epsilon^2}$. In this case, the rate of convergence ϵ_n can again be characterized as the smallest solution of the entropy inequality

$$\sup_{\epsilon > \epsilon_n} \log N(\xi\epsilon, \{\theta : \epsilon < d_n(\theta, \theta_0) \leq 2\epsilon\}, d_n) \leq n\epsilon_n^2 \tag{2.14}$$

and meeting the condition on concentration probability in a Kullback–Leibler neighborhood of the joint density

$$\Pi\{\theta : K(p_{\theta_0}^n; p_\theta^n) \leq n\epsilon_n^2, \quad V(p_{\theta_0}^n; p_\theta^n) \leq n\epsilon_n^2\} \geq e^{-n\epsilon_n^2}. \tag{2.15}$$

Admittedly, the statement looks complicated, but substantial simplification is possible in the important special cases of independent, non-identically distributed (i.n.i.d.) variables and Markov processes. For i.n.i.d. variables, the Kullback–Leibler divergence measures can be replaced by the sum of individual divergence measures and the testing condition holds automatically if d_n is the root average squared Hellinger distance $d_n^2(\theta, \theta_0) = n^{-1} \sum_{i=1}^n d^2(p_{i,\theta}, p_{i,\theta_0})$, so entropies need to be evaluated when the distance is measured by d_n. For Markov processes, the root average squared Hellinger distance is given by $\int d^2(p_{\theta_1}(\cdot|x), p_{\theta_2}(\cdot|x)) \mathrm{d}\nu(x)$, where $p_\theta(\cdot|x)$ stands for the transition density and ν is a probability measure. Thus the square-root average squared Hellinger distance emerges as the canonical metric for i.n.i.d. or Markov models, playing the role of the Hellinger distance for i.i.d. observations. The non-i.i.d. extension is extremely useful in estimations involving nonparametric regression with fixed covariates, estimation of the spectral density of a time series using the Whittle likelihood and so on. Explicit convergence rates were obtained in Ghosal and van der Vaart (2007a) for these models and various choices of prior distributions. For instance, for a nonparametric normal regression model with nonrandom covariates, the rate $n^{-\alpha/(2\alpha+1)}$ is obtained using a suitable random B-spline series prior when the true regression function is α-smooth.

2.6 Adaptation and model selection

2.6.1 Motivation and description

Given the level of smoothness, we have seen in the last section that the minimax rate of convergence may be achieved using a suitable prior distribution. However, it is important to know the level of smoothness to construct the appropriate prior. For instance, in constructing a prior for α-smooth densities using an exponential family based on splines, the number of elements in the B-spline basis was chosen depending on the value of α. In particular, this implies that different priors are needed for different smoothness levels. In practice, the smoothness level of the target class is rarely known, so a prior appropriate for a hypothesized class will give suboptimal rate at a true density if it is actually smoother or coarser than the wrongly targeted class. Therefore the question arises whether we can actually achieve the optimal rate by using a prior constructed without using the actual knowledge of smoothness. If such a prior exists, then the posterior is called *rate adaptive* or simply *adaptive* in short.

In classical statistics, estimators with optimal mean squared error have been constructed in the adaptive framework, that is, without knowing the correct smoothness level. The property can be considered as an *oracle property*, in the sense that the lack of knowledge of the smoothness does not diminish the performance of the estimator compared to the oracle, which uses the extra information on smoothness. Although in classical statistics, even the constant appearing in the limiting distribution may be matched with that of the oracle estimator, such a goal will not be achieved in the Bayesian framework since the best constants in our nonadaptive posterior convergence theorems are also unknown.

From the Bayesian point of view, it is natural to treat the unknown level of smoothness α as a parameter and put a prior distribution on it. Given the value of α, a prior for the class of densities may be obtained as before, using the spline based exponential family. Thus the resulting two-level hierarchical prior is actually a mixture of the spline series prior constructed before. Then the natural question is whether the resulting posterior distribution will converge at the right rate for all smoothness levels, or at least for a rich class of smoothness levels.

It can be seen that there is a strong connection between adaptation and the posterior probability attached to various smoothness levels. In the hierarchy of the models indexed by a smoothness parameter, classes of densities are nested, so a coarser level of smoothness spans a bigger class, increasing the size of the support of the prior. This corruption of support leads to larger entropy estimates slowing down the rate, unless the additional portion can be handled separately. Thus adaptation requires showing that the posterior probability of obtaining a coarser model converges to zero. The treatment of smoother models is somewhat different. Assuming that the correct smoothness level is chosen by the prior with positive probability, the contribution of the smoother priors in the mixture can be ignored for the purpose of lower bounding the prior concentration, at the expense of incorporating an irrelevant constant, which is eventually ignored for the purpose of calculating the rate. Clearly, smoother priors do not contribute to increasing the size of the support of the mixture prior. Thus, if the posterior asymptotically ignores models coarser than the true one, adaptation is expected to take place.

2.6.2 *Infinite-dimensional normal models*

The natural strategy for adaptation works in several models. We begin with one of the simplest cases, the infinite-dimensional normal model: $X_i \overset{\text{ind}}{\sim} \text{Nor}(\theta_i, n^{-1})$, $\boldsymbol{\theta} = (\theta_1, \theta_2, \ldots) \in \ell_2$, considered in Belitser and Ghosal (2003). The "smoothness" of $\boldsymbol{\theta}$ is measured by the behavior of the tail of the sequence $\theta_1, \theta_2, \ldots$, and is defined as the largest q for which $\sum_{i=1}^{\infty} i^{2q}\theta_i^2 < \infty$. It is well kown that the minimax rate of estimation of $\boldsymbol{\theta}$ on the subspace $\sum_{i=1}^{\infty} i^{2q}\theta_i^2 < \infty$ is $n^{-q/(2q+1)}$. The minimax rate is attained by the Bayes estimator with respect to the prior $\theta_i \sim \text{Nor}(0, i^{-(2q+1)})$, which we denote by Π_q. This assertion follows by direct calculations of posterior mean and variance and bounding posterior probabilities of deviations by Chebyshev's inequality. However, the prior is clearly dependent on the unknown smoothness level q. We consider only countably many possible values of q and suppose that they do not accumulate at any point other than 0 or ∞. Hence we can arrange the possible values of q in an increasing double sequence $\ldots, q_{-1}, q_0, q_1, \ldots$, where q_0 is the true value of smoothness. We attach positive weights λ_q to each possible value of q and mix to form the hierarchical prior $\sum_q \lambda_q \Pi_q$. By using explicit properties of normal likelihood, an upper bound for the posterior probability of each q can be obtained, leading to exponential-type decay $\Pi(q < q_0|X_1, X_2, \ldots) \le \mathrm{e}^{-cn^{\delta}}$ with $\delta > 0$. Here, the fact that the q values are separated by a positive distance plays a crucial role. Thus the role of $q < q_0$ is asymptotically negligible. To treat the case $q > q_0$, again due to positive separation, it is easily shown by Chebyshev's inequality that the contribution of Π_q,

$q > q_0$, to the posterior probability of $C^c := \{\sum_{i=1}^{\infty} i^{2q_0}\theta_i^2 > B\}$ is negligible for large B. Thus what matters eventually is the contribution of Π_{q_0}, for which the correct rate $n^{-q_0/(2q_0+1)}$ is already in force, and that of Π_q, $q > q_0$, restricted to C. The entropy of C grows as ϵ^{-1/q_0}, while the rate of concentration of the mixture prior, due to the presence of the component Π_{q_0} in the convex combination, is at least $\lambda_{q_0} \exp(-c\epsilon^{-1/q_0})$. The last assertion is a consequence of tail behavior of random $\boldsymbol{\theta}$ from Π_{q_0} and for large N, a lower bound for the probability of a ball $\{\sum_{i=1}^{N}(\theta_i - \theta_{i0})^2 \leq \delta\}$ of the form e^{-cN}. Therefore the natural Bayesian strategy for adaptation works well in the infinite-dimensional normal model.

2.6.3 General theory of Bayesian adaptation

For countably many competing abstract models indexed by α, say, a general result was obtained in Ghosal, Lember and van der Vaart (2008) based on some earlier results by Ghosal, Lember and van der Vaart (2003), Huang (2004) and Lember and van der Vaart (2007). We allow the index set A for α and the prior λ_α to depend on n, but we do not make n explicit in the notation. We work with the Hellinger distance for definiteness. Let $\epsilon_{n,\alpha}$ be the usual optimal rate in model α and β stand for the "true value" of α in the sense that the βth model is the best approximating model for the true density p_0. Let $B_\beta(\epsilon)$ be the Kullback–Leibler neighborhood in the true model. For simplicity, we do not include sieves in the condition by assuming that the required entropy condition holds in the original model, but the result will be easily modified to the general case assuming that sieves have exponentially small prior probability $\mathrm{e}^{-cn\epsilon_{n,\beta}^2}$. Essentially four basic conditions ensure adaptation.

(i) For each model α, the local entropy condition $\log N(\epsilon/3, C_\alpha(\epsilon), d) \leq E_\alpha \epsilon_{n,\alpha}^2$ for all $\epsilon > \epsilon_{n,\alpha}$ holds, where $C_\alpha(\epsilon) = \{p : d(p, p_0) \leq 2\epsilon\}$ and E_α is a constant free of n.

(ii) For coarser models $\alpha < \beta$ and any positive integer j, the net prior probabilities of $C_\alpha(j\epsilon_{n,\alpha})$ are comparable to that of $B_\beta(\epsilon_{n,\beta})$ by the condition

$$\frac{\lambda_\alpha \Pi_\alpha(C_\alpha(j\epsilon_{n,\alpha}))}{\lambda_\beta \Pi_\beta(B_\beta(\epsilon_{n,\beta}))} \leq \mu_\alpha \mathrm{e}^{Lj^2 n\epsilon_{n,\alpha}^2}.$$

(iii) For finer models $\alpha \geq \beta$, the net prior probabilities of $C_\alpha(j\epsilon_{n,\beta})$ are comparable to that of $B_\beta(\epsilon_{n,\beta})$ by the condition

$$\frac{\lambda_\alpha \Pi_\alpha(C_\alpha(j\epsilon_{n,\beta}))}{\lambda_\beta \Pi_\beta(B_\beta(\epsilon_{n,\beta}))} \leq \mu_\alpha \mathrm{e}^{Lj^2 n\epsilon_{n,\beta}^2}.$$

(iv) For a sufficiently large B, the total prior mass of a $B\epsilon_{n,\alpha}$-ball in coarser models compared to the concentration rate in the true model is significantly small in that

$$\sum_{\alpha<\beta} \frac{\lambda_\alpha \Pi_\alpha(C_\alpha(B\epsilon_{n,\alpha}))}{\lambda_\beta \Pi_\beta(B_\beta(\epsilon_{n,\beta}))} = o(\mathrm{e}^{-2n\epsilon_{n,\beta}^2});$$

here the constants μ_α satisfy $\sum \sqrt{\mu_\alpha} \leq \mathrm{e}^{n\epsilon_{n,\beta}^2}$.

In the above, we restricted attention to nonparametric models. If parametric models with $n^{-1/2}$-convergence rates are involved, then slight modifications of the conditions are necessary, as argued in Section 2.5.3, to avoid an undesirable logarithmic factor. In the adaptive context, the prior concentration level in ϵ-balls then should be given by a power of ϵ, say ϵ^D and the right-hand side of condition (iv) should be replaced by $o(n^{-3D})$.

It may be seen that in order to satisfy the required conditions, one may control the weight sequence λ_α suitably, in particular, depending on n. The choice $\lambda_\alpha \propto \mu_\alpha \mathrm{e}^{-cn\epsilon_{n,\alpha}^2}$ is sometimes fruitful.

2.6.4 Density estimation using splines

The above general result applies directly in the context of density estimation in Hölder α-classes with log-spline priors as discussed in the last section. Interestingly, a simple hierarchical prior obtained by using a single prior on α leads to the optimal rate $n^{-\alpha/(2\alpha+1)}$ only up to a logarithmic factor, as in Ghosal, Lember and van der Vaart (2003). The logarithmic factor was removed by using very special weights in Huang (2004), who also treated a similar problem for nonparametric regression using a wavelet basis. This result requires restricting to finitely many competing smoothness classes. For such a case, the same result can be obtained from the general theory of Ghosal, Lember and van der Vaart (2008) by using a sequence of weights $\lambda_\alpha \propto \prod_{\gamma<\alpha}(C\epsilon_{n,\gamma})^{J_{n,\gamma}}$ where, as before, $J_{n,\alpha} \sim n^{1/(1+2\alpha)}$ is the dimension of the optimal spline approximation model. Another possibility is to consider a discrete uniform prior on $\epsilon_{n,\alpha}$-nets in Hölder classes with $\lambda_\alpha \propto \mu_\alpha \mathrm{e}^{-cn\epsilon_{n,\alpha}^2}$.

As discussed in Section 2.6.1 and exhibited in the infinite normal model, it is to be expected that the posterior probability of models coarser than the true one combined together should be negligible. On the other hand, the posterior probability of the complement of a large multiple of $\epsilon_{n,\beta}$-balls in smoother models is small. In particular, if the true density lies outside the closure of these models, then the posterior probability of smoother models also converges to zero.

2.6.5 Bayes factor consistency

The results become much more transparent when we consider just a pair of competing models. Then adaptation is equivalent to the *consistency of the Bayes factor* of the smoother model relative to the coarser model – the Bayes factor goes to infinity if the true density belongs to the smoother model while the Bayes factor goes to zero otherwise. Since we allow the true density to lie outside the models, we interpret Bayes factor consistency in the following generalized sense: the Bayes factor converges to zero if the true density stays away from the smoother models by an amount larger than its convergence rate, while the Bayes factor converges to infinity if the true density is within the convergence rate of the smoother model.

The asymptotic behavior of Bayes factors has been studied by many authors in the parametric case, but only a few results are available in the nonparametric case. When the smaller model is a singleton and the prior in the larger model satisfies the Kullback–Leibler property, then Dass and Lee (2004) showed Bayes factor consistency. Their proof uses Doob's consistency theorem and is heavily dependent on the assumption that there is a positive prior mass at the true density. Hence it is extremely difficult to generalize this result to composite null hypotheses. Another result was obtained by Walker, Damien and Lenk (2004), who showed that if the Kullback–Leibler property holds in one model and does not hold in another model, then the Bayes factor shows the same kind of dichotomy. This result helps only if the two models separate well, but such a situation is rare. What is commonly observed is nested families with differing convergence rates. In such a case, consistency of Bayes factors can be obtained from the calculation done in our result on adaptation. We can state the result roughly as follows.

Let $\epsilon_{n,1} > \epsilon_{n,2}$ be the two possible rates in the respective competing models and suppose, for simplicity, the models are given equal weight. Assume that the prior of the $\epsilon_{n,j}$-size Kullback–Leibler neighborhood around the true density in model j is at least $e^{-n\epsilon_{n,j}^2}$, $j = 1, 2$, and further the prior probability of a large multiple of an $\epsilon_{n,1}$-size Hellinger ball in model 1 is at most $O(e^{-3n\epsilon_{n,2}^2})$. Then the Bayes factor is consistent.

The result applies readily to some goodness-of-fit tests for parametric models against nonparametric alternatives; see Ghosal, Lember and van der Vaart (2008) for details.

2.7 Bernshteĭn–von Mises theorems

2.7.1 Parametric Bernshteĭn–von Mises theorems

The convergence rate of a posterior distribution asserts concentration at the true value of the parameter at a certain speed, but does not tell us the asymptotic shape of the posterior distribution. For smooth parametric families, a remarkable result, popularly known as the *Bernshteĭn–von Mises theorem*, says that the posterior distribution asymptotically looks like a normal distribution centered at the MLE with variance equal to the inverse of the Fisher information; see van der Vaart (1998) for example. As a consequence, the posterior distribution of $\sqrt{n}(\theta - \hat{\theta}_n)$ conditioned on the sample, where θ is the (random) parameter and $\hat{\theta}_n$ is the MLE of θ, approximately coincides with its frequentist distribution under the parameter value θ, where $\hat{\theta}_n$ carries the randomness. This is a remarkable, and very mysterious result, since the interpretations of randomness in the two situations are quite different and the two quantities involved in the process are obtained from very different principles. From an application point of view, the importance of the Bernshteĭn–von Mises theorem lies in its ability to construct approximately valid confidence sets using Bayesian methods. This is very useful especially in complex problems since the sampling distribution is often hard to compute while the samples from the posterior distribution can be obtained relatively easily using various computational devices. The Bernshteĭn–von Mises phenomenon extends beyond smooth families; see Ghosal, Ghosh and Samanta (1995) for a precise description and a necessary and sufficient condition leading to the asymptotic matching in the first order. The prior, which needs only to have positive and continuous density at the true value of the parameter, plays a relatively minor role in the whole development.

2.7.2 Nonparametric Bernshteĭn–von Mises theorems

For infinite-dimensional cases, only very few results are available so far. Here, the validity of the phenomenon depends not only on the model, but also heavily on the prior. For estimating a c.d.f. F with a Dirichlet process prior, Lo (1983) showed that the posterior of $\sqrt{n}(F - \mathbb{F}_n)$ converges weakly to the F_0-Brownian bridge process a.s. under F_0, where $\mathbb{F}_n$ stands for the empirical c.d.f. On the other hand, the well known Donsker theorem tells us that $\sqrt{n}(\mathbb{F}_n - F_0)$ converges weakly to the F_0-*Brownian bridge*. Thus by the symmetry of the Brownian bridge, again we observe that the two limiting distributions coincide, leading to Bayesian matching. The result is extended in the multidimensional case in Lo (1986). Both results use the explicit structure of the Dirichlet process posterior and hence are difficult to extend to other priors. Recently, the result was substantially generalized to a wide class of Lévy process priors for cumulative hazard functions by Kim and Lee (2004) using fine properties of Poisson random measures and corresponding conjugacy results. The result concludes that the posterior distribution of $\sqrt{n}(H - \mathbb{H}_n)$, where H is the cumulative hazard function and $\mathbb{H}_n$ is the Nelson–Aalen estimator, converges weakly to a Brownian motion process a.s. under the true cumulative hazard H_0. From classical survival analysis and the martingale central limit theorem, it is known that the same Brownian motion process appears as the limit of $\sqrt{n}(\mathbb{H}_n - H_0)$ when H_0 is the true cumulative hazard function.

2.7.3 Semiparametric Bernshteĭn–von Mises theorems

For semiparametric problems, often only the parametric part θ is of interest. Thus the marginal posterior distribution of θ is especially of interest. One expects that, in spite of the possibly slower convergence rate of the posterior distribution of the nonparametric part η, the convergence rate for θ is still $n^{-1/2}$. Indeed, one can hope that the Bernshteĭn–von Mises phenomenon holds in the sense that the marginal posterior distribution of $\sqrt{n}(\theta - \hat{\theta}_n)$ is asymptotically normal and asymptotically coincides with the frequentist distribution of $\sqrt{n}(\hat{\theta}_n - \theta_0)$ a.s. under the distribution generated by (θ_0, η_0) for most pairs of true values (θ_0, η_0) of (θ, η). A major barrier in deriving such a result is the fact that even the marginal posterior of θ is obtained through the posterior of the whole parameter (θ, η), making it difficult to study unless explicit expressions are available. A semiparametric Bernshteĭn–von Mises theorem was obtained in Shen (2002), but it seems that the conditions are somewhat difficult to verify. More transparent results with mild and verifiable conditions are highly desirable. For the specific case of the Cox proportional hazard model, a semiparametric Bernshteĭn–von Mises theorem was obtained by Kim (2006).

2.7.4 Nonexistence of Bernshteĭn–von Mises theorems

All the Bernshteĭn von Mises results obtained thus far in infinite-dimensional models appear to be associated with convergence rate $n^{-1/2}$. In principle, there is no reason why a Bernshteĭn–von Mises theorem should be restricted to the $n^{-1/2}$-convergence domain. However, certain negative results lead to the suspicion that the Bernshteĭn–von Mises theorem may not hold for slower convergence. The first result of this kind was obtained in Cox (1993), where the problem of estimating the signal in a white noise model was considered, and the signal was assumed to lie in a *Sobolev space*. The prior was obtained by expanding the function in a suitable series and putting independent normal priors on the coefficients. It was observed that the coverage probability of credible intervals constructed from the posterior distribution can converge to an arbitrarily small number with increasing sample size. Thus not only does the Bernshteĭn–von Mises theorem fail, but it fails in the worst possible way. The main problem seems to be the optimal trade-off used in smoothing, making the order of the bias the same as that of the standard deviation. For the infinite-dimensional normal model discussed in Section 2.6.2, it was also shown in Freedman (1999) that the frequentist and the Bayesian distributions of the L_2-norm of the difference of the Bayes estimator and the parameter differ by an amount equal to the scale of interest. This again leads to the conclusion that the frequentist coverage probability of a Bayesian credible set for the parameter can be infinitesimally small. This is disappointing since a confidence set must now be found using only frequentist methds and hence is harder to obtain. Interestingly, if a (sequence of) functional depends only on the first p_n coordinates where $p_n/\sqrt{n} \to 0$, then the Bernshteĭn–von Mises theorem holds for that functional. Indeed, the posterior distribution of the entire p_n-vector consisting of the first p_n-coordinates centered by the MLE is approximated by the p_n-dimensional normal distribution which approximates the distribution of the MLE; see Ghosal (2000) for details.

2.8 Concluding remarks

The nonparametric Bayesian approach to solving problems is rapidly gaining popularity among practitioners as theoretical properties become increasingly better understood and computational hurdles are being removed. Nowadays, many new Bayesian nonparametric methods for complex models arising in biomedical, geostatistical, environmental, econometric and many other applications are being proposed. The purpose of this chapter is to give a brief outline of the fundamental basis of prior construction and large sample behavior of the posterior distribution. Naturally, it has not been possible to cite every relevant paper here.

In this chapter, we have reviewed the properties of the Dirichlet process, other priors constructed using the Dirichlet process and asymptotic properties of the posterior distribution for nonparametric and semiparametric problems. We discussed a naive construction of the Dirichlet process, indicated measure theoretic difficulties associated with the approach and subsequently rectified the problem by working with a suitable countable generator. We also discussed a method of construction using a gamma process. We then discussed basic properties of the Dirichlet process such as expressions of prior mean and variance, interpretation of the parameters and posterior conjugacy with implications such as explicit expressions for posterior expectation and variance, and some peculiar consequences of the presence of point mass in the base measure for the posterior Dirichlet process. We further discussed the discreteness property, characterization of support, self-similarity, convergence, tail behavior and mutual singularity of the Dirichlet process. The joint distribution of samples drawn from a random probability measure obtained from a Dirichlet process was described by the Blackwell–MacQueen generalized Pólya urn scheme and its relation to MCMC sampling, clustering and distribution of distinct observations was discussed. We described Sethuraman's stick-breaking representation of a Dirichlet process and its potential role in constructing new processes and computation involving Dirichlet processes.

The Dirichlet process leads to various new processes through some common operations. We discussed the role of a mixture of Dirichlet processes in eliciting a prior distribution. We then thoroughly discussed the importance of kernel smoothing applied to a Dirichlet process leading to Dirichlet mixtures. Computational techniques through MCMC sampling using the generalized Pólya urn structure were briefly outlined. We also mentioned the role of symmetrization and conditioning operations especially in semiparametric applications leading to respectively the symmetrized Dirichlet and conditioned Dirichlet processes.

We discussed consistency and its implications thoroughly. We started with a very general theorem by Doob which concludes consistency at almost all parameter values with respect to the prior measure. However, null sets could be huge and we discussed some prominent examples of inconsistency. In fact, we mentioned that this inconsistency behavior is more common than consistency in a topological sense if the prior distribution is completely arbitrarily chosen. We then discussed the tail-freeness property and its role in providing consistency. In particular, consistency follows for Pólya tree priors. We then presented the general formulation of Schwartz's theory of consistency and the role of the condition of prior positivity in Kullback–Leibler neighborhoods. We argued that a size condition on the model described by the existence of uniformly exponentially consistent tests plays a key role, which further reduces to entropy conditions for commonly used metrics. This is because a desired test can be constructed by covering the space with small balls guided by the bounds for metric entropy, finding appropriate tests against these balls and combining these basic tests. The role of sieves in compactifying a noncompact space is shown to be extremely important. The general result on consistency was applied to density estimation using Dirichlet mixtures or Gaussian process priors leading to very explicit conditions for consistency for these commonly used priors. We also argued that Schwartz's theory is an appropriate tool for studying posterior consistency in semiparametric problems. Further we indicated how Schwartz's theory can be extended to the case of independent nonidentically distributed and some dependent observations, with applications to estimating binary regression, spectral density and transition density using commonly used priors.

We next studied convergence rates of posterior distribution. We discussed the theory developed in Ghosal, Ghosh and van der Vaart (2000) refining Schwartz's theory for consistency. It was again seen that the concentration of priors in Kullback–Leibler type neighborhoods and the growth rate of entropy functions determine the rate. We explicitly constructed priors using bracketing approximations or exponential families based on splines to achieve optimal rates. We further discussed the convergence rate of Dirichlet mixture priors and Gaussian priors under various setups. Extensions to misspecified or non-i.i.d. models were briefly indicated with applications.

The issue of adaptation was considered next. It was argued that mixing "optimal priors" by putting a prior distribution on the indexing parameter is a prudent strategy. We described conditions under which this natural Bayesian strategy works, and in particular showed that the adaptation takes place in an infinite-dimensional normal model and class of log-spline densities. We also discussed the connection between adaptation and model selection. Under the same conditions, we argued that the posterior probability of the true model increases to one leading to the consistency of the Bayes factor when only two models are present. Finally, we discussed the importance of the Bernshteĭn–von Mises theorem and mentioned some instances where the theorem holds true, and where it fails.

Although a lot of development has taken place in the last ten years, we still need to know much more about finer properties of the posterior distribution in various applications. Asymptotics can play a key role in separating a desirable approach from an undesirable one. It will be of great interest to compose a catalog of priors appropriate for applications having desirable posterior convergence properties.

References

Amewou-Atisso, M., Ghosal, S., Ghosh, J. K. and Ramamoorthi, R. V. (2003). Posterior consistency for semiparametric regression problems. *Bernoulli*, **9**, 291–312.

Antoniak, C. (1974). Mixtures of Dirichlet processes with application to Bayesian nonparametric problems. *Annals of Statistics*, **2**, 1152–74.

Barron, A. R. (1999). Information-theoretic characterization of Bayes performance and the choice of priors in parametric and nonparametric problems. In *Bayesian Statistics 6*, ed. J. M. Bernardo et al., 27–52. Oxford: Oxford University Press.

Barron, A. R., Schervish, M. and Wasserman, L. (1999). The consistency of posterior distributions in nonparametric problems. *Annals of Statistics*, **27**, 536–61.

Belitser, E. N. and Ghosal, S. (2003). Adaptive Bayesian inference on the mean of an infinite-dimensional normal distribution. *Annals of Statistics*, **31**, 536–59.

Blackwell, D. and MacQueen, J. B. (1973). Ferguson distributions via Pólya urn schemes. *Annals of Statistics*, **1**, 353–55.

Choi, T. and Schervish, M. (2007). Posterior consistency in nonparametric Bayesian problems using Gaussian process prior. *Journal of Multivariate Analysis*, **98**, 1969–87.

Choudhuri, N., Ghosal, S. and Roy, A. (2004). Bayesian estimation of the spectral density of a time series. *Journal of the American Statistical Association*, **99**, 1050–59.

Coram, M. and Lalley, S. P. (2006). Consistency of Bayes estimators of a binary regression. *Annals of Statistics*, **34**, 1233–69.

Cox, D. D. (1993). An analysis of Bayesian inference for nonparametric regression. *Annals of Statistics*, **21**, 903–23.

Dalal, S. R. (1979). Dirichlet invariant processes and applications to nonparametric estimation of symmetric distribution functions. *Stochastic Processes and Their Applications*, **9**, 99–107.

Dass, S. C. and Lee, J. (2004). A note on the consistency of Bayes factors for testing point null versus nonparametric alternatives. *Journal of Statistical Planning and Inference*, **119**, 143–52.

Diaconis, P. and Freedman, D. (1986). On the consistency of Bayes estimates (with discussion). *Annals of Statistics*, **14**, 1–67.

Diaconis, P. and Freedman, D. (1993). Nonparametric binary regression: a Bayesian approach. *Annals of Statistics*, **21**, 2108–37.

Doob, J. L. (1948). Application of the theory of martingales. In *Le Calcul des Probabilités et ses Applications*, Colloques Internationaux du Centre National de la Recherche Scientifique, 13, 23–37. Paris: CNRS.

Doss, H. (1985a). Bayesian nonparametric estimation of the median. I: Computation of the estimates. *Annals of Statistics*, **13**, 1432–44.

Doss, H. (1985b). Bayesian nonparametric estimation of the median. II: Asymptotic properties of the estimates. *Annals of Statistics*, **13**, 1445–64.

Doss, H. and Sellke, T. (1982). The tails of probabilities chosen from a Dirichlet process. *Annals of Statistics*, **10**, 1302–05.

Escobar, M. and West, M. (1995). Bayesian density estimation and inference using mixtures. *Journal of the American Statistical Association*, **90**, 577–88.

Ferguson, T. S. (1973). A Bayesian analysis of some nonparametric problems. *Annals of Statistics*, **1**, 209–30.

Ferguson, T. S. (1983). Bayesian density estimation by mixtures of normal distributions. In *Recent Advances in Statistics*, ed. M. Rizvi, J. Rustagi and D. Siegmund, 287–302. New York: Academic Press.

Freedman, D. (1963). On the asymptotic distribution of Bayes estimates in the discrete case. I. *Annals of Mathematical Statistics*, **34**, 1386–403.

Freedman, D. (1999). On the Bernstein–von Mises theorem with infinite-dimensional parameters. *Annals of Statistics*, **27**, 1119–40.

Ghosal, S. (2000). Asymptotic normality of posterior distributions for exponential families with many parameters. *Journal of Multivariate Analysis*, **74**, 49–69.

Ghosal, S., Ghosh, J. K. and Ramamoorthi, R. V. (1999a). Posterior consistency of Dirichlet mixtures in density estimation. *Annals of Statistics*, **27**, 143–58.

Ghosal, S., Ghosh, J. K. and Ramamoorthi, R. V. (1999b). Consistent semiparametric Bayesian inference about a location parameter. *Journal of Statistical Planning and Inference*, **77**, 181–93.

Ghosal, S., Ghosh, J. K. and Samanta, T. (1995). On convergence of posterior distributions. *Annals of Statistics*, **23**, 2145–52.

Ghosal, S., Ghosh, J. K. and van der Vaart, A. W. (2000). Convergence rates of posterior distributions. *Annals of Statistics*, **28**, 500–31.

Ghosal, S., Lember, J. and van der Vaart, A. W. (2003). On Bayesian adaptation. *Acta Applicanadae Mathematica*, **79**, 165–75.

Ghosal, S., Lember, J. and van der Vaart, A. W. (2008). Nonparametric Bayesian model selection and averaging. *Electronic Journal of Statistics*, **2**, 63–89.

Ghosal, S. and Roy, A. (2006). Posterior consistency of Gaussian process prior for nonparametric binary regression. *Annals of Statistics*, **34**, 2413–29.

Ghosal, S. and Tang, Y. (2006). Bayesian consistency for Markov processes. *Sankhyā*, **68**, 227–39.

Ghosal, S. and van der Vaart, A. W. (2001). Entropies and rates of convergence for maximum likelihood and Bayes estimation for mixture of normal densities. *Annals of Statistics*, **29**, 1233–63.

Ghosal, S. and van der Vaart, A. W. (2007a). Convergence of posterior distributions for non iid observations. *Annals of Statistics*, **35**, 192–223.

Ghosal, S. and van der Vaart, A. W. (2007b). Posterior convergence of Dirichlet mixtures at smooth densities. *Annals of Statistics*, **29**, 697–723.

Ghosal, S. and van der Vaart, A. W. (2010). *Fundamentals of Nonparametric Bayesian Inference.* Cambridge: Cambridge University Press, to appear.

Ghosh, J. K. and Ramamoorthi, R. V. (2003). *Bayesian Nonparamterics.* New York: Springer-Verlag.

Hjort, N. L. and Ongaro, A. (2005). Exact inference for Dirichlet means. *Statistical Inference for Stochastic Processes*, **8**, 227–54.

Huang, T. Z. (2004). Convergence rates for posterior distribution and adaptive estimation. *Annals of Statistics*, **32**, 1556–93.

Kim, Y. (2006). The Bernstein–von Mises theorem for the proportional hazard model. *Annals of Statistics*, **34**, 1678–700.

Kim, Y. and Lee, J. (2001). On posterior consistency of survival models. *Annals of Statistics*, **29**, 666–86.

Kim, Y. and Lee, J. (2004). A Bernstein–von Mises theorem in the nonparametric right-censoring model. *Annals of Statistics*, **32**, 1492–512.

Kleijn, B. and van der Vaart, A. W. (2006). Misspecification in infinite-dimensional Bayesian statistics. *Annals of Statistics*, **34**, 837–77.

Korwar, R. and Hollander, M. (1973). Contributions to the theory of Dirichlet processes. *Annals of Statistics*, **1**, 706–11.

Kraft, C. H. (1964). A class of distribution function processes which have derivatives. *Journal of Applied Probability*, **1**, 385–88.

Lavine, M. (1992). Some aspects of Polya tree distributions for statistical modeling. *Annals of Statistics*, **20**, 1222–35.

Lember, J. and van der Vaart, A. W. (2007). On universal Bayesian adaptation. *Statistics and Decisions*, **25**, 127–52.

Lijoi, A., Prünster, I. and Walker, S. G. (2005). On consistency of nonparametric normal mixtures for Bayesian density estimation. *Journal of the American Statistical Association*, **100**, 1292–96.

Lo, A. Y. (1983). Weak convergence for Dirichlet processes. *Sankhyā, Series A*, **45**, 105–11.

Lo, A. Y. (1984). On a class of Bayesian nonparametric estimates. I: Density estimates. *Annals of Statistics*, **12**, 351–57.

Lo, A. Y. (1986). A remark on the limiting posterior distribution of the multiparameter Dirichlet process. *Sankhyā, Series A*, **48**, 247–49.

Regazzini, E., Guglielmi, A. and Di Nunno, G. (2002). Theory and numerical analysis for exact distributions of functionals of a Dirichlet process. *Annals of Statistics*, **30**, 1376–1411.

Schwartz, L. (1965). On Bayes procedures. *Probability Theory and Related Fields*, **4**, 10–26.

Sethuraman, J. (1994). A constructive definition of Dirichlet priors. *Statistica Sinica*, **4**, 639–50.

Shen, X. (2002). Asymptotic normality of semiparametric and nonparametric posterior distributions. *Journal of the American Statistical Association*, **97**, 222–35.

Shen, X. and Wasserman, L. (2001). Rates of convergence of posterior distributions. *Annals of Statistics*, **29**, 687–714.

Tang, Y. and Ghosal, S. (2007). Posterior consistency of Dirichlet mixtures for estimating a transition density. *Journal of Statistical Planning and Inference*, **137**, 1711–26.

Teh, Y. W., Jordan, M. I., Beal, M. and Blei, D. M. (2006). Hierarchical Dirichlet processes. *Journal of the American Statistical Association*, **101**, 1566–81.

Tokdar, S. T. (2006). Posterior consistency of Dirichlet location-scale mixture of normal density estimation and regression. *Sankhyā*, **67**, 90–110.

Tokdar, S. T. and Ghosh, J. K. (2007). Posterior consistency of Gaussian process priors in density estimation. *Journal of Statistical Planning and Inference*, **137**, 34–42.

van der Vaart, A. W. (1998). *Asymptotic Statistics*. Cambridge: Cambridge University Press.

van der Vaart, A. W. and van Zanten, H. (2007). Bayesian inference with rescaled Gaussian process prior. *Electronic Journal of Statistics*, **1**, 433–48.

van der Vaart, A. W. and van Zanten, H. (2008). Rates of contraction of posterior distributions based on Gaussian process priors. *Annals of Statistics*, **36**, 1435–63.

Walker, S. G. (2004). New approaches to Bayesian consistency. *Annals of Statistics*, **32**, 2028–43.

Walker, S. G., Damien, P. and Lenk, P. (2004). On priors with a Kullback–Leibler property. *Journal of the American Statistical Association*, **99**, 404–8.

Walker, S. G. and Hjort, N. L. (2001). On Bayesian consistency. *Journal of the Royal Statistical Society, Series B*, **63**, 811–21.

Walker, S. G., Lijoi, A. and Prünster, I. (2005). Data tracking and the understanding of Bayesian consistency. *Biometrika*, **92**, 765–78.

Walker, S. G., Lijoi, A. and Prünster, I. (2007). On rates of convergence of posterior distributions in infinite-dimensional models. *Annals of Statistics*, **35**, 738–46.

Wu, Y. and Ghosal, S. (2008). Kullback–Leibler property of kernel mixture priors in Bayesian density estimation. *Electronic Journal of Statistics*, **2**, 298–331.

that the sequence $X^{(\infty)}$ is exchangeable if and only if it is a mixture of sequences of independent and identically distributed (i.i.d.) random variables.

Theorem 3.1 (de Finetti, 1937) *The sequence $X^{(\infty)}$ is exchangeable if and only if there exists a probability measure Q on the space $\mathcal{P}_{\mathbb{X}}$ of all probability measures on $\mathbb{X}$ such that, for any $n \geq 1$ and $A = A_1 \times \cdots \times A_n \times \mathbb{X}^\infty$, one has*

$$\mathbb{P}\left[X^{(\infty)} \in A\right] = \int_{\mathcal{P}_{\mathbb{X}}} \prod_{i=1}^{n} p(A_i)\, Q(\mathrm{d}p)$$

where $A_i \in \mathscr{X}$ for any $i = 1, \ldots, n$ and $\mathbb{X}^\infty = \mathbb{X} \times \mathbb{X} \times \cdots$.

In the statement of the theorem, the space $\mathcal{P}_{\mathbb{X}}$ is equipped with the topology of weak convergence which makes it a complete and separable metric space. The probability Q is also termed the *de Finetti measure* of the sequence $X^{(\infty)}$. We will not linger on technical details on exchangeability and its connections with other dependence properties for sequences of observations. The interested reader can refer to the exhaustive and stimulating treatments of Aldous (1985) and Kallenberg (2005).

The exchangeability assumption is usually formulated in terms of conditional independence and identity in distribution, i.e.

$$\begin{aligned} X_i \mid \tilde{p} &\overset{\text{i.i.d.}}{\sim} \tilde{p} \qquad i \geq 1 \\ \tilde{p} &\sim Q. \end{aligned} \tag{3.1}$$

Hence, $\tilde{p}^n = \prod_{i=1}^n \tilde{p}$ represents the conditional p.d. of $(X_1, \ldots, X_n)$, given $\tilde{p}$. Here $\tilde{p}$ is some random probability measure defined on $(\Omega, \mathscr{F}, \mathbb{P})$ and taking values in $\mathcal{P}_{\mathbb{X}}$: its distribution Q takes on the interpretation of prior distribution for Bayesian inference. Whenever Q degenerates on a finite-dimensional subspace of $\mathcal{P}_{\mathbb{X}}$, the inferential problem is usually called *parametric*. On the other hand, when the support of Q is infinite dimensional then one typically speaks of a *nonparametric* inferential problem.

In the following sections we focus our attention on various families of priors Q: some of them are well known and occur in many applications of Bayesian nonparametric statistics whereas some others have recently appeared in the literature and witness the great vitality of this area of research. We will describe specific classes of priors which are tailored for different applied statistical problems: each of them generalizes the Dirichlet process in a different direction, thus obtaining more modeling flexibility with respect to some specific feature of the prior process. This last point can be appreciated when considering the predictive structure implied by the Dirichlet process, which actually overlooks some important features of the data. Indeed, it is well known that, in a model of the type (3.1), the family of predictive distributions induced by a Dirichlet process, with baseline measure α, is

$$\mathbb{P}\left[X_{n+1} \in A \mid X_1, \ldots, X_n\right]$$
$$= \frac{\alpha(\mathbb{X})}{\alpha(\mathbb{X})+n} P_0(A) + \frac{n}{\alpha(\mathbb{X})+n} \frac{1}{n} \sum_{j=1}^{k} n_j \delta_{X_j^*}(A) \qquad \forall A \in \mathscr{X}$$

where δ_x denotes a point mass at $x \in \mathbb{X}$, $P_0 = \alpha/\alpha(\mathbb{X})$ and the X_j^* with frequency n_j denote the $k \leq n$ distinct observations within the sample. The previous expression implies that X_{n+1} will be a new observation X_{k+1}^* with probability $\alpha(\mathbb{X})/[\alpha(\mathbb{X})+n]$, whereas it will coincide with any of the previous observations with probability $n/[\alpha(\mathbb{X}) + n]$. Since these probability masses depend neither on the number of clusters into which the data are grouped nor on their frequencies, an important piece of information for prediction is neglected. It is quite complicated to obtain a tractable generalization of the Dirichlet process incorporating dependence on both the number of clusters and the frequencies; however, dependence on the number of clusters is achievable and the two-parameter Poisson–Dirichlet process, with $\sigma \in (0, 1)$ and $\theta > -\sigma$, represents a remarkable example. Details will be provided later, but here we anticipate that the predictive distribution implies that X_{n+1} will be a new value X_{k+1}^* with probability $[\theta + \sigma k]/[\theta + n]$, whereas X_{n+1} will coincide with a previously recorded observation with probability $[n - \sigma k]/[\theta + n]$. Hence, the probability of obtaining new values is monotonically increasing in k and the value of σ can be used to tune the strength of the dependence on k.

The analysis of general classes of priors implies that, in most cases and in contrast to what happens for the Dirichlet process, one has to work with nonconjugate models. This should not be a big concern, since conjugacy corresponds to mere mathematical convenience: from a conceptual point of view, there is no justification for requiring conjugacy. On the contrary, one may argue that conjugacy constrains the posterior to having the same structure as the prior which, in a nonparametric setup, may represent a limitation to the desired flexibility. So it is definitely worth exploring the potential of random probability measures which do not have this feature and it will be seen that, even if conjugacy fails, one can find many alternatives to the Dirichlet process which preserve mathematical tractability. Since most of these general classes of priors are obtained as suitable transformations of completely random measures, in the next subsection we provide a concise digression on this topic.

3.1.2 A concise account of completely random measures

We start with the definition of completely random measure, a concept introduced in Kingman (1967). Denote, first, by $\mathcal{M}_{\mathbb{X}}$ the space of boundedly finite measures

A detailed treatment of this subject can be found in the superb book by Kingman (1993).

If $\tilde{\mu}$ is defined on $\mathbb{X} = \mathbb{R}$, one can also consider the càdlàg random distribution function induced by $\tilde{\mu}$, namely $\{\tilde{\mu}((-\infty, x]) : x \in \mathbb{R}\}$. Such a random function defines an *increasing additive process*, that is a process whose increments are nonnegative, independent and possibly not stationary. See Sato (1999) for an exhaustive account. To indicate such processes we will also use the term *independent increments processes*, whereas in the Bayesian literature they are more frequently referred to as Lévy processes: this terminology is not completely appropriate since, in probability theory, the notion of Lévy process is associated to processes with independent and stationary increments. We rely on CRMs in most of our exposition since they represent an elegant, yet simple, tool for defining nonparametric priors. Moreover, one can easily realize that posterior inferences are achieved by virtue of the simple structure featured by CRMs conditional on the data. Indeed, in most of the examples we will illustrate, a posteriori a CRM turns out to be the sum of two independent components: (i) a CRM with no fixed points of discontinuity and whose Lévy intensity is obtained by applying an updating rule to the prior Lévy intensity; (ii) a sum of random jumps. These jumps occur at: (a) the a priori fixed points of discontinuities with updated jump distribution; (b) the new fixed points of discontinuity corresponding to the observations with jump distribution determined by the Lévy intensity of the CRM. Given this common structure, the specific updating of the involved quantities depends on the specific transformation of the CRM that has been adopted for defining the prior.

Finally, note that, without loss of generality, one can a priori consider CRMs with no fixed points of discontinuity which implies $\tilde{\mu} = \tilde{\mu}_c$. In the sequel we adopt this assumption when specifying some of the nonparametric priors we deal with and it will be pointed out how fixed points of discontinuity arise when evaluating the posterior distribution, given a sample $X_1, \ldots, X_n$.

3.2 Models for survival analysis

Survival analysis has been the focus of many contributions to Bayesian nonparametric theory and practice. Indeed, many statistical problems arising in the framework of survival analysis require function estimation and, hence, they are ideally suited for a nonparametric treatment. Moreover, this represents an area where the interest in generalizations of the Dirichlet process has emerged with particular emphasis. The main reason for this is due to the particular nature of survival data which are governed by some censoring mechanism. The breakthrough in the treatment of these issues in a Bayesian nonparametric setup can be traced back to Doksum (1974) where the notion of neutral-to-the-right (NTR) random probability is introduced. The law of a NTR process can be used as a prior for the distribution function of survival times and the main advantage of Doksum's definition is that NTR priors are conjugate (in a sense to be made precise later), even when right-censored observations are present. While this enables one to model a random distribution function for the survival times, a different approach yields priors for cumulative hazards and hazard rates. This has been pursued in a number of papers such as Dykstra and Laud (1981), Lo and Weng (1989), Hjort (1990), Kim (1999), Nieto-Barajas and Walker (2004) and James (2005). All the proposals we are going to examine arise as suitable transformations of CRMs.

3.2.1 Neutral-to-the-right priors

A simple and useful approach for defining a prior on the space of distribution functions on $\mathbb{R}^+$ has been devised by Doksum (1974) who introduces the notion of neutral-to-the-right prior.

Definition 3.5 A random distribution function $\tilde{F}$ on $\mathbb{R}^+$ is neutral-to-the-right (NTR) if, for any $0 \leq t_1 < t_2 < \cdots < t_k < \infty$ and any $k \geq 1$, the random variables

$$\tilde{F}(t_1), \quad \frac{\tilde{F}(t_2) - \tilde{F}(t_1)}{1 - \tilde{F}(t_1)}, \quad \ldots \quad , \frac{\tilde{F}(t_k) - \tilde{F}(t_{k-1})}{1 - \tilde{F}(t_{k-1})}$$

are independent.

The concept of *neutrality* has been introduced in Connor and Mosimann (1969) and it designates a random vector $(\tilde{p}_1, \ldots, \tilde{p}_{k+1})$ of proportions with $\sum_{i=1}^{k+1} \tilde{p}_i = 1$ such that $\tilde{p}_1$ is independent from $\tilde{p}_2/(1 - \tilde{p}_1)$, $(\tilde{p}_1, \tilde{p}_2)$ is independent from $\tilde{p}_3/(1 - \tilde{p}_1 - \tilde{p}_2)$ and so on. This can be seen as a method for randomly splitting the unit interval and, as will be shown in Section 3.3.4, it is also exploited in order to define the so-called stick-breaking priors. In the definition above, one has $\tilde{p}_i = \tilde{F}(t_i) - \tilde{F}(t_{i-1})$ for any $i = 1, \ldots, k$, where $\tilde{F}(t_0) = 0$.

We recall the connection between NTR priors and CRMs on $\mathbb{R}^+$ which has been pointed out by Doksum (1974).

Theorem 3.6 (Doksum, 1974) *A random distribution function* $\tilde{F} = \{\tilde{F}(t) : t \geq 0\}$ *is NTR if and only if it has the same p.d. of the process* $\{1 - \mathrm{e}^{-\tilde{\mu}((0,t])} : t \geq 0\}$, *for some CRM* $\tilde{\mu}$ *on* $\mathbb{X} = \mathbb{R}^+$ *such that* $\mathbb{P}[\lim_{t\to\infty} \tilde{\mu}((0, t]) = \infty] = 1$.

By virtue of this result one can characterize both the prior and, as we shall see, the posterior distribution of $\tilde{F}$ in terms of the Lévy intensity ν associated to $\tilde{\mu}$. For instance, one can evaluate the prior guess at the shape of $\tilde{F}$ since

$$\mathbb{E}[\tilde{F}(t)] = 1 - \mathbb{E}\left[\mathrm{e}^{-\tilde{\mu}((0,t])}\right] = 1 - \mathrm{e}^{-\int_{(0,t]}\int_{\mathbb{R}^+}[1-\mathrm{e}^{-s}]\,\rho_x(\mathrm{d}s)\,\alpha(\mathrm{d}x)}.$$

Another feature which makes NTR priors attractive for applications is their conjugacy property.

Theorem 3.7 (Doksum, 1974) *If $\tilde{F}$ is* NTR($\tilde{\mu}$), *then the posterior distribution of $\tilde{F}$, given the data $X_1, \ldots, X_n$, is* NTR($\tilde{\mu}^*$) *where $\tilde{\mu}^*$ is a CRM with fixed points of discontinuity.*

In light of the previous result it is worth remarking that, in a Bayesian nonparametric setup, the term "conjugacy" is used with slightly different meanings. For this reason, we introduce here a distinction between *parametric conjugacy* and *structural conjugacy*. The former occurs when the p.d. of the posterior process is the same as the p.d. of the prior process with updated parameters: for instance, the posterior distribution of the Dirichlet process with parameter measure α, given uncensored data, is still a Dirichlet process with updated parameter measure $\alpha + \sum_{i=1}^n \delta_{X_i}$. The latter, namely structural conjugacy, identifies a model where the posterior process has the same structure of the prior process in the sense that they both belong to the same general class of random probability measures. Hence, Theorem 3.7 establishes that NTR priors are structurally conjugate: the posterior of a NTR($\tilde{\mu}$) process is still NTR. Note that structural conjugacy does not necessarily imply parametric conjugacy: the posterior CRM $\tilde{\mu}^*$ characterizing the NTR process is not necessarily of the same type as the prior. On the other hand, parametric conjugacy of a specific prior implies structural conjugacy.

An explicit description of the posterior CRM $\tilde{\mu}^*$ has been provided in Ferguson (1974). Denote by $\bar{\Lambda}(x) := \sum_{i=1}^n \delta_{X_i}([x, \infty))$ the number of individuals still alive right before x, i.e. the so-called *at risk process*. Moreover, $X_1^*, \ldots, X_k^*$ represent the k distinct observations among $X_1, \ldots, X_n$, with $1 \leq k \leq n$. As mentioned before, we suppose, for notational simplicity, that $\tilde{\mu}$ does not have a priori fixed points of discontinuity.

Theorem 3.8 (Ferguson, 1974) *If $\tilde{F}$ is* NTR($\tilde{\mu}$) *and $\tilde{\mu}$ has Lévy intensity* (3.4), *then*

$$\tilde{\mu}^* \stackrel{d}{=} \tilde{\mu}_c^* + \sum_{i=1}^k J_i \, \delta_{X_i^*} \tag{3.8}$$

where $\tilde{\mu}_c^*$ is independent from $J_1, \ldots, J_k$ and the J_i are mutually independent. Moreover, the Lévy intensity of the CRM $\tilde{\mu}_c^*$ is updated as

$$\nu^*(\mathrm{d}s, \mathrm{d}x) = \mathrm{e}^{-\bar{\Lambda}(x) s} \, \rho_x(\mathrm{d}s) \, \alpha(\mathrm{d}x).$$

One can also determine an expression for the p.d. of the jumps J_i at the distinct observations. To this end, consider the distinct observations in an increasing order $X_{(1)}^* < \cdots < X_{(k)}^*$. Moreover, let $n_i = \sum_{j=1}^n \delta_{X_j}(\{X_{(i)}^*\})$ be the frequency of the ith ordered observation in the sample: in terms of the at risk process one has

$n_i = \bar{\Lambda}(X^*_{(i)}) - \bar{\Lambda}(X^*_{(i+1)})$ for any $i = 1, \ldots, k$ with the proviso that $X^*_{(k+1)} = \infty$. The p.d. of J_i is given by

$$G_i(\mathrm{d}s) = \frac{(1 - \mathrm{e}^{-s})^{n_i} \, \mathrm{e}^{-s\,\bar{n}_{i+1}} \, \rho_{X^*_{(i)}}(\mathrm{d}s)}{\int_{\mathbb{R}^+} (1 - \mathrm{e}^{-v})^{n_i} \, \mathrm{e}^{-v\,\bar{n}_{i+1}} \, \rho_{X^*_{(i)}}(\mathrm{d}v)}$$

where, for the sake of simplicity, we have set $\bar{n}_i := \bar{\Lambda}(X^*_{(i)}) = \sum_{j=i}^k n_j$. If ν is homogeneous, then $\rho_{X^*_{(i)}} = \rho$ and the distribution of J_i does not depend on the location where the jump occurs.

The above posterior characterization does not take into account the possibility that the data are subject to a censoring mechanism according to which not all observations are exact. In particular, in survival analysis, in reliability and in other models for the time elapsing up to a terminal event, a typical situation is represented by right-censoring. For example, when studying the survival of a patient subject to a treatment in a hospital, the observation is right-censored if her/his survival time cannot be recorded after she/he leaves the hospital. Formally, right-censoring can be described as follows. Suppose $c_1, \ldots, c_n$ are n censoring times which can be either random or nonrandom. For ease of exposition we assume that the c_i are deterministic. To each survival time X_i associate $\Delta_i = \mathbb{1}_{(0,c_i]}(X_i)$ and set $T_i = \min\{X_i, c_i\}$. Clearly $\Delta_i = 1$ if X_i is observed exactly, and $\Delta_i = 0$ if X_i is right censored and the observed data are then given by $(T_1, \Delta_1), \ldots, (T_n, \Delta_n)$. Supposing there are $k \le n$ distinct observations among $\{T_1, \ldots, T_n\}$, we record them in an increasing order as $T^*_{(1)} < \cdots < T^*_{(k)}$. Correspondingly, define

$$n_i^c := \sum_{\{j:\,\Delta_j = 0\}} \delta_{T_j}(\{T^*_{(i)}\}) \quad \text{and} \quad n_i := \sum_{\{j:\,\Delta_j = 1\}} \delta_{T_j}(\{T^*_{(i)}\}) \tag{3.9}$$

as the number of right-censored and exact observations, respectively, occurring at $T^*_{(i)}$ for any $i = 1, \ldots, k$. Finally, set $\bar{n}_i^c = \sum_{j=i}^k n_j^c$ and $\bar{n}_i = \sum_{j=i}^k n_j$.

Theorem 3.9 (Ferguson and Phadia, 1979) *Suppose $\tilde{F}$ is* NTR$(\tilde{\mu})$ *where $\tilde{\mu}$ has no fixed jump points. Then the posterior distribution of $\tilde{F}$, given $(T_1, \Delta_1), \ldots, (T_n, \Delta_n)$, is* NTR$(\tilde{\mu}^*)$ *with*

$$\tilde{\mu}^* \stackrel{d}{=} \tilde{\mu}^*_c + \sum_{\{i:\,n_i \ge 1\}} J_i \, \delta_{T^*_{(i)}}. \tag{3.10}$$

Hence, the posterior distribution of $\tilde{F}$ preserves the same structure of the uncensored case and the jumps occur only at the exact observations, i.e. those distinct observations for which n_i is positive. In (3.10) $\tilde{\mu}^*_c$ is a CRM without fixed points of discontinuity and it is independent from the jumps J_i. Its Lévy measure coincides with

$$\nu^*(\mathrm{d}s, \mathrm{d}x) = \mathrm{e}^{-\bar{\Lambda}(x)s}\, \rho_x(\mathrm{d}s)\, \alpha(\mathrm{d}x)$$

where $\bar{\Lambda}(x) = \sum_{i=1}^{n} \delta_{T_i}([x, \infty))$ is the at risk process based on both exact and censored observations.

Moreover, the p.d. of the jump J_i occurring at each exact distinct observation, i.e. $T^*_{(i)}$ with $n_i \geq 1$, is given by

$$G_i(\mathrm{d}s) = \frac{(1 - \mathrm{e}^{-s})^{n_i}\, \mathrm{e}^{-(\bar{n}_{i+1} + \tilde{n}_i^c)s}\, \rho_{T^*_{(i)}}(\mathrm{d}s)}{\int_{\mathbb{R}^+} (1 - \mathrm{e}^{-v})^{n_i}\, \mathrm{e}^{-(\bar{n}_{i+1} + \tilde{n}_i^c)v}\, \rho_{T^*_{(i)}}(\mathrm{d}v)}.$$

Also in this case, if $\rho_{T^*_{(i)}} = \rho$ the distribution of J_i does not depend on the location at which the jump occurs. We close this subsection with a detailed description of two important examples of NTR priors.

Example 3.10 (The Dirichlet process) One might wonder whether the Dirichlet process defined by Ferguson (1973) is also a NTR prior. This amounts to asking oneself whether there exists a CRM $\tilde{\mu}$ such that the random distribution function $\tilde{F}$ defined by $\tilde{F}(t) \stackrel{d}{=} 1 - \mathrm{e}^{-\tilde{\mu}((0,t])}$ for any $t > 0$ is generated by a Dirichlet process prior with parameter measure α on $\mathbb{R}^+$. The answer to such a question is affirmative. Indeed, if $\tilde{\mu}$ is a CRM whose Lévy intensity is defined by

$$\nu(\mathrm{d}s, \mathrm{d}x) = \frac{\mathrm{e}^{-s\, \alpha((x,\infty))}}{1 - \mathrm{e}^{-s}}\, \alpha(\mathrm{d}x)\, \mathrm{d}s$$

then $\tilde{F} \stackrel{d}{=} 1 - \mathrm{e}^{-\tilde{\mu}}$ is a Dirichlet process with parameter measure α, see Ferguson (1974). One can, then, apply results from Ferguson and Phadia (1979) in order to characterize the posterior distribution of a Dirichlet random distribution function given right-censored data. It is to be mentioned that such an analysis was originally developed by Susarla and Van Ryzin (1976) without resorting to the notion of NTR prior. They show that the Dirichlet process features the property of parametric conjugacy if the observations are all exact, whereas it does not in the presence of right-censored data. Indeed, Blum and Susarla (1977) characterize the posterior distribution of a Dirichlet process given right-censored data as a mixture of Dirichlet processes in the sense of Antoniak (1974). In the present setting, a simple application of Theorem 3.9 allows us to recover the results in Susarla and Van Ryzin (1976). Moreover, Theorem 3.9 implies that the Dirichlet process, in the presence of right-censored observations, is structurally conjugate when seen as a member of the class of NTR priors. The posterior distribution of the Dirichlet random distribution function $\tilde{F}$ is NTR($\tilde{\mu}^*$) with $\tilde{\mu}^*$ as in (3.10). The Lévy intensity of $\tilde{\mu}_c^*$ coincides with

$$\nu^*(ds, dx) = \frac{e^{-\{\alpha((x,\infty))+\bar{\Lambda}(x)\}\, s}}{1 - e^{-s}}\, \alpha(dx)\, ds$$

and the distribution of the jump J_i at each exact distinct observation (i.e. $T^*_{(i)}$ with $n_i \geq 1$) coincides with the distribution of the random variable $-\log(B_i)$ where $B_i \sim \text{Beta}(\alpha((T^*_{(i)}, \infty)) + \bar{n}_{i+1} + \tilde{n}_i^c;\ n_i)$. Note that if the observations are all exact, then $\tilde{F}$ given the data is a Dirichlet process with parameter measure $\alpha + \sum_{i=1}^n \delta_{X_i}$ which coincides with the well-known result proved by Ferguson (1973). □

Example 3.11 (The beta-Stacy process) Having pointed out the lack of parametric conjugacy of the Dirichlet process in a typical inferential problem for survival analysis, one might wonder whether, conditionally on a sample featuring right-censored data, there exists a NTR process prior which shares both structural and parametric conjugacy. The problem was successfully faced in Walker and Muliere (1997), where the authors define the beta-Stacy NTR prior. Its description can be provided in terms of the Lévy intensity of $\tilde{\mu}$ where, as usual, we are supposing that a priori $\tilde{\mu}$ does not have fixed jump points. To this end, suppose that α is some probability measure on $\mathbb{R}^+$ which is absolutely continuous with respect to the Lebesgue measure and $c : \mathbb{R}^+ \to \mathbb{R}^+$ some piecewise continuous function. Use the notation F_α to denote the distribution function corresponding to α, i.e. $F_\alpha(x) = \alpha((0, x])$ for any x. A beta-Stacy process $\tilde{F}$ with parameters α and c is NTR($\tilde{\mu}$) if $\tilde{\mu}$ is a CRM whose Lévy intensity is defined by

$$\nu(ds, dx) = \frac{e^{-s\, c(x)\, \alpha((x,\infty))}}{1 - e^{-s}}\, c(x)\, ds\, \alpha(dx). \tag{3.11}$$

Note that one obtains $\mathbb{E}[\tilde{F}] = F_\alpha$ and that the Dirichlet process arises when $c(x) \equiv c$. It is to be said, however, that the definition originally provided in Walker and Muliere (1997) is more general and it allows possible choices of parameter measures α having point masses. Here, for ease of exposition we confine ourselves to this simplified case.

Theorem 3.12 (Walker and Muliere, 1997) *Let $\tilde{F}$ be a beta-Stacy process with parameters α and c satisfying the conditions given above. Then $\tilde{F}$, given $(T_1, \Delta_1), \ldots, (T_n, \Delta_n)$, is still a beta-Stacy process with updated parameters*

$$\alpha^*((0, t]) = 1 - \prod_{x \in [0,t]} \left\{1 - \frac{c(x)\, dF_\alpha(x) + d\Phi(x)}{c(x)\alpha([x, \infty)) + \bar{\Lambda}(x)}\right\}$$

$$c^*(x) = \frac{c(x)\alpha([x, \infty)) + \bar{\Lambda}(x) - \sum_{i=1}^n \delta_{T_i}(\{x\})\delta_{\Delta_i}(\{1\})}{\alpha^*([x, \infty))}$$

where $\Phi(x) = \sum_{i=1}^{n} \delta_{T_i}((0,x])\, \delta_{\Delta_i}(\{1\})$ *is the counting process for the uncensored observations.*

In the previous statement $\prod_{x\in[0,t]}$ denotes the *product integral*, a quite standard operator in the survival analysis literature. If $l_m = \max_{i=1,\dots,m} |x_i - x_{i-1}|$, the following definition holds true

$$\prod_{x\in[a,b]} \{1 + \mathrm{d}Y(x)\} := \lim_{l_m \to 0} \prod_j \left\{1 + Y(x_j) - Y(x_{j-1})\right\}$$

where the limit is taken over all partitions of $[a,b]$ into intervals determined by the points $a = x_0 < x_1 < \cdots < x_m = b$ and these partitions get finer and finer as $m \to \infty$. See Gill and Johansen (1990) for a survey of applications of product integrals to survival analysis. Finally, the Bayes estimator of $\tilde{F}$, under squared loss, coincides with the distribution function F_{α^*} associated to α^*. Interestingly, if the function c goes to 0 (pointwise) then F_{α^*} converges to the Kaplan–Meier estimator.

Remark 3.13 An appealing feature of NTR processes is that they allow for quite a rich prior specification in terms of the parameters of the Lévy intensity: in addition to the prior guess at the shape $\mathbb{E}[\tilde{F}]$, it is often also possible to assign a functional form to $\mathrm{Var}[\tilde{F}]$, whereas in the Dirichlet case, after selecting $\mathbb{E}[\tilde{F}]$, one is left with a single constant parameter to fix. A few details on this can be found in Walker and Muliere (1997) and Walker and Damien (1998).

Remark 3.14 The posterior characterizations in Theorems 3.8 and 3.9 may not seem particularly appealing at first glance. However, they reveal explicitly the posterior structure and constitute the fundamental element for devising a sampling strategy for achieving posterior inferences. Indeed, relying on some algorithm for simulating the trajectories of independent increment processes $\{\tilde{\mu}((0,x]) : x \geq 0\}$, thanks to Theorems 3.8 and 3.9 a full Bayesian analysis can be carried out. This allows us to derive Bayes estimates such as $\mathbb{E}\left[\tilde{F}(t) \mid \text{data}\right]$ or any other posterior quantity of statistical interest, see for example Ferguson and Klass (1972), Damien, Laud and Smith (1995), Walker and Damien (1998, 2000) and Wolpert and Ickstadt (1998).

3.2.2 Priors for cumulative hazards: the beta process

An alternative approach to inference for survival analysis, due to Hjort (1990), consists in assessing a prior for the cumulative hazard defined as

$$\tilde{H}_x = \tilde{H}_x(\tilde{F}) = \int_0^x \frac{\mathrm{d}\tilde{F}(v)}{1 - \tilde{F}(v^-)} \tag{3.12}$$

where $F(v^-) = \lim_{z \downarrow 0} F(v - z)$ and the integrand is the *hazard rate*, i.e. the conditional probability of observing a death/failure/event at time v given that the individual is still alive (or the system is still functioning or the event has not yet occurred) at v. From (3.12) one has the following product integral representation of $\tilde{F}$ in terms of the cumulative hazard $\tilde{H}_x$:

$$\tilde{F}(t) = 1 - \prod_{x \in [0,t]} \left\{1 - \mathrm{d}\tilde{H}_x\right\}. \tag{3.13}$$

Hence assessing a prior for the distribution function $\tilde{F}$ is the same as specifying a prior for $\tilde{H} = \{\tilde{H}_x : x \geq 0\}$ or for the hazard rate. The relation (3.13) between $\tilde{F}$ and $\tilde{H}$ suggests that the prior for $\tilde{H}$ should be such that

$$0 \leq \tilde{H}_x - \tilde{H}_{x-} \leq 1 \qquad \forall x \tag{3.14}$$

almost surely.

The main idea is, then, to model $\tilde{H}$ as a CRM $\tilde{\mu}$ by setting $x \mapsto \tilde{H}_x := \tilde{\mu}((0, x])$. However, due to (3.14), such a CRM must have all jumps of size less than 1. As shown in Hjort (1990), this happens if and only if the jump part of the Lévy intensity ν is concentrated on $[0, 1]$, i.e.

$$\rho_x((1, \infty)) = 0 \qquad \forall x > 0. \tag{3.15}$$

Within this context, Hjort's beta process prior stands, in terms of relevance, as the analog of the Dirichlet process for modeling probability distributions. Let, again, $c : \mathbb{R}^+ \to \mathbb{R}^+$ be a piecewise continuous function and H_0 be the baseline cumulative hazard which, for simplicity, we assume to be absolutely continuous. Consider now the beta CRM $\tilde{\mu}$ on $\mathbb{R}^+$ which is characterized by the Lévy intensity

$$\nu(\mathrm{d}s,\ \mathrm{d}x) = c(x)\, s^{-1}\, (1-s)^{c(x)-1}\, \mathrm{d}s\, \mathrm{d}H_{0,x}$$

for any $x \geq 0$ and $0 < s < 1$. Then, the beta process is defined as $\tilde{H} = \{\tilde{\mu}((0, x]) : x \geq 0\}$. In symbols, we write $\tilde{H} \sim \text{Beta}(c, H_0)$. Note that $\mathbb{E}[\tilde{H}_x] = H_{0,x}$. The relation between modeling the cumulative hazard with a CRM and specifying a NTR prior for the distribution function is clarified by the following theorem.

Theorem 3.15 (Hjort, 1990) *A random distribution function $\tilde{F}$ is* NTR($\tilde{\mu}$) *for some CRM $\tilde{\mu}$ if and only if the corresponding cumulative hazard $\tilde{H}(\tilde{F}) = \{\tilde{H}_x(\tilde{F}) : x \geq 0\}$ is an independent increments process with Lévy intensity satisfying condition* (3.15).

For an interesting illustration of further connections between priors for cumulative hazards and NTR processes see Dey, Erickson and Ramamoorthi (2003).

In analogy with NTR processes, a posterior characterization in terms of an updated CRM with fixed points of discontinuity corresponding to the exact observations can be given for general CRM cumulative hazards, see Hjort (1990). For brevity, here we focus on the beta process. Indeed, an important aspect of the beta process, which makes it appealing for applications to survival analysis, is its parametric conjugacy with respect to right-censoring. Recall that $\Phi(x) = \sum_{i=1}^n \delta_{T_i}((0,x])\delta_{\Delta_i}(\{1\})$ is the number of uncensored observations occurring up to time x and $\bar{\Lambda}(x) = \sum_{i=1}^n \delta_{T_i}([x,\infty))$ is the at risk process. One then has the following theorem.

Theorem 3.16 (Hjort, 1990) *Let $(T_1, \Delta_1), \ldots, (T_n, \Delta_n)$ be a sample of survival times. If $\tilde{H} \sim \text{Beta}(c, H_0)$ then*

$$\tilde{H} \mid \text{data} \sim \text{Beta}\left(c + \bar{\Lambda}, \int \frac{\mathrm{d}\Phi + c\,\mathrm{d}H_0}{c + \bar{\Lambda}}\right). \tag{3.16}$$

It follows immediately that the Bayes estimators of $\tilde{H}$ and $\tilde{F}$, with respect to a squared loss function, are

$$\hat{H}_x = \int_0^x \frac{c\,\mathrm{d}H_0 + \mathrm{d}\Phi}{c + \bar{\Lambda}}$$

and

$$\hat{F}(t) = 1 - \prod_{[0,t]} \left\{1 - \frac{c\,\mathrm{d}H_0 + \mathrm{d}\Phi}{c + \bar{\Lambda}}\right\}$$

respectively. Again, if we let the function c tend to zero, one obtains in the limit the Nelson–Aalen and the Kaplan–Meier estimators for $\tilde{H}$ and $\tilde{F}$, respectively.

In order to highlight the underlying posterior structure, Theorem 3.16 can be reformulated as follows. Suppose there are $k \le n$ distinct values among $\{T_1, \ldots, T_n\}$ so that the data can be equivalently represented as $(T_{(1)}^*, n_1^c, n_1), \ldots, (T_{(k)}^*, n_k^c, n_k)$ with n_i^c and n_i defined as in (3.9). If $\tilde{H} \sim \text{Beta}(c, H_0)$, then one has

$$\tilde{H} \mid \text{data} \stackrel{d}{=} \tilde{H}^* + \sum_{\{i:\, n_i \ge 1\}} J_i\, \delta_{T_{(i)}^*}, \tag{3.17}$$

where $\tilde{H}^* \stackrel{d}{=} \{\tilde{\mu}^*((0,x]) : x \ge 0\}$ and $\tilde{\mu}^*$ is a beta CRM with updated Lévy intensity

$$\nu(\mathrm{d}s, \mathrm{d}x) = s^{-1}(1-s)^{c(x)+\bar{\Lambda}(x)-1} c(x)\,\mathrm{d}H_{0,x}. \tag{3.18}$$

The random jump at each distinct exact observation (i.e. $T_{(i)}^*$ with $n_i \ge 1$) has the following distribution:

$$J_i \sim \text{Beta}\left([c(T_{(i)}^*) + \bar{\Lambda}(T_{(i)}^*)]\mathrm{d}H_{0,T_{(i)}^*}^*;\ [c(T_{(i)}^*) + \bar{\Lambda}(T_{(i)}^*)]\left\{1 - \mathrm{d}H_{0,T_{(i)}^*}^*\right\}\right),$$

where $\mathrm{d}H^*_{0,x} = [\mathrm{d}H_{0,x} + \mathrm{d}\Phi(x)]/[c(x) + \bar{\Lambda}(x)]$. These jumps can be merged with the updated beta CRM in (3.18) yielding the posterior representation in (3.16): note that the posterior baseline hazard in (3.16) is not continuous anymore. This sets up an analogy with what happens in the updating of the Dirichlet process, to be clarified in Section 3.3.1.

Remark 3.17 Recently, an interesting Bayesian nonparametric approach for dealing with factorial models with unbounded number of factors was introduced in Griffiths and Ghahramani (2006). The marginal process, termed the Indian buffet process, represents the analog of the Blackwell–MacQueen, or Chinese restaurant, process for the Dirichlet model. As shown in Thibaux and Jordan (2007), the de Finetti measure of the Indian buffet process is a beta process defined on a bounded space $\mathbb{X}$. Specifically, the Indian buffet process is an i.i.d. mixture of suitably defined Bernoulli processes with mixing measure the beta process. Such developments show how classes of random measures can become important also for completely different applications than the ones they were designed for. This witnesses the importance of studying general classes of random measures independently of possible immediate applications.

Two interesting extensions of Hjort's pioneering work can be found in Kim (1999) and James (2006a). The model adopted in Kim (1999) allows for more general censoring schemes. Let $\mathcal{N}_i = \{\mathcal{N}_{i,x} : x \geq 0\}$, for $i = 1, \ldots, n$, be counting processes where $\mathcal{N}_{i,x}$ denotes the number of events (these being, for instance, deaths or failures) observed up to time x for the ith counting process. Moreover, let the process $Y_i = \{Y_{i,x} : x \geq 0\}$ be the cumulative intensity associated to $\mathcal{N}_i$, thus entailing that $\mathcal{N}_i - Y_i$ is a martingale with respect to some filtration $(\mathscr{F}_x)_{x \geq 0}$. If the cumulative intensity can be represented as

$$Y_{i,x} = \int_0^x Z_{i,s}\, \mathrm{d}\tilde{H}_s \tag{3.19}$$

where $Z_i = \{Z_{i,x} : x \geq 0\}$ is an $(\mathscr{F}_x)_{x \geq 0}$ adapted process, then we have a *multiplicative intensity model*, a general class of models introduced in Aalen (1978). Moreover, if survival times $X_1, \ldots, X_n$ are subject to right censoring, with $c_1, \ldots, c_n$ denoting the n (possibly random) censoring times and $a \wedge b := \min\{a, b\}$, then $\mathcal{N}_{i,x} = \mathbb{1}_{(0, x \wedge c_i]}(X_i)$ is equal to 1 if the ith observation is both uncensored and not greater than x. In this case the process Z_i is such that $Z_{i,x} = \mathbb{1}_{(0,T_i]}(x)$ where $T_i = X_i \wedge c_i$ is the possibly right-censored survival time (or time to failure or time to an event) for the ith individual. On the other hand, when data are both left and right censored with left and right censoring times denoted by $\boldsymbol{e} = (e_1, \ldots, e_n)$ and on $\boldsymbol{c} = (c_1, \ldots, c_n)$, respectively, both independent from the X_i, one is led to consider $\mathcal{N}_{i,x} = \mathbb{1}_{(e_i, c_i \wedge x]}(X_i)$. Hence, conditional on $\boldsymbol{e}$ and on $\boldsymbol{c}$, $\mathcal{N}_i$ is a counting process governed by a multiplicative intensity model (3.19) with $Z_{i,x} = \mathbb{1}_{(e_i,c_i]}(x)$, where e_i denotes an entrance time and c_i a censoring time. The main result proved in Kim (1999) is structural conjugacy of $\tilde{H} = \{\tilde{H}_x : x \geq 0\}$ in (3.19). Specifically, if $\tilde{H}$ is a process with independent increments, then $\tilde{H}|$data is again a process with independent increments and fixed points of discontinuity in correspondence to the exact observation with random jumps expressed in terms of the Lévy intensity. For the case of right-censored observations with $\tilde{H}$ generated by a beta process, Hjort's result is recovered.

In James (2006a), the author proposes a new family of priors named *spatial neutral-to-the-right* processes. This turns out to be useful when one is interested in modeling survival times X coupled with variables Y which take values in a general space. Typically, Y can be considered as a spatial component. A spatial neutral-to-the-right process is a random probability measure associated to a cumulative hazard at y defined by

$$\tilde{H}_t(\mathrm{d}y) = \int_{(0,t]} \tilde{\mu}(\mathrm{d}x, \mathrm{d}y)$$

where $\tilde{\mu}$ is some CRM on $\mathbb{R}^+ \times \mathbb{Y}$ and $\mathbb{Y}$ is some complete and separable metric space. Hence, by (3.7), $\tilde{\mu}(\mathrm{d}x, \mathrm{d}y) = \int_{[0,1]} s\, N(\mathrm{d}s, \mathrm{d}x, \mathrm{d}y)$ where N is a Poisson random measure on $[0, 1] \times \mathbb{R}^+ \times \mathbb{Y}$ whose intensity is

$$\nu(\mathrm{d}s, \mathrm{d}x, \mathrm{d}y) = \rho_x(\mathrm{d}s)\mathrm{d}A_0(x, y).$$

In accordance with the previous notation, ρ_x is, for any x in $\mathbb{R}^+$, a measure on $[0, 1]$ and A_0 is some hazard measure on $\mathbb{R}^+ \times \mathbb{Y}$ which plays the role of baseline hazard. Correspondingly, one has

$$\tilde{S}(t^-) = 1 - \tilde{F}(t^-) = \exp\left\{\int_{[0,1]\times(0,t)\times\mathbb{Y}} \log(1 - s)\, N(\mathrm{d}s, \mathrm{d}x, \mathrm{d}y)\right\}$$

and $\tilde{p}(\mathrm{d}x, \mathrm{d}y) = \tilde{S}(x^-)\, \tilde{\mu}(\mathrm{d}x,\ \mathrm{d}y)$ is the random probability measure on $\mathbb{R}^+ \times \mathbb{Y}$ whose law acts as a prior for the distribution of (X, Y). James (2006a) shows also that the posterior distribution of $\tilde{p}$, given a sample of exchangeable observations $(X_1, Y_1), \ldots, (X_n, Y_n)$, arises as the sum of two independent components: one has a form similar to the prior, the only difference being an updating of $\tilde{S}$ and $\tilde{\mu}$, and the other is given by fixed points of discontinuity corresponding to the distinct observations. The analysis provided by James (2006a) also offers an algorithm for sampling from the marginal distribution of the observations, which represents an analog of the Blackwell–MacQueen urn scheme for these more general priors. Finally, as pointed out in James (2006a), there are some nice connections between this area of research in Bayesian nonparametrics and the theory of regenerative composition structures in combinatorics, see Gnedin and Pitman (2005b).

3.2.3 Priors for hazard rates

A number of papers have focused on the issue of specifying a prior for the hazard rate, instead of the cumulative hazard. For simplicity we assume that the data are generated by a p.d. on $\mathbb{R}^+$ which is absolutely continuous with respect to the Lebesgue measure. Then, the hazard rate is $h(x) = F'(x)/[1 - F(x^-)]$ and a prior for it can be defined in terms of a mixture with respect to a CRM. Let $k(\cdot \,|\, \cdot)$ be some kernel on $\mathbb{R}^+ \times \mathbb{Y}$, i.e. k is bimeasurable and for any bounded $B \in \mathscr{B}(\mathbb{R}^+)$ one has $\int_B k(x\,|\,y)\,\mathrm{d}x < \infty$ for any $y \in \mathbb{Y}$. Then, a prior for the hazard rate coincides with the p.d. of the random hazard rate defined by

$$\tilde{h}(x) = \int_{\mathbb{Y}} k(x\,|\,y)\,\tilde{\mu}(\mathrm{d}y) \tag{3.20}$$

where $\tilde{\mu}$ is a CRM on $(\mathbb{Y}, \mathscr{Y})$. The corresponding cumulative hazard is clearly given by $\tilde{H}_x = \int_0^x \tilde{h}(s)\mathrm{d}s$. From (3.20), provided $\tilde{H}_x \to \infty$ for $x \to \infty$ almost surely, one can define a random density function $\tilde{f}$ as

$$\tilde{f}(x) = \tilde{h}(x)\,\mathrm{e}^{-\tilde{H}_x}$$

where $\tilde{S}(x) = \exp(-\tilde{H}_x)$ is the survival function at x. Such models are often referred to as life-testing models. The random hazard $\tilde{h}$ in (3.20) can also be used to define the intensity rate of a counting process $\mathcal{N}_i = \{\mathcal{N}_{i,x} : x \geq 0\}$ as $Z_{i,x}\,\tilde{h}(x)$ where $Z_i = \{Z_{i,x} : x \geq 0\}$ is a process satisfying the same conditions pointed out in Kim (1999). Various specific models proposed in the literature fit within this framework according to the choices of k, $\tilde{\mu}$ and Z_i. For example, Dykstra and Laud (1981) consider the case where $k(x\,|\,y) \equiv \mathbb{1}_{(0,x]}(y)\beta(x)$ for some measurable and nonnegative function β, $Z_{i,x} = \mathbb{1}_{(0,T_i]}(x)$ and $\tilde{\mu}$ is a gamma process characterized by the Lévy intensity (3.5).

The random hazard $\tilde{h} = \{\tilde{h}(x) : x \geq 0\}$ corresponding to the mixing kernel described above is termed an *extended gamma process* with parameters α and β in Dykstra and Laud (1981) and is again an independent increment process with nonhomogeneous Lévy intensity

$$\nu(\mathrm{d}s, \mathrm{d}x) = \frac{\mathrm{e}^{-\beta(x)^{-1}s}}{s}\,\mathrm{d}s\,\alpha(\mathrm{d}x). \tag{3.21}$$

Lo and Weng (1989) consider $\tilde{h}$ in (3.20) with a generic kernel k and process Z_i, and with $\tilde{\mu}$ an extended gamma process, or weighted gamma process in their terminology. Due to linearity of the relation in (3.20), a characterization of the posterior distribution of $\tilde{\mu}$ given the data would easily yield a posterior representation of the hazard rate $\tilde{h}$. In order to determine a posterior characterization of $\tilde{\mu}$, it is convenient to interpret the variable y in the kernel $k(\cdot\,|\,y)$ as a latent variable: hence the posterior distribution of $\tilde{\mu}$ arises by mixing the conditional distribution of $\tilde{\mu}$, given the data and the latent, with respect to the posterior distribution of the latent variables, given the data. Such a strategy is pursued in James (2005) where the author achieves an explicit posterior characterization for general multiplicative intensity models with mixture random hazards (3.20) driven by a generic CRM $\tilde{\mu}$. For brevity, here we focus on the simple life-testing model case with exact observations denoted by $\boldsymbol{X} = (X_1, \ldots, X_n)$. The likelihood function is then given by

$$\mathscr{L}(\mu; \boldsymbol{x}) = \mathrm{e}^{-\int_{\mathbb{Y}} K_n(y)\mu(\mathrm{d}y)} \prod_{i=1}^{n} \int_{\mathbb{Y}} k(x_i|y)\mu(\mathrm{d}y), \tag{3.22}$$

where $K_n(y) = \sum_{i=1}^{n} \int_0^{x_i} k(s|y)\mathrm{d}s$. Now, augmenting the likelihood with respect to the latent variables $\boldsymbol{y} = (y_1, \ldots, y_n)$, (3.22) reduces to

$$\begin{aligned}\mathscr{L}(\mu; \boldsymbol{x}, \boldsymbol{y}) &= \mathrm{e}^{-\int_{\mathbb{Y}} K_n(y)\mu(\mathrm{d}y)} \prod_{i=1}^{n} k(x_i|y_i)\mu(\mathrm{d}x_i)\\ &= \mathrm{e}^{-\int_{\mathbb{Y}} K_n(y)\mu(\mathrm{d}y)} \prod_{j=1}^{k} \mu(\mathrm{d}y_j^*)^{n_j} \prod_{i \in C_j} k(x_i|y_j^*),\end{aligned}$$

where $\boldsymbol{y}^* = (y_1^*, \ldots, y_k^*)$ denotes the vector of the $k \leq n$ distinct latent variables, n_j is the frequency of y_j^* and $C_j = \{r : y_r = y_j^*\}$. We are now in a position to state the posterior characterization of the mixture hazard rate.

Theorem 3.18 (James, 2005) *Let $\tilde{h}$ be a random hazard rate as defined in* (3.20). *Then, given $\boldsymbol{X}$ and $\boldsymbol{Y}$, the posterior distribution of $\tilde{\mu}$ coincides with*

$$\tilde{\mu}^* \stackrel{d}{=} \tilde{\mu}_c^* + \sum_{i=1}^{k} J_i\, \delta_{Y_j^*} \tag{3.23}$$

where $\tilde{\mu}_c^*$ is a CRM with intensity measure

$$\nu^*(\mathrm{d}s, \mathrm{d}y) = \mathrm{e}^{-s\, K_n(y)}\, \rho_y(\mathrm{d}\, s)\, \alpha(\mathrm{d}y), \tag{3.24}$$

the jumps J_i $(i = 1, \ldots, k)$ are mutually independent, independent from $\tilde{\mu}_c^*$ and their distribution can be described in terms of the Lévy intensity of $\tilde{\mu}$.

Hence, we have again the posterior structure of an updated CRM with fixed points of discontinuity, the only difference being that in such a mixture setup one has to deal with both latent variables and observables. Moreover, the p.d. of the jumps J_i corresponding to the latents Y_i^* is

$$G_i(\mathrm{d}s) \propto s^{n_i} \mathrm{e}^{-s\, K_n(y_i^*)} \rho_{y_i^*}(\mathrm{d}s).$$

To complete the description the distribution of the latent variables $\boldsymbol{Y}$ conditionally on the data is needed. Setting $\tau_{n_j}(u|y) = \int_{\mathbb{R}^+} s^{n_j} e^{-us} \rho_y(ds)$ for any $u > 0$, one has

$$f(dy_1^*, .., dy_k^* | \boldsymbol{X}) = \frac{\prod_{j=1}^k \tau_{n_j}(K_n(y_j^*)|y_j^*) \prod_{i \in C_j} k(x_i, y_j^*) \alpha(dy_j^*)}{\sum_{k=1}^n \sum_{\boldsymbol{n} \in \Delta_{n,k}} \prod_{j=1}^k \int_{\mathbb{Y}} \tau_{n_j}(K_n(y)|y) \prod_{i \in C_j} k(x_i, y) \alpha(dx)} \tag{3.25}$$

for any $k \in \{1, \ldots, n\}$ and $\boldsymbol{n} := (n_1, \ldots, n_k) \in \Delta_{n,k} := \{(n_1, \ldots, n_k) : n_j \geq 1, \sum_{j=1}^k n_j = n\}$. We also recall that an alternative posterior characterization, valid when modeling decreasing hazard rate functions, has been provided in Ho (2006) and it is formulated in terms of $\boldsymbol{S}$-paths. In the light of Theorem 3.18, the distribution of $\tilde{\mu}$, given $\boldsymbol{X}$, can in principle be evaluated exactly by integrating (3.23) with respect to (3.25). Performing such an integration is a very difficult task since one needs to average with respect to all possible partitions of the integers $\{1, \ldots, n\}$. Nonetheless, the posterior representation is crucial for devising suitable simulation algorithms such as those provided in Nieto-Barajas and Walker (2004) and Ishwaran and James (2004). The latter paper contains also a wealth of applications, which highlight the power of the mixture model approach to multiplicative intensity models.

A variation of the use of weighted gamma or of beta processes for modeling hazards is suggested in Nieto-Barajas and Walker (2002). Consider a sequence $(t_n)_{n \geq 1}$ of ordered points, $0 < t_1 < t_2 < \cdots$, and set λ_k to be the hazard in the interval $(t_{k-1}, t_k]$. A first attempt to model the different hazard rates might be based on independence of the λ_k: this is done in Walker and Mallick (1997) where the λ_k are taken to be independent gamma random variables. Alternatively, a discrete version of Hjort's model implies that, given a set of failure or death times $\{t_1, t_2, \ldots\}$, the hazard rates $\pi_k = \mathbb{P}[T = t_k \mid T \geq t_k]$ are independent beta-distributed random variables. However, in both cases it seems sensible to assume dependence among the λ_k or among the π_k. The simplest form of dependence one might introduce is Markovian and this is pursued in Nieto-Barajas and Walker (2002). Hence, if θ_k is the parameter of interest, one may set $\mathbb{E}[\theta_{k+1}|\theta_1, \ldots, \theta_k] = f(\theta_k)$ for some function f. This assumption gives rise to what the authors name *Markov gamma and beta* processes. The most interesting feature is that, conditionally on a latent variable, the hazard rates have a very simple structure which naturally yields a MCMC simulation scheme for posterior inferences. An early contribution to this approach is due to Arjas and Gasbarra (1994).

3.3 General classes of discrete nonparametric priors

In this section we will describe in some detail a few recent probabilistic models that are natural candidates for defining nonparametric priors Q which select discrete distributions with probability 1. There are essentially two ways of exploiting such priors: (a) they can be used to model directly the data when these are generated by a discrete distribution; (b) they are introduced as basic building blocks in hierarchical mixtures if the data arise from a continuous distribution. The latter use will be detailed in Section 3.4.1.

3.3.1 Normalized random measures with independent increments

Among the various generalizations of the Dirichlet process, the one we will illustrate in the present section is inspired by a construction of the Dirichlet process provided in Ferguson (1973). Indeed, a Dirichlet process on a complete and separable metric space, $\mathbb{X}$, can also be obtained by normalizing the increments of a gamma CRM $\tilde{\gamma}$ with parameter α as described in Example 3.3: the random probability measure $\tilde{p} = \tilde{\gamma}/\tilde{\gamma}(\mathbb{X})$ has the same distribution as the Dirichlet process on $\mathbb{X}$ with parameter measure α. Given this, one might naturally wonder whether a full Bayesian analysis can be performed if in the above normalization the gamma process is replaced by any CRM with a generic Lévy intensity $\nu(\mathrm{d}s, \mathrm{d}x) = \rho_x(\mathrm{d}s)\,\alpha(\mathrm{d}x)$. Though Bayesians have seldom considered "normalization" as a tool for defining random probability measures, this idea has been exploited and applied in a variety of contexts not closely related to Bayesian inference such as storage problems, computer science, population genetics, ecology, statistical physics, combinatorics, number theory and excursions of stochastic processes; see Pitman (2006) and references therein. Some important theoretical insight on the properties of normalized random measures was first given in Kingman (1975), where a random discrete distribution generated by the σ-stable subordinator is considered. Further developments can be found in Perman, Pitman and Yor (1992), where a description of the atoms of random probability measures, obtained by normalizing increasing processes with independent and stationary increments, in terms of a stick-breaking procedure, is provided. From a Bayesian perspective, the idea of normalization was taken up again in Regazzini, Lijoi and Prünster (2003), where a normalized random measure with independent increments is introduced as a random probability measure on $\mathbb{R}$ obtained by normalizing a suitably time-changed increasing process with independent but not necessarily stationary increments. A definition stated in terms of CRMs is as follows.

Definition 3.19 Let $\tilde{\mu}$ be a CRM on $\mathbb{X}$ such that $0 < \tilde{\mu}(\mathbb{X}) < \infty$ almost surely. Then, the random probability measure $\tilde{p} = \tilde{\mu}/\tilde{\mu}(\mathbb{X})$ is termed a normalized random measure with independent increments (NRMI).

Both finiteness and positiveness of $\tilde{\mu}(\mathbb{X})$ are clearly required for the normalization to be well defined and it is natural to express such conditions in terms of the Lévy intensity of the CRM. Indeed, it is enough to have $\rho_x(\mathbb{R}^+) = \infty$ for every x and $0 < \alpha(\mathbb{X}) < \infty$. The former is equivalent to requiring that the CRM $\tilde{\mu}$ has infinitely many jumps on any bounded set: in this case $\tilde{\mu}$ is also called an *infinite activity* process. The previous conditions can also be strengthened to necessary and sufficient conditions but we do not pursue this here.

In the following we will speak of *homogeneous (nonhomogeneous)* NRMIs, if the underlying CRM (or, equivalently, the Lévy intensity (3.4)) is homogeneous (nonhomogeneous).

Example 3.20 (The σ-stable NRMI) Suppose $\sigma \in (0, 1)$ and let $\tilde{\mu}_\sigma$ be the σ-stable CRM examined in Example 3.4 with Lévy intensity (3.6). If α in (3.6) is finite, the required positivity and finiteness conditions are satisfied. One can, then, define a random probability measure $\tilde{p} = \tilde{\mu}_\sigma/\tilde{\mu}_\sigma(\mathbb{X})$ which takes on the name of normalized σ-stable process with parameter σ. This random probability measure was introduced in Kingman (1975) in relation to optimal storage problems. The possibility of application in Bayesian nonparametric inference was originally pointed out by A. F. M. Smith in the discussion to Kingman (1975).

Example 3.21 (The generalized gamma NRMI) Consider now a generalized gamma CRM (Brix, 1999) which is characterized by a Lévy intensity of the form

$$\nu(\mathrm{d}s, \mathrm{d}x) = \frac{\sigma}{\Gamma(1-\sigma)}\, s^{-1-\sigma}\, \mathrm{e}^{-\tau s}\mathrm{d}s\, \alpha(\mathrm{d}x), \tag{3.26}$$

where $\sigma \in (0, 1)$ and $\tau > 0$. Let us denote it by $\tilde{\mu}_{\sigma,\tau}$. Note that if $\tau = 0$ then $\tilde{\mu}_{\sigma,0}$ coincides with the σ-stable CRM $\tilde{\mu}_\sigma$, whereas if $\sigma \to 0$ the gamma CRM (3.5) is obtained. If α in (3.26) is finite, we have $0 < \tilde{\mu}_{\sigma,\tau}(\mathbb{X}) < \infty$ almost surely and a NRMI $\tilde{p} = \tilde{\mu}_{\sigma,\tau}/\tilde{\mu}_{\sigma,\tau}(\mathbb{X})$, which is termed the normalized generalized gamma process. See Pitman (2003) for a discussion on its representation as a Poisson–Kingman model, a class of random distributions described in Section 3.3.3. The special case of $\sigma = 1/2$, corresponding to the normalized inverse Gaussian process, was examined in Lijoi, Mena and Prünster (2005) who also provide an expression for the family of finite-dimensional distributions of $\tilde{p}$.

Example 3.22 (The extended gamma NRMI) A nonhomogeneous NRMI arises by considering the extended gamma process of Dykstra and Laud (1981) characterized by the Lévy intensity (3.21). If the function $\beta : \mathbb{X} \to \mathbb{R}^+$ is such that $\int_{\mathbb{X}} \log[1 + \beta(x)]\, \alpha(\mathrm{d}x) < \infty$, then the corresponding NRMI is well defined and will be termed an extended gamma NRMI.

These examples, together with others one could think of by simply providing a Lévy intensity, suggest that NRMIs identify a very large class of priors and one might then wonder whether they are amenable of practical use for inferential purposes. A first thing to remark is that, apart from the Dirichlet process, NMRIs are not structurally conjugate, see James, Lijoi and Prünster (2006). Nonetheless one can still provide a posterior characterization of NRMIs in the form of a mixture representation. In the sequel, we will always work with a NRMI, whose underlying Lévy intensity has a non-atomic α in (3.4). Suppose that the data are exchangeable according to model (3.1) where Q is the probability distribution of a NRMI. Since NRMIs are almost surely discrete, data can display ties and we denote by $X_1^*, \ldots, X_k^*$ the k distinct observations, with frequencies $n_1, \ldots, n_k$, present within the sample $\boldsymbol{X} = (X_1, \ldots, X_n)$. Before stating the posterior characterization, we introduce the key latent variable. For any $n \geq 1$, let U_n be a positive random variable whose density function, conditional on the sample $\boldsymbol{X}$, is

$$q_{\boldsymbol{X}}(u) \propto u^{n-1} \, \mathrm{e}^{-\psi(u)} \prod_{j=1}^{k} \tau_{n_j}(u|X_j^*), \tag{3.27}$$

where ψ is the *Laplace exponent* of $\tilde{\mu}$, i.e.

$$\psi(u) = \int_{\mathbb{X}} \int_{\mathbb{R}^+} (1 - \mathrm{e}^{-u\,v}) \rho_x(\mathrm{d}v) \, \alpha(\mathrm{d}x)$$

and, for any $m \geq 1$, $\tau_m(u|x) := \int_{\mathbb{R}^+} s^m \, \mathrm{e}^{-us} \, \rho_x(\mathrm{d}s)$. The following result states that the posterior distribution of $\tilde{\mu}$ and of $\tilde{p}$, given a sample $\boldsymbol{X}$, is a mixture of NRMIs with fixed points of discontinuity in correspondence to the observations and the mixing density is $q_{\boldsymbol{X}}$ in (3.27).

Theorem 3.23 (James, Lijoi and Prünster, 2005, 2009) *If $\tilde{p}$ is a NRMI obtained by normalizing $\tilde{\mu}$, then*

$$\tilde{\mu} \,|\, (\boldsymbol{X}, U_n) \overset{d}{=} \tilde{\mu}_{U_n} + \sum_{i=1}^{k} J_i^{(U_n)} \, \delta_{X_i^*}, \tag{3.28}$$

where $\tilde{\mu}_{U_n}$ is a CRM with Lévy intensity $\nu^{(U_n)}(\mathrm{d}s, \mathrm{d}x) = \mathrm{e}^{-U_n s} \, \rho_x(\mathrm{d}s) \, \alpha(\mathrm{d}x)$, the nonnegative jumps $J_i^{(U_n)}$ are mutually independent and independent from $\tilde{\mu}_{U_n}$ with density function $f_i(s) \propto s^{n_i} \mathrm{e}^{-U_n s} \rho_{X_i^*}(\mathrm{d}s)$. Moreover,

$$\tilde{p} \,|\, (\boldsymbol{X}, \, U_n) \overset{d}{=} w \frac{\tilde{\mu}_{U_n}}{\tilde{\mu}_{U_n}(\mathbb{X})} + (1 - w) \frac{\sum_{i=1}^{k} J_i^{(U_n)} \, \delta_{X_i^*}}{\sum_{r=1}^{k} J_r^{(U_n)}} \tag{3.29}$$

where $w = \tilde{\mu}_{U_n}(\mathbb{X}) / [\tilde{\mu}_{U_n}(\mathbb{X}) + \sum_{i=1}^{k} J_i^{(U_n)}]$.

The above result displays the same posterior structure, namely CRM with fixed points of discontinuity, that has already occurred on several occasions in Section 3.2. Here the only difference is that such a representation holds conditionally on a suitable latent variable, which makes it slightly more elaborate. This is due to the fact that the structural conjugacy property is not satisfied. Nonetheless, NRMIs give rise to more manageable predictive structures than, for instance, NTR processes, see also James, Lijoi and Prünster (2009).

Since the Dirichlet process is a special case of NRMI, it is interesting to see how the posterior representation of Ferguson (1973) is recovered. Indeed, if $\tilde{\mu}$ is a gamma CRM with parameter measure α on $\mathbb{X}$ such that $\alpha(\mathbb{X}) = \theta \in (0, \infty)$, then $\tilde{\mu}_{U_n} + \sum_{i=1}^k J_i^{(U_n)} \delta_{X_i^*}$ is a gamma CRM with Lévy intensity

$$\nu^{(U_n)}(\mathrm{d}s, \mathrm{d}x) = \frac{\mathrm{e}^{-(1+U_n)s}}{s} \,\mathrm{d}s\, \alpha_n^*(\mathrm{d}x) \tag{3.30}$$

where $\alpha_n^* = \alpha + \sum_{i=1}^k n_i\, \delta_{X_i^*}$. However, since the CRM characterized by (3.30) is to be normalized, we can, without loss of generality, set the scale parameter $1 + U_n$ in (3.30) equal to 1. The random probability in (3.29) turns out to be a Dirichlet process with parameter α_n^* and its distribution does not depend on U_n. Note also the analogy with the posterior updating of the beta process sketched after Theorem 3.16.

In analogy with NTR processes, the availability of a posterior representation is essential for the implementation of sampling algorithms in order to simulate the trajectories of the posterior CRM. A possible algorithm suggested by the representation (3.28) is

1. sample U_n from q_X,
2. sample the jump $J_i^{(U_n)}$ at X_i^* from the density $f_i(s) \propto s^{n_i} \mathrm{e}^{-U_n s} \rho_{X_i^*}(\mathrm{d}s)$,
3. simulate a realization of $\tilde{\mu}_{U_n}$ with Lévy measure $\nu^{(U_n)}(\mathrm{d}x, \mathrm{d}s) = \mathrm{e}^{-U_n s}$ $\rho_x(\mathrm{d}s)\, \alpha(\mathrm{d}x)$ via the Ferguson and Klass algorithm; see Ferguson and Klass (1972) and Walker and Damien (2000).

For an application of this computational technique see Nieto-Barajas and Prünster (2009).

Example 3.24 (The generalized gamma NRMI) Consider the normalized generalized gamma process defined in Example 3.21. The (posterior) distribution of $\tilde{\mu}$, given U_n and $\boldsymbol{X}$, coincides in distribution with the CRM $\tilde{\mu}_{U_n} + \sum_{i=1}^k J_i^{(U_n)} \delta_{X_i^*}$ where $\tilde{\mu}_{U_n}$ is a generalized gamma CRM with Lévy intensity $\nu^{(U_n)}(\mathrm{d}s, \mathrm{d}x) = \frac{\sigma}{\Gamma(1-\sigma)}\, s^{-1-\sigma}\, \mathrm{e}^{-(U_n+1)s} \mathrm{d}s\, \alpha(\mathrm{d}x)$, the fixed points of discontinuity coincide with the distinct observations X_i^* and the ith jump $J_i^{(U_n)}$ is Gamma $(U_n + 1, n_i - \sigma)$ distributed, for $i = 1, \ldots, k$. Finally, the density function of U_n, conditional on $\boldsymbol{X}$, is $q_{\boldsymbol{X}}(u) \propto u^{n-1}\,(1+u)^{k\sigma - n}\, \mathrm{e}^{-\alpha(\mathbb{X})(1+u)^\sigma}$.

3.3.2 Exchangeable partition probability function

The nature of the realizations of NRMIs and, in general, of discrete random probability measures, quite naturally leads to analysis of the partition structures among the observations that they generate. Indeed, given n observations $X_1, \ldots, X_n$ generated from model (3.1), discreteness of $\tilde{p}$ implies that there might be ties within the data, i.e. $\mathbb{P}[X_i = X_j] > 0$ for $i \neq j$. Correspondingly, define Ψ_n to be a random partition of the integers $\{1, \ldots, n\}$ such that any two integers i and j belong to the same set in Ψ_n if and only if $X_i = X_j$. Let $k \in \{1, \ldots, n\}$ and suppose $\{C_1, \ldots, C_k\}$ is a partition of $\{1, \ldots, n\}$ into k sets C_i. Hence, $\{C_1, \ldots, C_k\}$ is a possible realization of Ψ_n. A common and sensible specification for the probability distribution of Ψ_n consists in assuming that it depends on the frequencies of each set in the partition. To illustrate this point, recall that

$$\Delta_{n,k} := \left\{ (n_1, \ldots, n_k) : \ n_i \geq 1, \ \sum_{i=1}^{k} n_i = n \right\}.$$

For $n_i = \text{card}(C_i)$, then $(n_1, \ldots, n_k) \in \Delta_{n,k}$ and

$$\mathbb{P}[\Psi_n = \{C_1, \ldots, C_k\}] = \Pi_k^{(n)}(n_1, \ldots, n_k). \tag{3.31}$$

A useful and intuitive metaphor is that of species sampling: one is not much interested in the realizations of the X_i, which stand as species labels thus being arbitrary, but rather in the probability of observing k distinct species with frequencies $(n_1, \ldots, n_k)$ in $n \geq k$ draws from a population.

Definition 3.25 Let $(X_n)_{n \geq 1}$ be an exchangeable sequence. Then, $\{\Pi_k^{(n)} : 1 \leq k \leq n, \ n \geq 1\}$ with $\Pi_k^{(n)}$ defined in (3.31) is termed an exchangeable partition probability function (EPPF).

Indeed, the EPPF defines an important tool which was introduced in Pitman (1995) and it determines the distribution of a random partition of $\mathbb{N}$. It is worth noting that the fundamental contributions J. Pitman has given to this area of research have been deeply influenced by, and appear as natural developments of, some earlier relevant work on random partitions by J. F. C. Kingman, see, for example, Kingman (1978, 1982).

From the above definition it follows that, for any $n \geq k \geq 1$ and any $(n_1, \ldots, n_k) \in \Delta_{n,k}$, $\Pi_k^{(n)}$ is a symmetric function of its arguments and it satisfies the addition rule $\Pi_k^{(n)}(n_1, \ldots, n_k) = \Pi_{k+1}^{(n+1)}(n_1, \ldots, n_k, 1) + \sum_{j=1}^{k} \Pi_k^{(n+1)}(n_1, \ldots, n_j + 1, \ldots, n_k)$. On the other hand, as shown in Pitman (1995), every non-negative symmetric function satisfying the addition rule is the EPPF of some exchangeable sequence, see Pitman (1995, 2006) for a thorough and useful analysis of EPPFs.

The availability of the EPPF yields, as a by-product, the system of predictive distributions induced by Q. Indeed, suppose Q in model (3.1) coincides with a discrete nonparametric prior and $\{\Pi_k^{(n)} : 1 \le k \le n,\ n \ge 1\}$ is the associated EPPF. If the sample $\boldsymbol{X} = (X_1, \ldots, X_n)$ contains k distinct values $X_1^*, \ldots, X_k^*$ and n_j of them are equal to X_j^* one has

$$\mathbb{P}[X_{n+1} = \text{ new} \mid \boldsymbol{X}] = \frac{\Pi_{k+1}^{(n+1)}(n_1, \ldots, n_k, 1)}{\Pi_k^{(n)}(n_1, \ldots, n_k)},$$

$$\mathbb{P}[X_{n+1} = X_j^* \mid \boldsymbol{X}] = \frac{\Pi_k^{(n+1)}(n_1, \ldots, n_j + 1, \ldots, n_k)}{\Pi_k^{(n)}(n_1, \ldots, n_k)}.$$

If $\tilde{p}$ is a NRMI (with nonatomic parameter measure α), the associated EPPF is

$$\Pi_k^{(n)}(n_1, \ldots, n_k) = \frac{1}{\Gamma(n)} \int_0^\infty u^{n-1}\, \mathrm{e}^{-\psi(u)} \left\{ \prod_{j=1}^k \int_{\mathbb{X}} \tau_{n_j}(u|x)\, \alpha(\mathrm{d}x) \right\} \mathrm{d}u \tag{3.32}$$

and from it one can deduce the system of predictive distributions of X_{n+1}, given $\boldsymbol{X}$,

$$\mathbb{P}\left[X_{n+1} \in \mathrm{d}x_{n+1} \mid \boldsymbol{X}\right] = w_k^{(n)}\, P_0(\mathrm{d}x_{n+1}) + \frac{1}{n} \sum_{j=1}^k w_{j,k}^{(n)}\, \delta_{X_j^*}(\mathrm{d}x_{n+1}) \tag{3.33}$$

where $P_0 = \alpha/\alpha(\mathbb{X})$ and

$$w_k^{(n)} = \frac{1}{n} \int_0^{+\infty} u\, \tau_1(u|x_{n+1})\, q_X(u)\, \mathrm{d}u, \tag{3.34}$$

$$w_{j,k}^{(n)} = \int_0^{+\infty} u\, \frac{\tau_{n_j+1}(u|X_j^*)}{\tau_{n_j}(u|X_j^*)}\, q_X(u)\, \mathrm{d}u. \tag{3.35}$$

In the homogeneous case, i.e. $\rho_x = \rho$, the previous formulae reduce to those given in Pitman (2003). Closed form expressions are derivable for some specific NRMI. For example, if $\tilde{p}$ is the σ-stable NRMI, then $w_k^{(n)} = k\sigma/n$ and $w_{j,k}^{(n)} = (n_j - \sigma)$, see Pitman (1996). On the other hand, if $\tilde{p}$ is the normalized generalized gamma process, one has (3.33) with

$$w_k^{(n)} = \frac{\sigma}{n} \frac{\sum_{i=0}^n \binom{n}{i}(-1)^i\, \beta^{i/\sigma}\, \Gamma\left(k+1-\frac{i}{\sigma}; \beta\right)}{\sum_{i=0}^{n-1} \binom{n-1}{i}(-1)^i\, \beta^{i/\sigma}\, \Gamma\left(k-\frac{i}{\sigma}; \beta\right)},$$

$$w_{j,k}^{(n)} = \frac{\sum_{i=0}^n \binom{n}{i}(-1)^i\, \beta^{i/\sigma}\, \Gamma\left(k-\frac{i}{\sigma}; \beta\right)}{\sum_{i=0}^{n-1} \binom{n-1}{i}(-1)^i\, \beta^{i/\sigma}\, \Gamma\left(k-\frac{i}{\sigma}; \beta\right)}\, (n_j - \sigma),$$

see Lijoi, Mena and Prünster (2007a). The availability of closed form expressions of the predictive distributions is essential for the implementation of Blackwell–MacQueen-type sampling schemes, which are a key tool for drawing inference in complex mixture models. Nonetheless, even when no closed form expressions are available, drawing samples from the predictive is still possible by conditioning on the latent variable U_n. Indeed, one has

$$\mathbb{P}[X_{n+1} \in \mathrm{d}x_{n+1} \mid \boldsymbol{X}, U_n = u] \propto \kappa_1(u)\, \tau_1(u|x_{n+1})\, P_0(\mathrm{d}x_{n+1}) + \sum_{j=1}^{k} \frac{\tau_{n_j+1}(u|X_j^*)}{\tau_{n_j}(u|X_j^*)}\, \delta_{X_j^*}(\mathrm{d}x_{n+1})$$

where $\kappa_1(u) = \int_{\mathbb{X}} \tau_1(u|x)\, \alpha(\mathrm{d}x)$. From this one can implement an analog of the Blackwell–MacQueen urn scheme in order to draw a sample $X_1, \ldots, X_n$ from $\tilde{p}$. Let $m(\mathrm{d}x|u) \propto \tau_1(u|x)\, \alpha(\mathrm{d}x)$ and, for any $i \geq 2$, set $m(\mathrm{d}x_i|x_1, \ldots, x_{i-1}, u) = \mathbb{P}[X_i \in \mathrm{d}x_i|X_1, \ldots, X_{i-1}, U_{i-1} = u]$. Moreover, set U_0 to be a positive random variable whose density function is given by $q_0(u) \propto \mathrm{e}^{-\psi(u)} \int_{\mathbb{X}} \tau_1(u|x)\, \alpha(\mathrm{d}x)$. The sampling scheme can be described as follows.

1. Sample U_0 from q_0.
2. Sample X_1 from $m(\mathrm{d}x|U_0)$.
3. At step i

   (a) sample U_{i-1} from $q_{\boldsymbol{X}_{i-1}}(u)$, where $\boldsymbol{X}_{i-1} = (X_1, \ldots, X_{i-1})$;
   (b) generate ξ_i from $f_i(\xi) \propto \tau_1(U_{i-1}|\xi)\, P_0(\mathrm{d}\xi)$;
   (c) sample X_i from $m(\mathrm{d}x|\boldsymbol{X}_{i-1}, U_{i-1})$, which implies

   $$X_i = \begin{cases} \xi_i & \text{prob} \propto \kappa_1(U_{i-1}) \\ X^*_{j,i-1} & \text{prob} \propto \tau_{n_{j,i-1}+1}(U_{i-1}|X^*_{j,i-1})/\tau_{n_{j,i-1}+1}(U_{i-1}|X^*_{j,i-1}) \end{cases}$$

   where $X^*_{j,i-1}$ is the jth distinct value among $X_1, \ldots, X_{i-1}$ and $n_{j,i-1}$ is the cardinality of the set $\{X_s : X_s = X^*_{j,i-1},\ s = 1, \ldots, i-1\}$.

3.3.3 Poisson–Kingman models and Gibbs-type priors

Consider a discrete random probability measure $\tilde{p} = \sum_{i\geq 1} \tilde{p}_i \delta_{X_i}$ where the locations X_i are i.i.d. from a nonatomic probability measure P_0 on $\mathbb{X}$. Furthermore, suppose the locations are independent from the weights $\tilde{p}_i$. The specification of $\tilde{p}$ is completed by assigning a distribution for the weights. Pitman (2003) identifies a method for achieving this goal: he derives laws, which are termed Poisson–Kingman distributions, for sequences of ranked random probability masses $\tilde{p}_i$. To be more specific, consider a homogeneous CRM $\tilde{\mu}$ whose intensity $\nu(\mathrm{d}s, \mathrm{d}x) = \rho(\mathrm{d}s)\alpha(\mathrm{d}x)$ is such that $\rho(\mathbb{R}^+) = \infty$ and $\alpha = P_0$ is a nonatomic probability measure. Denote by $J_{(1)} \geq J_{(2)} \geq \cdots$ the ranked jumps of the CRM, set $T = \sum_{i \geq 1} J_{(i)}$ and assume that the p.d. of the total mass T is absolutely continuous with respect to the Lebesgue measure on $\mathbb{R}$. Next, define

$$\tilde{p}_{(i)} = \frac{J_{(i)}}{T} \tag{3.36}$$

for any $i = 1, 2, \ldots$ and denote by $S^* = \{(p_1, p_2, \ldots) : p_1 \geq p_2 \geq \cdots \geq 0, \sum_{i \geq 1} p_i = 1\}$ the set of all sequences of ordered nonnegative real numbers that sum up to 1.

Definition 3.26 Let $P_{\rho,t}$ be the conditional distribution of the sequence $(\tilde{p}_{(i)})_{i \geq 1}$ of ranked random probabilities generated by a CRM through (3.36), given $T = t$. Let η be a probability distribution on $\mathbb{R}^+$. The distribution

$$\int_{\mathbb{R}^+} P_{\rho,t}\, \eta(\mathrm{d}t)$$

on S^* is termed a Poisson–Kingman distribution with Lévy intensity ρ and mixing distribution η. It is denoted by $\mathrm{PK}(\rho, \eta)$.

If η coincides with the p.d. of T, we use the notation $\mathrm{PK}(\rho)$ to indicate the corresponding random probability with masses in S^*. The discrete random probability measure $\tilde{p} = \sum_{i \geq 1} \tilde{p}_{(i)} \delta_{X_i}$, where the $\tilde{p}_{(i)}$ follow a $\mathrm{PK}(\rho, \eta)$ distribution, is termed a $\mathrm{PK}(\rho, \eta)$ *random probability measure*. It is important to remark that $\mathrm{PK}(\rho)$ random probability measures are equivalent to homogeneous NRMIs defined in Section 3.3.1. Pitman (2003) derives an expression for the EPPF of a general $\mathrm{PK}(\rho, \eta)$ model but it is difficult to evaluate. However, in the special case of a $\mathrm{PK}(\rho)$ model it reduces to the simple expression implied by (3.32) when the dependence on the locations of the jumps is removed. Although the potential use of general $\mathrm{PK}(\rho, \eta)$ random probability measures for statistical inference is quite limited, their theoretical importance can be traced back to two main reasons: (i) the two-parameter Poisson–Dirichlet process is a $\mathrm{PK}(\rho, \eta)$ model, whereas it is not a NRMI; (ii) $\mathrm{PK}(\rho, \eta)$ models generate the class of Gibbs-type random probability measure which possess a conceptually very appealing predictive structure. Both examples involve $\mathrm{PK}(\rho, \eta)$ models based on the σ-stable CRM.

Example 3.27 (The two-parameter Poisson–Dirichlet process) One of the main reasons of interest for the class of $\mathrm{PK}(\rho, \eta)$ priors is due to the fact that it contains, as a special case, the two-parameter Poisson–Dirichlet process, introduced in Perman, Pitman and Yor (1992). This process and the distribution of the ranked probabilities, termed the two-parameter Poisson–Dirichlet distribution, were further studied in the remarkable papers by Pitman (1995) and Pitman and Yor (1997a). Its name is also explained by the fact that it can be seen as a natural extension of the

From (3.38), it follows that the predictive distributions induced by a Gibbs-type prior are of the form

$$\mathbb{P}\left[X_{n+1} \in \mathrm{d}x \mid \boldsymbol{X}\right] = \frac{V_{n+1,k+1}}{V_{n,k}} P_0(\mathrm{d}x) + \frac{V_{n+1,k}}{V_{n,k}} \sum_{j=1}^{k} (n_j - \sigma)\, \delta_{X_j^*}(\mathrm{d}x). \quad (3.39)$$

The structure of (3.39) provides some insight into the inferential implications of the use of Gibbs-type priors. Indeed, the prediction rule can be seen as resulting from a two-step procedure: the $(n+1)$th observation X_{n+1} is either "new" (i.e. not coinciding with any of the previously observed X_i^*) or "old" with probability depending on n and k but not on the frequencies n_i. Given X_{n+1} is "new", it is sampled from P_0. Given X_{n+1} is "old" (namely X_{n+1} is equal to one of the already sampled X_i^*), it will coincide with a particular X_j^* with probability $(n_j - \sigma)/(n - k\sigma)$. By comparing the predictive distributions (3.39) with those arising from the models dealt with so far, one immediately sees that the PD(σ, θ) process (hence, a fortiori the Dirichlet process) and the normalized generalized gamma process belong to the class of Gibbs priors. Considered as a member of this general class, the Dirichlet process is the only prior for which the probability of sampling a "new" or "old" observation does not depend on the number of distinct ones present in the sample. On the other hand, one may argue that it is desirable to have prediction rules for which the assignment to "new" or "old" depends also on the frequencies n_i. However, this would remarkably increase the mathematical complexity and so Gibbs priors appear to represent a good compromise between tractability and richness of the predictive structure. An investigation of the predictive structure arising from Gibbs-type priors can be found in Lijoi, Prünster and Walker (2008a).

An important issue regarding the class of Gibbs-type priors is the characterization of its members. In other terms, one might wonder which random probability measures induce an EPPF of the form (3.38). An answer has been successfully provided by Gnedin and Pitman (2005a). Let ρ_σ be the jump part of the intensity of a σ-stable CRM and consider PK(ρ_σ, η) random probability measures with arbitrary mixing distribution η: for brevity we refer to them as the σ-stable PK models. Then, $\tilde{p}$ is a Gibbs-type prior with $\sigma \in (0, 1)$ if and only if it is a σ-stable PK model. Hence, the corresponding $V_{n,k}$, which specify the prior completely, are of the form

$$V_{n,k} = \int_{\mathbb{R}^+} \frac{\sigma^k t^{-n}}{\Gamma(n - k\sigma)\, f_\sigma(t)} \int_0^t s^{n-k\sigma-1} f_\sigma(t-s)\, \mathrm{d}s\, \eta(\mathrm{d}t),$$

where f_σ denotes, as before, the σ-stable density. Moreover, $\tilde{p}$ is a Gibbs-type prior with $\sigma = 0$ if and only if it is a mixture, with respect to the parameter $\theta = \alpha(\mathbb{X})$, of a Dirichlet process. See Pitman (2003, 2006) and Gnedin and Pitman (2005a) for more details and interesting connections to combinatorics. Finally, the only NRMI, which is also of Gibbs type with $\sigma \in (0, 1)$, is the normalized generalized gamma process (Lijoi, Prünster and Walker, 2008b).

3.3.4 Species sampling models

Species sampling models, introduced and studied in Pitman (1996), are a very general class of discrete random probability measures $\tilde{p} = \sum_{j \geq 1} \tilde{p}_j \, \delta_{X_j}$ in which the weights $\tilde{p}_j$ are independent of the locations X_j. Such a generality provides some insight on the structural properties of these random probability measures; however, for possible uses in concrete applications, a distribution for the weights $\tilde{p}_j$ has to be specified. Indeed, homogeneous NRMI and Poisson–Kingman models belong to this class and can be seen as completely specified species sampling models. On the other hand, NTR and nonhomogeneous NRMI do not fall within this framework.

Definition 3.29 Let $(\tilde{p}_j)_{j \geq 1}$ be a sequence of nonnegative random weights such that $\sum_{j \geq 1} \tilde{p}_j \leq 1$ and suppose that $(\xi_n)_{n \geq 1}$ is a sequence of i.i.d. random variables with nonatomic p.d. P_0. Moreover, let the ξ_i be independent from the $\tilde{p}_j$. Then, the random probability measure

$$\tilde{p} = \sum_{j \geq 1} \tilde{p}_j \, \delta_{\xi_j} + \left(1 - \sum_{j \geq 1} \tilde{p}_j \right) P_0$$

is a species sampling model.

Accordingly, a sequence of random variables $(X_n)_{n \geq 1}$, which is conditionally i.i.d. given a species sampling model, is said to be a *species sampling sequence*. Moreover, if in the previous definition one has $\sum_{j \geq 1} \tilde{p}_j = 1$, almost surely, then the model is termed *proper*. We will focus on this specific case and provide a description of a few well-known species sampling models.

The use of the terminology species sampling is not arbitrary. Indeed, discrete nonparametric priors are not well suited for modeling directly data generated by a continuous distribution (in such cases they are used at a latent level within a hierarchical mixture). However, as already noted in Pitman (1996), when the data come from a discrete distribution, as happens for species sampling problems in ecology, biology and population genetics, it is natural to assign a discrete nonparametric prior to the unknown proportions. More precisely, suppose that a population consists of an ideally infinite number of species: one can think of $\tilde{p}_i$ as the proportion of the ith species in the population and ξ_i is the label assigned to species i. Since the labels ξ_i are generated by a nonatomic distribution they are almost surely distinct: hence, distinct species will have distinct labels attached. The following characterization provides a formal description of the family of predictive distributions induced by a species sampling model.

Theorem 3.30 (Pitman, 1996) *Let* $(\xi_n)_{n\ge1}$ *be a sequence of i.i.d. random variables with p.d.* P_0. *Then* $(X_n)_{n\ge1}$ *is a species sampling sequence if and only if there exists a collection of weights* $\{p_{j,n}(n_1,\ldots,n_k): \ 1\le j\le k,\ 1\le k\le n,\ n\ge1\}$ *such that* $X_1=\xi_1$ *and, for any* $n\ge1$,

$$X_{n+1}\mid(X_1,\ldots,X_n)=\begin{cases}\xi_{n+1} & \text{with prob } p_{k_n+1,n}(n_1,\ldots,n_{k_n},1)\\ X^*_{n,j} & \text{with prob } p_{k_n,n}(n_1,\ldots,n_j+1,\ldots,n_{k_n})\end{cases}$$

where k_n *is the number of distinct values* $X^*_{n,1},\ldots,X^*_{n,k_n}$ *among the conditioning observations.*

The main issue with the statement above lies in the fact that it guarantees the existence of the predictive weights $p_{j,n}(n_1,\ldots,n_k)$, but it does not provide any hint on their form. As mentioned earlier, in order to evaluate the predictive distribution it is necessary to assign a p.d. to the weights $\tilde p_j$. An alternative to the normalization procedure used for NRMI and PK models, is represented by the *stick-breaking* mechanism which generates species sampling models with stick-breaking weights. Let $(V_i)_{i\ge1}$ be a sequence of independent random variables taking values in $[0,1]$ and set

$$\tilde p_1=V_1,\qquad \tilde p_i=V_i\prod_{j=1}^{i-1}(1-V_j)\quad i\ge2.$$

These random weights define a proper species sampling model if and only if $\sum_{i\ge1}\mathbb{E}\left[\log(1-V_i)\right]=-\infty$, see Ishwaran and James (2001). The rationale of the construction is apparent. Suppose one has a unit length stick and breaks it into two bits of length V_1 and $1-V_1$. The first bit represents $\tilde p_1$ and in order to obtain $\tilde p_2$ it is enough to split the remaining part, of length $1-V_1$, into two parts having respective lengths $V_2(1-V_1)$ and $(1-V_2)(1-V_1)$. The former will coincide with $\tilde p_2$ and the latter will be split to generate $\tilde p_3$, and so on. The Dirichlet process with parameter measure α represents a special case, which corresponds to the Sethuraman (1994) series representation: if $\alpha(\mathbb{X})=\theta$, then the V_i are i.i.d. with $\text{Beta}(1,\theta)$ distribution. Another nonparametric prior which admits a stick-breaking construction is the $\text{PD}(\sigma,\theta)$ process. If in the stick-breaking construction one takes independent V_i such that

$$V_i\sim\text{Beta}(\theta+i\sigma,\ 1-\sigma),$$

the resulting $\tilde p$ is a $\text{PD}(\sigma,\theta)$ process, see Pitman (1995). Moreover, Teh, Görür and Ghahramani (2007) derived a simple and interesting construction of the beta process, which is based on a variation of the stick-breaking scheme described above.

Remark 3.31 There has recently been a growing interest for stick-breaking priors as a tool for specifying priors within regression problems. Based on an initial idea set forth by MacEachern (1999, 2000, 2001) who introduced the so-called dependent Dirichlet process, many subsequent papers have provided variants of the stick-breaking construction so to allow either the random masses $\tilde{p}_j$ or the random locations X_i to depend on a set of covariates $z \in \mathbb{R}^d$. In this respect, stick-breaking priors are particularly useful, since they allow us to introduce dependence in a relatively simple way. This leads to a family of random probability measures $\{\tilde{p}_z : z \in \mathbb{R}^d\}$ where

$$\tilde{p}_z = \sum_{j \geq 1} \tilde{p}_{j,z} \, \delta_{X_{j,z}}.$$

A natural device for incorporating dependence on z into the $\tilde{p}_j$ is to let the variables V_i depend on $z \in \mathbb{R}^d$: for example one might have $V_{i,z} \sim \text{Beta}(a_z, b_z)$. As for the dependence of the locations on z, the most natural approach is to take the $X_{i,z}$ i.i.d. with distribution $P_{0,z}$. Anyhow, we will not enter the technical details related to these priors: these, and other interesting proposals, are described extensively in Chapter 7.

Turning attention back to the PD(σ, θ) process as a species sampling model, the weights $p_{j,n}$ defining the predictive distribution induced by $\tilde{p}$ are known. Indeed, if $\xi_1, \ldots, \xi_n$ are i.i.d. random variables with distribution P_0, then $X_1 = \xi_1$ and, for any $i \geq 2$, one has

$$X_{n+1} \mid (X_1, \ldots, X_n) = \begin{cases} \xi_{n+1} & \text{with prob } (\theta + \sigma k_n)/(\theta + n) \\ X^*_{n,j} & \text{with prob } (n_{n,j} - \sigma)/(\theta + n) \end{cases}$$

with $X^*_{n,j}$ being the jth of the k_n distinct species observed among $X_1, \ldots, X_n$ and $n_{n,j}$ is the number of times the jth species $X^*_{n,j}$ has been observed. Besides the characterization in terms of predictive distributions, Pitman (1996) has also provided a representation of the posterior distribution of a PD(σ, θ) process $\tilde{p}$, given the data $\boldsymbol{X}$. Suppose $\mathbb{E}[\tilde{p}] = P_0$ and let $\boldsymbol{X} = (X_1, \ldots, X_n)$ be such that it contains $k \leq n$ distinct values $X^*_1, \ldots, X^*_k$, with respective frequencies $n_1, \ldots, n_k$. Then

$$\tilde{p} \mid \boldsymbol{X} \stackrel{d}{=} \sum_{j=1}^{k} p^*_j \, \delta_{X^*_j} + \left(1 - \sum_{j=1}^{k} p^*_j\right) \tilde{p}^{(k)} \tag{3.40}$$

where $\tilde{p}^{(k)}$ is a PD$(\sigma, \theta + k\sigma)$ such that $\mathbb{E}[\tilde{p}^{(k)}] = P_0$ and $(p^*_1, \ldots, p^*_k) \sim \text{Dir}(n_1 - \sigma, \ldots, n_k - \sigma, \theta + k\sigma)$. The posterior distribution of a PD(σ, θ) process can also be described in terms of a mixture with respect to a latent random variable, thus replicating the structure already encountered for NRMI. Let $\boldsymbol{X}$ be, as usual, the set of n data with $k \leq n$ distinct values $X^*_1, \ldots, X^*_k$ and let U_k be a positive random variable with density

$$q_{\sigma,\theta,k}(u) = \frac{\sigma}{\Gamma(k+\theta/\sigma)}\, u^{\theta+k\sigma-1}\, \mathrm{e}^{-u^{\sigma}}.$$

It can be shown that the distribution of a PD(σ, θ) process, conditional on the data $\boldsymbol{X}$ and on U_k, coincides with the distribution of a normalized CRM

$$\tilde{\mu}_{U_k} + \sum_{i=1}^{k} J_i^{(U_k)}\, \delta_{X_i^*}$$

where $\tilde{\mu}_{U_k}$ is a generalized gamma process with $\rho_x^{(U_k)}(\mathrm{d}s) = \rho^{(U_k)}(\mathrm{d}s) = \frac{\sigma}{\Gamma(1-\sigma)}\, s^{-1-\sigma}\, \mathrm{e}^{-U_k s}\, \mathrm{d}s$. The jumps $J_i^{(U_k)}$ at the observations X_i^* are independent gamma random variables with $\mathbb{E}[J_i^{(U_k)}] = (n_i - \sigma)/U_k$. Moreover, the jumps $J_i^{(U_k)}$ and the random measure $\tilde{\mu}_{U_k}$ are, conditional on U_k, independent. This characterization shows quite nicely the relation between the posterior behavior of the PD(σ, θ) process and of the generalized gamma NRMI, detailed in Example 3.24. Finally, note that the posterior representation in (3.40) is easily recovered by integrating out U_k.

Remark 3.32 Species prediction problems based on these models have been considered by Lijoi, Mena and Prünster (2007b). Specifically, they assume that data are directed by a Gibbs-type prior. Conditionally on $X_1, \ldots, X_n$, exact evaluations are derived for the following quantities: the p.d. of the number of new species that will be detected among the observations $X_{n+1}, \ldots, X_{n+m}$; the probability that the observation X_{n+m+1} will show a new species. Various applications, for example gene discovery prediction in genomics, illustrate nicely how discrete nonparametric priors can be successfully used to model directly the data, if these present ties. In this context the need for predictive structures, which exhibit a more flexible clustering mechanism than the one induced by the Dirichlet process, becomes apparent.

3.4 Models for density estimation

Up to now we have mainly focused on nonparametric priors, which select almost surely discrete probability measures. Due to the nonparametric nature of the models, it is clear that the set of such discrete distributions is not dominated by a fixed σ-finite measure. In the present section we illustrate two different approaches for defining priors whose realizations yield, almost surely, p.d.s admitting a density function with respect to some σ-finite measure λ on $\mathbb{X}$. The results we are going to describe are useful, for example, when one wants to model directly data generated by a continuous distribution on $\mathbb{X} = \mathbb{R}$.

3.4.1 Mixture models

An important and general device for defining a prior on densities was first suggested by Lo (1984). The basic idea consists in introducing a sequence of exchangeable latent variables $(\theta_n)_{n\geq 1}$ governed by some discrete random probability measure $\tilde{p}$ on Θ, a Polish space endowed with the Borel σ-field, which is convoluted with a suitable kernel k. To be more precise, k is a jointly measurable application from $\mathbb{X} \times \Theta$ to $\mathbb{R}^+$ and, given the dominating measure λ, the application $C \mapsto \int_C k(x,\theta)\lambda(\mathrm{d}x)$ defines a probability measure on $\mathbb{X}$ for any $\theta \in \Theta$. Hence, for any θ, $k(\,\cdot\,,\theta)$ is a density function on $\mathbb{X}$ with respect to λ. A hierarchical mixture model can, then, be defined as follows

$$
\begin{aligned}
X_i \mid \theta_i, \tilde{p} &\overset{\text{ind}}{\sim} k(\,\cdot\,,\theta_i)\\
\theta_i \mid \tilde{p} &\overset{\text{iid}}{\sim} \tilde{p}\\
\tilde{p} &\sim Q.
\end{aligned}
$$

This is the same as saying that, given the random density

$$
x \mapsto \tilde{f}(x) = \int_\Theta k(x,\theta)\,\tilde{p}(\mathrm{d}\theta) = \sum_{j\geq 1} k(x,\theta_j)\,\tilde{p}_j, \tag{3.41}
$$

the observations X_i are independent and identically distributed and the common p.d. has density function $\tilde{f}$. In (3.41), the $\tilde{p}_j$ are the probability masses associated to the discrete mixing distribution $\tilde{p}$. The original formulation of the model provided by Lo (1984) sets $\tilde{p}$ to coincide with a Dirichlet process: hence it takes on the name of *mixture of Dirichlet process* whose acronym MDP is commonly employed in the Bayesian literature. It is apparent that one can replace the Dirichlet process in (3.41) with any of the discrete random probability measures examined in Section 3.3. As for the choice of the kernels the most widely used is represented by the Gaussian kernel: in this case, if the nonparametric prior is assigned to both mean and variance, then $\tilde{p}$ is defined on $\Theta = \mathbb{R} \times \mathbb{R}^+$. Such an approach to density estimation yields, as a by-product, a natural framework for investigating the clustering structure within the observed data. Indeed, given the discreteness of $\tilde{p}$, there can be ties among the latent variables in the sense that $\mathbb{P}[\theta_i = \theta_j] > 0$ for any $i \neq j$. Possible coincidences among the θ_i induce a partition structure within the observations. Suppose, for instance, that there are $k \leq n$ distinct values $\theta_1^*, \ldots, \theta_k^*$ among $\theta_1, \ldots, \theta_n$ and let $C_j := \{i : \theta_i = \theta_j^*\}$ for $j = 1, \ldots, k$. According to such a definition, any two different indices i and l belong to the same group C_j if and only if $\theta_i = \theta_l = \theta_j^*$. Hence, the C_j describe a clustering scheme for the observations X_i: any two observations X_i and X_l belong to the same cluster if and only if $i, l \in I_j$ for some j. In particular, the number of distinct values θ_i^* among the latent θ_i identifies the number of clusters into which the n observations can be partitioned.

by the addition of a further acceleration step. Once the number $k^{(t)}$ of distinct latents has been sampled according to the scheme above, one proceeds to re-sampling the values of the $k^{(t)}$ distinct latent variables from their marginal distribution. In other words, given $k^{(t)}$ and the vector $\boldsymbol{\theta}^{(t)} = (\theta^*_{1,t}, \ldots, \theta^*_{k^{(t)},t})$, one re-samples the labels of $\boldsymbol{\theta}^{(t)}$ from the distribution

$$\mathbb{P}\left[\theta^*_{1,t} \in \mathrm{d}\theta_1, \ldots, \theta^*_{k^{(t)},t} \in \mathrm{d}\theta_{k^{(t)}} \,\middle|\, \boldsymbol{X}, \boldsymbol{\theta}^{(t)}\right] \propto \prod_{j=1}^{k^{(t)}} \prod_{i \in C_{j,t}} k(X_i, \theta_j)\, P_0(\mathrm{d}\theta_j)$$

where the $C_{j,t}$ are sets of indices denoting the membership to each of the $k^{(t)}$ clusters at iteration t. Such an additional sampling step was suggested in MacEachern (1994) and Bush and MacEachern (1996), see also Ishwaran and James (2001). Another difficulty arises for nonconjugate models where it is not possible to sample from $P_{0,i}(\mathrm{d}\theta_i)$ and evaluate exactly the weights $q^*_{i,0}$. A variant to the sampler in this case was proposed by MacEachern and Müller (1998), Neal (2000) and Jain and Neal (2007). Note that, even if these remedies where devised for the MDP, they work for any mixture of random probability measure. □

Remark 3.34 According to a terminology adopted in Papaspiliopoulos and Roberts (2008), the previous Gibbs sampling scheme can be seen as a marginal method in the sense that it exploits the integration with respect to the underlying $\tilde{p}$. The alternative family of algorithms is termed conditional methods: these rely on the simulation of the whole model and, hence, of the latent random probability measure as well. The simulation of $\tilde{p}$ can be achieved either by resorting to the Ferguson and Klass (1972) algorithm or by applying MCMC methods tailored for stick-breaking priors; see Ishwaran and James (2001, 2003b), Walker (2007) and Papaspiliopoulos and Roberts (2008). Here we do not pursue this point and refer the interested reader to the above mentioned articles. In particular, Papaspiliopoulos and Roberts (2008) discuss a comparison between the two methods. It is important to stress that both approaches require an analytic knowledge of the posterior behavior of the latent random probability measure: for marginal methods the key ingredient is represented by the predictive distributions, whereas for conditional methods a posterior representation for $\tilde{p}$ is essential. □

We now describe a few examples where the EPPF is known and a full Bayesian analysis for density estimation and clustering can be carried out using marginal methods.

Example 3.35 (Mixture of the PD(σ, θ) process) These mixtures have been examined by Ishwaran and James (2001) and, within the more general framework of species sampling models, by Ishwaran and James (2003a). For a PD(σ, θ) process

$\tilde{p}$, equation (3.37) yields the following weights

$$q_{i,0}^* \propto (\theta + \sigma k_{i,n-1}) \int_\Theta k(X_i, \theta)\, P_0(\mathrm{d}\theta), \qquad q_{i,j}^* \propto (n_{i,j} - \sigma)\, k(X_i, \theta_{i,j}^*)$$

for any $j = 1, \ldots, k_{i,n-1}$. As expected, when $\sigma \to 0$ one obtains the weights corresponding to the Dirichlet process.

Example 3.36 (Mixture of the generalized gamma NRMI) If the mixing $\tilde{p}$ is a normalized generalized gamma process described in Example 3.21, one obtains a mixture discussed in Lijoi, Mena and Prünster (2007a). The Gibbs sampler is again implemented in a straightforward way since the EPPF is known: the weights $q_{i,j}^*$, for $j = 0, \ldots, k_{i,n-1}$, can be determined from the weights of the predictive, $w_{k_{i,n-1}}^{(n-1)}$ and $w_{j,k_{i,n-1}}^{(n-1)}$ as displayed in Section 3.3.2. In Lijoi, Mena and Prünster (2007a) it is observed that the parameter σ has a significant influence on the description of the clustering structure of the data. First of all, the prior distribution on the number of components of the mixture, induced by $\tilde{p}$, is quite flat if σ is not close to 0. This is in clear contrast to the highly peaked distribution corresponding to the Dirichlet case. Moreover, values of σ close to 1 tend to favor the formation of a large number of clusters most of which of size (frequency) $n_j = 1$. This phenomenon gives rise to a reinforcement mechanism driven by σ: the mass allocation, in the predictive distribution, is such that clusters of small size are penalized whereas those few groups with large frequencies are reinforced in the sense that it is much more likely that they will be re-observed. The role of σ suggests a slight modification of the Gibbs sampler above and one needs to consider the full conditional of σ as well. Hence, if it is supposed that the prior for σ is some density q on $[0, 1]$, one finds out that the conditional distribution of σ, given the data $\boldsymbol{X}$ and the latent variables $\boldsymbol{\theta}$, is

$$\begin{aligned} q\left(\sigma \mid \boldsymbol{X}, \boldsymbol{\theta}\right) &= q(\sigma \mid \boldsymbol{\theta}) \propto q(\sigma)\, \sigma^{k\ 1} \left(\prod_{j=1}^{k} (1-\sigma)_{n_j - 1} \right) \\ &\quad \times \sum_{i=0}^{n-1} \binom{n-1}{i} (-1)^i\, \beta^{i/\sigma}\, \Gamma\left(k - \frac{i}{\sigma}; \beta\right) \end{aligned}$$

where, again, $n_1, \ldots, n_k$ are the frequencies with which the $K_n = k$ distinct values among the θ_i are recorded. This strategy turns out to be very useful when inferring on the number of clusters featured by the data. It is apparent that similar comments about the role of σ apply to the PD(σ, θ) process as well.

We close this subsection with another interesting model of mixture introduced in Petrone (1999a,b): random Bernstein polynomials.

Example 3.37 (Random Bernstein polynomials) A popular example of a nonparametric mixture model for density estimation was introduced by Petrone (1999a,b). The definition of the prior is inspired by the use of Bernstein polynomials for the approximation of real functions. Indeed, it is well known that if F is a continuous function defined on [0, 1] then the polynomial of degree m defined by

$$B_m^F(x) = \sum_{j=0}^{m} F\left(\frac{j}{m}\right)\binom{m}{j} x^j(1-x)^{k-j} \tag{3.43}$$

converges, uniformly on [0, 1], to F as $m \to \infty$. The function B_m^F in (3.43) takes on the name of Bernstein polynomial on [0, 1]. It is clear that, when F is a distribution function on [0, 1], then B_m^F is a distribution function as well. Moreover, if the p.d. corresponding to F does not have a positive mass on $\{0\}$ and $\beta(x; a, b)$ denotes the density function of a beta random variable with parameters a and b, then

$$b_m^F(x) = \sum_{j=1}^{m} [F(j/m) - F((j-1)/m)]\ \beta(x; j, m-j+1) \tag{3.44}$$

for any $x \in [0, 1]$ is named a Bernstein density. If F has density f, it can be shown that $b_m^F \to f$ pointwise as $m \to \infty$. These preliminary remarks on approximation properties for Bernstein polynomials suggest that a prior on the space of densities on [0, 1] can be constructed by randomizing both the polynomial degree m and the weights of the mixture (3.44). In order to define properly a random Bernstein prior, let $\tilde{p}$ be, for instance, some NRMI generated by a CRM $\tilde{\mu}$ with Lévy intensity $\rho_x(\mathrm{d}s)\alpha(\mathrm{d}x)$ concentrated on $\mathbb{R}^+ \times [0, 1]$ and $\alpha([0, 1]) = a \in (0, \infty)$. Next, for any integer $m \geq 1$, introduce a discretization of α as follows:

$$\alpha^{(m)} = \sum_{j=1}^{m} \alpha_{j,m}\, \delta_{j/m}$$

where the weights $\alpha_{j,m}$ are nonnegative and such that $\sum_{j=1}^{m} \alpha_{j,m} = a$. One may note that the intensity $\nu^{(m)}(\mathrm{d}s, \mathrm{d}x) = \rho_x(\mathrm{d}s)\, \alpha^{(m)}(\mathrm{d}x)$ defines a NRMI $\tilde{p}_m$ which is still concentrated on $S_m := \{1/m, \ldots, (m-1)/m, 1\}$, i.e.

$$\tilde{p}_m = \sum_{j=1}^{m} \tilde{p}_{j,m}\, \delta_{j/m}$$

where $\tilde{p}_{j,m} = p(((j-1)/m,\ j/m])$, for any $j = 2, \ldots, m$, and $\tilde{p}_{1,m} = \tilde{p}([0, 1/m])$. Hence, if π is a prior on $\{1, 2, \ldots\}$, a Bernstein random polynomial prior is defined as the p.d. of the random density $\tilde{f}(x) = \sum_{m \geq 1} \pi(m)\, \tilde{f}_m(x)$, where

$$\tilde{f}_m(x) = \int_{[0,1]} \beta(x; my, m - my + 1)\, \tilde{p}_m(\mathrm{d}y) \tag{3.45}$$

is a mixture of the type (3.41). Conditional on m, $\tilde{f}_m$ defines a prior on the space of densities on $[0, 1]$. The previous definition can be given by introducing a vector of latent variables $\boldsymbol{Y} = (Y_1, \ldots, Y_n)$ and function $x \mapsto Z_m(x) = \sum_{j=1}^{m} j\, \mathbb{1}_{B_{j,m}}(x)$ where $B_{1,m} = [0, 1/m]$ and $B_{j,m} = ((j-1)/m, j/m]$ for any $j = 2, \ldots, m$. Hence, a Bernstein random polynomial prior can be defined through the following hierarchical mixture model

$$\begin{aligned} X_j \mid m, \tilde{p}, Y_j &\overset{\text{ind}}{\sim} \text{Beta}(Z_m(Y_j), m - Z_m(Y_j) + 1) \qquad j = 1, \ldots, n \\ Y_j \mid m, \tilde{p} &\overset{\text{i.i.d.}}{\sim} \tilde{p} \\ \tilde{p} \mid m &\sim Q \\ m &\sim \pi. \end{aligned}$$

The original definition provided in Petrone (1999a) involves a Dirichlet process, $\tilde{p}$, with parameter measure α and the author refers to it as a *Bernstein–Dirichlet prior* with parameters (π, α). The use of the Dirichlet process is very useful, especially when implementing the MCMC strategy devised in Petrone (1999a,b) since, conditional on m, the vector of weights $(\tilde{p}_{1,m}, \ldots, \tilde{p}_{m-1,m})$ in (3.45) turns out to be distributed according to an $(m-1)$-variate Dirichlet distribution with parameters $(\alpha_{1,m}, \ldots, \alpha_{m,m})$. Nonetheless, the posterior distribution of $(m, \tilde{p}_m)$, given $\boldsymbol{X} = (X_1, \ldots, X_n)$, is proportional to $\pi(m)\pi(p_{1,m}, \ldots, p_{m-1,m}) \prod_{i=1}^{n} \tilde{f}_m(X_i)$ which is analytically intractable since it consists of a product of mixtures. For example, it is impossible to evaluate the posterior distribution $\pi(m|X_1, \ldots, X_n)$ which is of great interest since it allows us to infer the number of components in the mixture and, hence, the number of clusters in the population. As for density estimation, the Bayesian estimate of $\tilde{f}$ with respect to a squared loss function is given by

$$\mathbb{E}[\tilde{f}(x) \mid X_1, \ldots, X_n] = \sum_{m \geq 1} \tilde{f}_m^*(x)\, \pi(m|X_1, \ldots, X_m)$$

with $\tilde{f}_m^*(x) = \sum_{j=1}^{m} \mathbb{E}[\tilde{p}_{j,m} \mid m, X_1, \ldots, X_n]\, \beta(x; j, m-j+1)$. This entails that the posterior estimate of $\tilde{f}$ is still a Bernstein random polynomial with updated weights, see Petrone (1999b).

Given the analytical difficulties we have just sketched, performing a full Bayesian analysis asks for the application of suitable computational schemes such as the MCMC algorithm devised in Petrone (1999b). The implementation of the algorithm is tailored to the Bernstein–Dirichlet process prior. It is assumed that the distribution function $x \mapsto F_0(x) = \alpha([0, x])/a$ is absolutely continuous with density f_0. Next, by making use of the latent variables $\boldsymbol{Y}$, a simple application of Bayes's theorem shows that

$$\pi(m|\boldsymbol{Y}, \boldsymbol{X}) \propto \pi(m) \prod_{i=1}^{m} \beta(X_i; Z_m(Y_j), m - Z_m(Y_j) + 1).$$

On the other hand, since $\tilde{p}$ is the Dirichlet process with parameter measure α, one has the following predictive structure for the latent variables

$$\pi(Y_j \mid m, \boldsymbol{Y}_{-j}, \boldsymbol{X}) \propto q(X_j, m)\, f_0(Y_j)\, \beta(X_j;\, Z_m(Y_j),\, m - Z_m(Y_j) + 1) + \sum_{i \neq j} q_i^*(X_j, m)\, \delta_{Y_i} \tag{3.46}$$

with $\boldsymbol{Y}_{-j}$ denoting the vector of latent variables obtained by deleting Y_j, the density $b_m^{F_0}$ defined as in (3.44) and

$$q(X_j, m) \propto a\, b_m^{F_0}(X_j), \qquad q_i^*(X_j, m) \propto \beta(X_j;\, Z_m(Y_i),\, m - Z_m(Y_i) + 1)$$

such that $q(X_j, m) + \sum_{i \neq j} q_i^*(X_j, m) = 1$. The predictive distribution in (3.46) implies that: (i) with probability $q(X_j, m)$ the value of Y_j is sampled from a density $f(y) \propto f_0(y)\, \beta(X_j;\, Z_m(y),\, m - Z_m(y) + 1)$ and (ii) with probability $q_i^*(X_j, m)$ the value of Y_j coincides with Y_i. Hence, one can apply the following Gibbs sampling algorithm in order to sample from the posterior distribution of $(m, \boldsymbol{Y}, \tilde{p}_{1,m}, \ldots, \tilde{p}_{m,m})$. Starting from initial values $(m^{(0)}, \boldsymbol{Y}^{(0)}, p_{1,m}^{(0)}, \ldots, p_{m,m}^{(0)})$, at iteration $t \geq 1$ one samples

1. $m^{(t)}$ from $\pi(m \mid \boldsymbol{Y}^{(t-1)}, \boldsymbol{X})$;
2. $Y_i^{(t)}$ from the predictive $\pi(Y_i \mid m^{(t)}, Y_1^{(t)}, \ldots, Y_{i-1}^{(t)}, Y_{i+1}^{(t-1)}, \ldots, Y_n^{(t-1)}, \boldsymbol{X})$ described in (3.46);
3. $(p_{1,m}^{(t)}, \ldots, p_{m,m}^{(t)})$ from an $(m-1)$-variate Dirichlet distribution with parameters $(\alpha_{1,m^{(t)}} + n_1, \ldots, \alpha_{m^{(t)},m^{(t)}} + n_{m^{(t)}})$, where n_j is the number of latent variables in $(Y_1^{(t)}, \ldots, Y_n^{(t)})$ in $B_{j,m^{(t)}}$.

For further details, see Petrone (1999a).

3.4.2 *Pólya trees*

Pólya trees are another example of priors which, under suitable conditions, are concentrated on absolutely continuous p.d.s with respect to the Lebesgue measure on $\mathbb{R}$. A first definition of Pólya trees can be found in Ferguson (1974) and a systematic treatment is provided by Lavine (1992, 1994) and Mauldin, Sudderth and Williams (1992). A useful preliminary concept is that of *tail-free prior* introduced by Freedman (1963). Let $\Gamma = \{\Gamma_k : k \geq 1\}$ be a nested tree of measurable partitions of $\mathbb{X}$. This means that Γ_{k+1} is a refinement of Γ_k, i.e. each set in Γ_{k+1} is the union of sets in Γ_k, and that $\cup_{k \geq 1} \Gamma_k$ generates $\mathscr{X}$, with $\mathscr{X}$ denoting the Borel σ-algebra of $\mathbb{X}$. One can, then, give the following definition.

Definition 3.38 A random probability measure $\tilde{p}$ on $\mathbb{X} \subset \mathbb{R}$ is tail-free with respect to Γ if there exist nonnegative random variables $\{V_{k,B} : k \geq 1,\ B \in \Gamma_k\}$ such that

(i) the families $\{V_{1,B} : B \in \Gamma_1\}, \{V_{2,B} : B \in \Gamma_2\}, \ldots$, are independent,
(ii) if $B_k \subset B_{k-1} \subset \cdots \subset B_1$, with $B_j \in \Gamma_j$, then $\tilde{p}(B_k) = \prod_{j=1}^{k} V_{j,B_j}$.

For tail-free processes a structural conjugacy property holds true: if $\tilde{p}$ is tail-free with respect to Γ, then $\tilde{p}$ given the data is still tail-free with respect to Γ.

Pólya trees can be recovered as special case of tail-free processes with the $V_{k,B}$ variables having a beta distribution. To illustrate the connection, consider the family Γ of partitions described as follows

$$\Gamma_1 = \{B_0, B_1\}, \quad \Gamma_2 = \{B_{00}, B_{01}, B_{10}, B_{11}\}, \quad \Gamma_3 = \{B_{000}, B_{001}, B_{010}, \ldots, B_{111}\}$$

and so on. In the above definition of the Γ_i we set $B_0 = B_{00} \cup B_{01}$, $B_1 = B_{10} \cup B_{11}$ and, given sets $B_{\boldsymbol{\varepsilon}0}$ and $B_{\boldsymbol{\varepsilon}1}$ in Γ_{k+1}, one has

$$B_{\boldsymbol{\varepsilon}0} \cup B_{\boldsymbol{\varepsilon}1} = B_{\boldsymbol{\varepsilon}}$$

for any $\boldsymbol{\varepsilon} = (\varepsilon_1, \ldots, \varepsilon_k) \in E^k = \{0, 1\}^k$. With this notation, the kth partition can be described as $\Gamma_k = \{B_{\boldsymbol{\varepsilon}} : \boldsymbol{\varepsilon} \in E^k\}$. Finally, let $E^* = \cup_{k \geq 1} E^k$ be the set of all sequences of zeros and ones and $\mathscr{A} = \{\alpha_{\boldsymbol{\varepsilon}} : \boldsymbol{\varepsilon} \in E^*\}$ a set of nonnegative real numbers.

Definition 3.39 A random probability measure $\tilde{p}$ is a Pólya tree process with respect to $\Gamma = \{\Gamma_k : k \geq 1\}$ and $\mathscr{A}$, in symbols $\tilde{p} \sim \mathrm{PT}(\mathscr{A}, \Gamma)$, if

(i) $\{\tilde{p}(B_{\boldsymbol{\varepsilon}0} | B_{\boldsymbol{\varepsilon}}) : \boldsymbol{\varepsilon} \in E^*\}$ is a collection of independent random variables,
(ii) $\tilde{p}(B_{\boldsymbol{\varepsilon}0} | B_{\boldsymbol{\varepsilon}}) \sim \mathrm{Beta}(\alpha_{\boldsymbol{\varepsilon}0}, \alpha_{\boldsymbol{\varepsilon}1})$.

The existence of a Pólya tree with respect to the parameters $\mathscr{A}$ is guaranteed by the validity of the following conditions expressed in terms of infinite products:

$$\frac{\alpha_{\boldsymbol{\varepsilon}0}}{\alpha_{\boldsymbol{\varepsilon}0} + \alpha_{\boldsymbol{\varepsilon}1}} \frac{\alpha_{\boldsymbol{\varepsilon}00}}{\alpha_{\boldsymbol{\varepsilon}00} + \alpha_{\boldsymbol{\varepsilon}01}} \cdots = 0, \qquad \frac{\alpha_1}{\alpha_0 + \alpha_1} \frac{\alpha_{11}}{\alpha_{10} + \alpha_{11}} \cdots = 0.$$

These ensure that the Pólya random probability measure is countably additive, almost surely. For a proof of this fact see, for example, Ghosh and Ramamoorthi (2003).

One of the most relevant properties of a Pólya tree prior $\mathrm{PT}(\mathscr{A}, \Gamma)$ is that, under a suitable condition on the parameters in $\mathscr{A}$, the realizations of $\tilde{p}$ are, almost surely, p.d.s that are absolutely continuous. In order to illustrate such a condition we confine ourselves to the case where $\mathbb{X} = [0, 1]$, the extension to the case $\mathbb{X} = \mathbb{R}$ being straightforward. Suppose that Γ is a sequence of dyadic partitions of $[0, 1]$, i.e. with $\boldsymbol{\varepsilon} \in E^k$ one has $B_{\boldsymbol{\varepsilon}} = (\sum_{j=1}^{k} \varepsilon_j 2^{-j}, \sum_{j=1}^{k} \varepsilon_j 2^{-j} + 2^{-k}]$. As noted in Ferguson (1974), using a result in Kraft (1964), one can show that if $\tilde{p} \sim \mathrm{PT}(\mathscr{A}, \Gamma)$ and the $\alpha_{\varepsilon_1 \cdots \varepsilon_k}$, seen as a function of the level on the partition tree, increase at a rate of at least k^2 or faster, then the p.d. of $\tilde{p}$ is concentrated on the set of probability measures that are absolutely continuous with respect to the Lebesgue measure.

heavily depend on the specific sequence of partitions Γ. In order to overcome the issue, Lavine (1992) suggests the use of mixtures of Pólya trees. This amounts to assuming the existence of random variables θ and ξ such that

$$\begin{aligned} \tilde{p} \mid (\theta, \xi) &\sim \mathrm{PT}(\mathscr{A}^{\theta}, \Gamma^{\xi}) \\ (\theta, \xi) &\sim \pi. \end{aligned}$$

If the prior π on the mixing parameters satisfies some suitable conditions, then the dependence on the partitions is smoothed out and the predictive densities can be continuous. A similar device is adopted in Paddock, Ruggeri, Lavine and West (2003) where the authors introduce a sequence of independent random variables which determine the end points partition elements in Γ_k, for any $k \geq 1$. Mixtures of Pólya trees are also used in Hanson and Johnson (2002) to model the regression error and the authors investigate applications to semiparametric accelerated failure time models.

3.5 Random means

The investigation of general classes of priors as developed in the previous sections is of great importance when it comes to study some quantities of statistical interest. Among these, here we devote some attention to random means, namely to linear functionals of random probability measures $\tilde{p}(f) = \int f \, \mathrm{d}\tilde{p}$, with f being some measurable function defined on $\mathbb{X}$. For instance, if the data are lifetimes, $\int x \, \tilde{p}(\mathrm{d}x)$ represents the random expected lifetime. The reason for focusing on this topic lies not only in the statistical issues that can be addressed in terms of means, but also because many of the results obtained for means of nonparametric priors do have important connections with seemingly unrelated research topics such as, for example, excursions of Bessel processes, the moment problem, special functions and combinatorics.

The first pioneering fundamental contributions to the study of means are due to D. M. Cifarelli and E. Regazzini. In their papers (Cifarelli and Regazzini, 1979a, 1979b, 1990) they provide useful insight into the problem and obtain closed form expressions for the p.d. of $\tilde{p}(f)$ when $\tilde{p}$ is a Dirichlet process. They first determine the remarkable identity for means of the Dirichlet process

$$\mathbb{E}\left[\frac{1}{\{1 + \mathrm{i}t\tilde{p}(f)\}^{\theta}}\right] = \exp\left\{-\int \log(1 + \mathrm{i}tf) \, \mathrm{d}\alpha\right\} \qquad \forall t \in \mathbb{R} \tag{3.48}$$

where f is any measurable function on $\mathbb{X}$ such that $\int \log(1 + |f|) \, \mathrm{d}\alpha < \infty$ and $\theta = \alpha(\mathbb{X}) \in (0, \infty)$. The left-hand side of (3.48) is the Stieltjes transform of order θ of the p.d., say $M_{\alpha,f}$, of the Dirichlet mean $\tilde{p}(f)$, while the right-hand side is the Laplace transform of $\int f \, \mathrm{d}\tilde{\gamma}$ where $\tilde{\gamma}$ is a gamma process with parameter measure α. Equation (3.48) has been termed the *Markov–Krein identity* because of its connections to the Markov moment problem, whereas it is named the *Cifarelli–Regazzini identity* in James (2006b). By resorting to (3.48), Cifarelli and Regazzini (1990) apply an inversion formula for Stieltjes transforms and obtain an expression for $M_{\alpha,f}$. For example, if $\theta = 1$, the density function corresponding to $M_{\alpha,f}$ coincides with

$$m_{\alpha,f}(x) = \frac{1}{\pi} \sin(\pi F_{\alpha^*}(x)) \exp\left\{-\mathrm{PV} \int_{\mathbb{R}} \log|y-x|\,\alpha^*(\mathrm{d}y)\right\}$$

where $\alpha^*(B) = \alpha(\{x \in \mathbb{R} : f(x) \in B\})$ is the image measure of α through f, F_{α^*} is the corresponding distribution function and $\mathrm{PV}\int$ means that the integral is a principal-value integral. In Diaconis and Kemperman (1996) one can find an interesting discussion with some applications of the formulae of Cifarelli and Regazzini (1990). Alternative expressions for $M_{\alpha,f}$ can be found in Regazzini, Guglielmi and Di Nunno (2002) where the authors rely on an inversion formula for characteristic functions due to Gurland (1948).

Since, in general, the exact analytic form of $M_{\alpha,f}$ is involved and difficult to evaluate, it is desirable to devise some convenient method to sample from $M_{\alpha,f}$ or to approximate it numerically. For example, Muliere and Tardella (1998) make use of the stick-breaking representation of the Dirichlet process and suggest an approximation based on a random stopping rule. In Regazzini, Guglielmi and Di Nunno (2002) one can find a numerical approximation of $M_{\alpha,f}$.

In Lijoi and Regazzini (2004) it is noted that when the baseline measure α is concentrated on a finite number of points, then the left-hand side of (3.48) coincides with the fourth Lauricella hypergeometric function, see Exton (1976). Such a connection has been exploited in order to provide an extension of (3.48) where the order of the Stieltjes transform does not need to coincide with the total mass of the baseline measure α. Other interesting characterizations of $M_{\alpha,f}$ can also be found in Hjort and Ongaro (2005). It is worth noting that Romik (2004, 2005) has recently pointed out how the p.d. $M_{\alpha,f}$ of a Dirichlet random mean coincides with the limiting distribution of a particular hook walk: it precisely represents the p.d. of the point where the hook walk intersects, on the plane, the graph of a continual Young diagram. Recall that a continual Young diagram is a positive increasing function g on some interval $[a, b]$ and it can be seen as the continuous analog of the Young diagram which is a graphic representation of a partition of an integer n. Romik (2004, 2005) has considered the problem of determining a formula for the baseline measure α (with support a bounded interval $[\xi_1, \xi_2]$) corresponding to a specified distribution $M_{\alpha,f}$ for the Dirichlet random mean. The solution he obtains is described by

$$F_\alpha(x) = \frac{1}{\pi} \operatorname{arccot}\left(\frac{1}{\pi\, m_{\alpha,f}(x)} \,\mathrm{PV}\int_{[\xi_1,\xi_2]} \frac{m_{\alpha,f}(y)}{y-x}\,\mathrm{d}y \right).$$

See also Cifarelli and Regazzini (1993) for an alternative representation of F_α and Hill and Monticino (1998) for an allied contribution.

There have also been recent contributions to the analysis of linear functionals of more general classes of priors of the type we have been presenting in this chapter. In Regazzini, Lijoi and Prünster (2003) the authors resort to Gurland's inversion formula for characteristic functions and provide an expression for the distribution function of linear functionals $\tilde{p}(f)$ of NRMIs. This approach can be naturally extended to cover means of the mixture of a Dirichlet process (Nieto-Barajas, Prünster and Walker, 2004). In Epifani, Lijoi and Prünster (2003) one can find an investigation of means of NTR priors which are connected to exponential functionals of Lévy processes: these are of great interest in the mathematical finance literature. The determination of the p.d. of a linear functional of a two-parameter Poisson–Dirichlet process is the focus of James, Lijoi and Prünster (2008). They rely on a representation of the Stieltjes transform of $\tilde{p}(f)$ as provided in Kerov (1998) and invert it. The formulae they obtain are of relevance also for the study of excursions of Bessel processes, which nicely highlights the connection of Bayes nonparametrics with other areas in strong development. Indeed, let $Y = \{Y_t, t \geq 0\}$ denote a real-valued process, such that: (i) the zero set $\mathcal{Z}$ of Y is the range of a σ-stable process and (ii) given $|Y|$, the signs of excursions of Y away from zero are chosen independently of each other to be positive with probability p and negative with probability $\bar{p} = 1 - p$. Examples of this kind of process are the Brownian motion ($\sigma = p = 1/2$), the skew Brownian motion ($\sigma = 1/2$ and $0 < p < 1$), the symmetrized Bessel process of dimension $2 - 2\sigma$, the skew Bessel process of dimension $2 - 2\sigma$. Then for any random time T which is a measurable function of $|Y|$,

$$A_T = \int_0^T \mathbb{1}_{(0,+\infty)}(Y_s)\,\mathrm{d}s \tag{3.49}$$

denotes the time spent positive by Y up to time T and A_T/T coincides in distribution with the distribution of $\tilde{p}(f)$ where $\tilde{p}$ is a PD(σ, σ) process and $f = \mathbb{1}_C$, the set C being such that $\alpha(C)/\theta = p$. See Pitman and Yor (1997b) for a detailed analysis. A recent review on means of random probability measures is provided in Lijoi and Prünster (2009).

3.6 Concluding remarks

In the present chapter we have provided an overview of the various classes of priors which generalize the Dirichlet process. As we have tried to highlight, most of them are suitable transformations of CRMs and they all share a common a posteriori structure. As far as the tools for deriving posterior representations are concerned, there are essentially two general techniques and both take the Laplace functional in (3.3) as starting point. The first one, set forth in James (2002) and developed and refined in subsequent papers, is termed Poisson partition calculus: the key idea consists in facing the problem at the level of the Poisson process underlying the CRM, according to (3.7), and then using Fubini-type arguments. The second approach, developed by the two authors of the present review and first outlined in Prünster (2002), tackles the problem directly at the CRM level, interprets observations as derivatives of the Laplace functional and then obtains the posterior representations as Radon–Nikodým derivatives.

A last remark concerns asymptotics, a research area under strong development which has been accounted for in Chapters 1 and 2. Among the asymptotic properties, consistency plays a predominant role. Despite the general validity of proper Bayesian Doob-style consistency, the "what if" or frequentist approach to consistency set forth by Diaconis and Freedman (1986) has recently gained great attention. The evaluation of a Bayesian procedure according to such a frequentist criterion is appropriate when one believes that data are i.i.d. from some "true" distribution P_0 and, nonetheless, assumes exchangeability as a tool which leads to a sensible rule for making predictions and for inductive reasoning. One is, then, interested in ascertaining whether the posterior distribution accumulates in suitable neighborhoods of P_0 as the sample size increases. A few examples of inconsistency provide a warning and suggest a careful treatment of this issue. Many sufficient conditions ensuring frequentist consistency are now available and results on rates of convergence have been derived as well. If one adheres to such a frequentist point of view, then one should choose, among priors for which consistency has been proved, the one featuring the fastest rate of convergence. When dealing with the discrete nonparametric priors examined in Sections 3.2 and 3.3 these considerations are clearly of interest: in fact, most of them, with the exceptions of the Dirichlet and the beta processes, are inconsistent if used to model directly continuous data. However, even an orthodox Bayesian who does not believe in the existence of a "true" P_0 and, hence, specifies priors regardless of frequentist asymptotic properties, would hardly use a discrete nonparametric prior on continuous data: this would mean assuming a model, which generates ties among observations with probability tending to 1 as the sample size diverges, for data which do not contain ties with probability 1. On the other hand, all the discrete priors we have been describing are consistent when exploited in situations they are structurally designed for. Specifically, they are consistent when used for modeling data arising from discrete distributions and, moreover, they are also consistent, under mild conditions, when exploited in a hierarchical mixture setup for continuous data. Thus, we have agreement of the two viewpoints on the models to use. Finally, note that rates of convergence do not seem to discriminate between different discrete priors in a mixture, since they are derived assuming i.i.d. data. In such cases we have to reverse the starting question and ask "what if the data are not i.i.d. but, indeed, exchangeable?" Then, the assessment of a prior should naturally be guided by considerations on the flexibility of the posterior and on the richness of the predictive structure, which also allow for a parsimonious model specification.

Acknowledgements Both authors wish to express their gratitude to Eugenio Regazzini who introduced them to the world of Bayesian statistics and has transmitted enthusiasm and skills of great help for the development of their own research. This research was partially supported by MIUR–Italy, grant 2008MK3AFZ.

References

Aalen, O. (1978). Nonparametric inference for a family of counting processes. *Annals of Statistics*, **6**, 701–26.

Aldous D. J. (1985). Exchangeability and related topics. In *École d'été de probabilités de Saint-Flour XIII*, Lecture Notes in Mathematics 1117, 1–198. Berlin: Springer.

Antoniak, C. E. (1974). Mixtures of Dirichlet processes with applications to Bayesian nonparametric problems. *Annals of Statistics*, **2**, 1152–74.

Arjas, E. and Gasbarra, D. (1994). Nonparametric Bayesian inference from right censored survival data using the Gibbs sampler. *Statistica Sinica*, **4**, 505–24.

Blackwell, D. (1973). Discreteness of Ferguson selections. *Annals of Statistics*, **1**, 356–58.

Blum, J. and Susarla, V. (1977). On the posterior distribution of a Dirichlet process given randomly right censored observations. *Stochastic Processes and Their Applications*, **5**, 207–11.

Brix, A. (1999). Generalized gamma measures and shot-noise Cox processes. *Advances in Applied Probability*, **31**, 929–53.

Bush, C. A. and MacEachern, S. N. (1996). A semiparametric Bayesian model for randomised block designs. *Biometrika*, **83**, 275–85.

Cifarelli, D. M. and Regazzini, E. (1979a). A general approach to Bayesian analysis of nonparametric problems. The associative mean values within the framework of the Dirichlet process: I (in Italian). *Rivista di Matematica per le Scienze Economiche e Sociali*, **2**, 39–52.

Cifarelli, D. M. and Regazzini, E. (1979b). A general approach to Bayesian analysis of nonparametric problems. The associative mean values within the framework of the Dirichlet process: II (in Italian). *Rivista di Matematica per le Scienze Economiche e Sociali*, **2**, 95–111.

Cifarelli, D. M. and Regazzini, E. (1990). Distribution functions of means of a Dirichlet process. *Annals of Statistics*, **18**, 429-442. (Correction in **22** (1994), 1633–34.)

Cifarelli, D. M. and Regazzini, E. (1993). Some remarks on the distribution functions of means of a Dirichlet process. *Technical Report 93.4*, IMATI–CNR, Milano.

Connor, R. J. and Mosimann, J. E. (1969). Concepts of independence for proportions with a generalization of the Dirichlet distribution. *Journal of the American Statistical Association*, **64**, 194–206

Daley, D. J. and Vere-Jones, D. (1988). *An Introduction to the Theory of Point Processes.* New York: Springer.

Damien, P., Laud, P. and Smith, A. F. M. (1995). Approximate random variate generation from infinitely divisible distributions with applications to Bayesian inference. *Journal of the Royal Statistical Society, Series B*, **57**, 547–63.

Dey, J., Erickson, R. V. and Ramamoorthi, R. V. (2003). Some aspects of neutral to the right priors. *International Statistical Review*, **71**, 383–401.

Diaconis, P. and Freedman, D. (1986). On the consistency of Bayes estimates. *Annals of Statistics*, **14**, 1–26.

Diaconis, P. and Kemperman, J. (1996). Some new tools for Dirichlet priors. In *Bayesian Statistics 5*, 97–106. New York: Oxford University Press.

Doksum, K. (1974). Tailfree and neutral random probabilities and their posterior distributions. *Annals of Probability*, **2**, 183–201.

Dykstra, R. L. and Laud, P. (1981). A Bayesian nonparametric approach to reliability. *Annals of Statistics*, **9**, 356–67.

Epifani, I., Lijoi, A. and Prünster, I. (2003). Exponential functionals and means of neutral-to-the-right priors. *Biometrika*, **90**, 791–808.

Escobar, M. D. (1988). Estimating the means of several normal populations by nonparametric estimation of the distribution of the means. *Ph.D. Dissertation*, Department of Statistics, Yale University.

Escobar, M. D. (1994). Estimating normal means with a Dirichlet process prior. *Journal of the American Statistical Association*, **89**, 268–77.

Escobar, M. D. and West, M. (1995). Bayesian density estimation and inference using mixtures. *Journal of the American Statistical Association*, **90**, 577–88.

Ewens, W. J. (1972). The sampling theory of selectively neutral alleles. *Theoretical Population Biology*, **3**, 87–112.

Exton, H. (1976). *Multiple hypergeometric Functions and Applications.* Chichester: Ellis Horwood.

Ferguson, T. S. (1973). A Bayesian analysis of some nonparametric problems. *Annals of Statistics*, **1**, 209–30.

Ferguson, T. S. (1974). Prior distributions on spaces of probability measures. *Annals of Statistics*, **2**, 615–29.

Ferguson, T. S. and Klass, M. J. (1972). A representation of independent increments processes without Gaussian components. *Annals of Mathematical Statistics*, **43**, 1634–43.

Ferguson, T. S. and Phadia, E. G. (1979). Bayesian nonparametric estimation based on censored data. *Annals of Statistics*, **7**, 163–86.

de Finetti, B. (1937). La prévision: ses lois logiques, ses sources subjectives. *Annales de l'nstitut Henri Poincaré*, **7**, 1–68.

Freedman, D. A. (1963). On the asymptotic behavior of Bayes' estimates in the discrete case. *Annals of Mathematical Statistics*, **34**, 1386–1403.

Ghosh, J. K. and Ramamoorthi, R. V. (2003). *Bayesian Nonparametrics*. New York: Springer.

Gill, R. D. and Johansen, S. (1990). A survey of product integration with a view towards survival analysis. *Annals of Statistics*, **18**, 1501–55.

Gnedin, A. and Pitman, J. (2005a). Exchangeable Gibbs partitions and Stirling triangles. *Zap. Nauchn. Sem. S.-Peterburg. Otdel. Mat. Inst. Steklov. (POMI)*, **325**, 83–102.

MacEachern, S. N. (1994). Estimating normal means with a conjugate style Dirichlet process prior. *Communications in Statistics: Simulation and Computation*, **23**, 727–41.

MacEachern, S. N. (1999). Dependent nonparametric processes. In *ASA Proceedings of the Section on Bayesian Statistical Science*, 50–5. Alexandria, Va.: American Statistical Association.

MacEachern, S. N. (2000). Dependent Dirichlet processes. *Technical Report*, Department of Statistics, Ohio State University.

MacEachern, S. N. (2001). Decision theoretic aspects of dependent nonparametric processes. In *Bayesian Methods with Applications to Science, Policy and Official Statistics*, ed. E. George, 551–60. Crete: International Society for Bayesian Analysis.

MacEachern, S. N. and Müller, P. (1998). Estimating mixture of Dirichlet process models. *Journal of Computational and Graphical Statistics*, **7**, 223–38.

Mauldin, R. D., Sudderth, W. D. and Williams, S. C. (1992). Pólya trees and random distributions. *Annals of Statistics*, **20**, 1203–21.

Muliere, P. and Tardella, L. (1998). Approximating distributions of random functionals of Ferguson–Dirichlet priors. *Canadian Journal of Statistics*, **26**, 283–97.

Muliere, P. and Walker, S. G. (1997). A Bayesian non-parametric approach to survival analysis using Pólya trees. *Scandinavian Journal of Statistics*, **24**, 331–40.

Müller, P. and Quintana, F. A. (2004). Nonparametric Bayesian data analysis. *Statistical Science*, **19**, 95–110.

Neal, R. M. (2000). Markov chain sampling methods for Dirichlet process mixture models. *Journal of Computational and Graphical Statistics*, **9**, 249–65.

Nieto-Barajas, L. E. and Prünster, I. (2009). A sensitivity analysis for Bayesian nonparametric density estimators. *Statistica Sinica*, **19**, 685–705.

Nieto-Barajas, L. E., Prünster, I. and Walker, S. G. (2004). Normalized random measures driven by increasing additive processes. *Annals of Statistics*, **32**, 2343–60.

Nieto-Barajas, L. E. and Walker, S. G. (2002). Markov beta and gamma processes for modelling hazard rates. *Scandinavian Journal of Statistics*, **29**, 413–24.

Nieto-Barajas, L. E. and Walker, S. G. (2004). Bayesian nonparametric survival analysis via Lévy driven Markov processes. *Statistica Sinica*, **14**, 1127–46.

Paddock, S. M, Ruggeri, F., Lavine, M. and West, M. (2003). Randomized Pólya tree models for nonparametric Bayesian inference. *Statistica Sinica*, **13**, 443–60.

Papaspiliopoulos, O. and Roberts, G. O. (2008). Retrospective Markov chain Monte Carlo methods for Dirichlet process hierarchical models. *Biometrika*, **95**, 169–86.

Perman, M., Pitman, J. and Yor, M. (1992). Size-biased sampling of Poisson point processes and excursions. *Probability Theory and Related Fields*, **92**, 21–39.

Petrone, S. (1999a). Random Bernstein polynomials. *Scandinavian Journal of Statistics*, **26**, 373–93.

Petrone, S. (1999b). Bayesian density estimation using Bernstein polynomials. *Canadian Journal of Statistics*, **27**, 105–26.

Pitman, J. (1995). Exchangeable and partially exchangeable random partitions. *Probability Theory and Related Fields*, **102**, 145–58.

Pitman, J. (1996). Some developments of the Blackwell–MacQueen urn scheme. In *Statistics, Probability and Game Theory. Papers in Honor of David Blackwell*, ed. T. S. Ferguson, L. S. Shapley and J. B. MacQueen, IMS Lecture Notes/Monographs 30, 245–67. Hayward, Calif.: Institute of Mathematical Statistics.

Pitman, J. (2003). Poisson–Kingman partitions. In *Statistics and Science: A Festschrift for Terry Speed*, ed. D. R. Goldstein, IMS Lecture Notes/Monographs 40, 1–34. Beachwood, Calif.: Institute of Mathematical Statistics.

Pitman, J. (2006). *Combinatorial Stochastic Processes.* Lectures from the 32nd Summer School on Probability Theory held in Saint-Flour, July 7–24, 2002. Lecture Notes in Mathematics 1875. Berlin: Springer.

Pitman, J. and Yor, M. (1997a). The two-parameter Poisson–Dirichlet distribution derived from a stable subordinator. *Annals of Probability*, **25**, 855–900.

Pitman, J., and Yor, M. (1997b). On the relative lengths of excursions derived from a stable subordinator. In *Séminaire de Probabilités XXXI*, ed. J. Azema, M. Emery, and M. Yor, Lecture Notes in Mathematics 1655, 287–305. Berlin: Springer.

Prünster, I. (2002). Random probability measures derived from increasing additive processes and their application to Bayesian statistics. *Ph.D. Thesis*, University of Pavia.

Regazzini, E. (2001). Foundations of Bayesian Statistics and Some Theory of Bayesian Nonparametric Methods. *Lecture Notes*, Stanford University.

Regazzini, E., Guglielmi, A. and Di Nunno, G. (2002). Theory and numerical analysis for exact distribution of functionals of a Dirichlet process. *Annals of Statistics*, **30**, 1376–411.

Regazzini, E., Lijoi, A. and Prünster, I. (2003). Distributional results for means of random measures with independent increments. *Annals of Statistics*, **31**, 560–85.

Romik, D. (2004). Explicit formulas for hook walks on continual Young diagrams. *Advances in Applied Mathematics*, **32**, 625–54.

Romik, D. (2005). Roots of the derivative of a polynomial. *American Mathematical Monthly*, **112**, 66–69.

Sato, K. (1999). *Lévy Processes and Infinitely Divisible Distributions.* Cambridge: Cambridge University Press.

Sethuraman, J. (1994). A constructive definition of the Dirichlet process prior. *Statistica Sinica*, **2**, 639–50.

Susarla, V. and Van Ryzin, J. (1976). Nonparametric Bayesian estimation of survival curves from incomplete observations. *Journal of the American Statistical Association*, **71**, 897–902.

Teh, Y. W., Görür, D. and Ghahramani, Z. (2007). Stick-breaking construction for the Indian buffet. In *Proceedings of the International Conference on Artifical Intelligence and Statistics,* Volume 11, ed. M. Meila and X. Shen, 556–63. Brookline, Mass.: Microtome.

Thibaux, R. and Jordan, M. I. (2007). Hierarchical beta processes and the Indian buffet process. *Proceedings of the International Conference on Artifical Intelligence and Statistics,* Volume 11, ed. M. Meila and X. Shen, Brookline, Mass.: Microtome.

Walker, S. G. (2007). Sampling the Dirichlet mixture model with slices. *Communications in Statistics: Simulation and Computation*, **36**, 45–54.

Walker, S. G. and Damien, P. (1998). A full Bayesian non–parametric analysis involving a neutral to the right process. *Scandinavian Journal of Statistics*, **25**, 669–80.

Walker, S. G. and Damien, P. (2000). Representation of Lévy processes without Gaussian components. *Biometrika*, **87**, 477–83.

Walker, S. G., Damien, P., Laud, P. W. and Smith, A. F. M. (1999). Bayesian nonparametric inference for random distributions and related functions. *Journal of the Royal Statistical Society, Series B*, **61**, 485–527.

where $h_0(s)$ and $c(s)$ are given positive functions. Then

$$\mathrm{E}\, H_m(t) = \sum_{j/m \le t} m^{-1} h_0(j/m) \to H_0(t) = \int_0^t h_0(s)\,\mathrm{d}s$$

and

$$\mathrm{Var}\, H_m(t) = \sum_{j/m \le t} \frac{m^{-1} h_0(j/m)\{1 - m^{-1} h_0(j/m)\}}{c(j/m) + 1} \to \int_0^t \frac{h_0(s)}{c(s) + 1}.$$

Hjort (1985, 1990) demonstrates that H_m indeed tends to a well-defined time-continuous independent increment process H, and that this limit has Lévy representation agreeing with what is laid out in Section 3.2. Thus a more direct view of the beta process is that its increments are independent and satisfy

$$\mathrm{d}H(s) \sim \mathrm{Beta}\big[c(s)\,\mathrm{d}H_0(s), c(s)\{1 - \mathrm{d}H_0(s)\}\big], \tag{4.2}$$

infinitesimally speaking. It is worth remarking that constructing or defining the beta process is a harder technical task than for example defining gamma processes, since the beta distribution has awkward convolution properties.

The main conjugacy theorem for the beta process, used as a prior for the cumulative hazard function for survival data, is given as Theorem 3.16. A good heuristic for appreciating that result is to combine the prior (4.2) for the $\mathrm{d}H(s)$ increment with the corresponding likelihood contribution

$$\{1 - \mathrm{d}H(s)\}^{Y(s) - \mathrm{d}N(s)} \mathrm{d}H(s)^{\mathrm{d}N(s)}, \tag{4.3}$$

with $\mathrm{d}N(s)$ being 1 if there is an observed death inside $[s, s + \mathrm{d}s]$ and 0 otherwise, and with $Y(s)$ counting the number of individuals at risk at time s. This leads (with some additional care and efforts to make the implied arguments following from (4.3) fully precise; see Hjort, 1990, Section 4) to H still being a beta process given the data, as formalized in Theorem 3.16. In particular we are led to proper Bayesian generalizations of the Nelson–Aalen estimator for the cumulative hazard and the Kaplan–Meier estimator for the survival function; these emerge as the noninformative limits as the $c(s)$ function tends to zero.

4.1.2 Transitions and Markov processes

In various situations, including applications in biostatistics, demographics and sociology, individuals do not merely belong to the "alive" and "dead" categories, as in standard survival analysis, but move between a number of states, say $1, \ldots, k$ – think of "employed," "unemployed but looking for work," "retired or permanently unemployed," for example. The hazard rates or forces of transition are defined infinitesimally via

$$\mathrm{d}H_{j,l}(s) = \Pr\{\text{moves to } l \text{ inside } [s, s+\mathrm{d}s] \mid \text{at } j \text{ at time } s\} \quad \text{for } j \neq l. \qquad (4.4)$$

There are $k(k-1)$ such cumulative hazard functions. For Bayes analysis we may let these be independent beta processes, say $H_{j,l} \sim \text{Beta}(c_j, H_{0,j,l})$. There is indeed a construction similar to but rather more elaborate than the one summarized above, as we now need to start with independent Dirichlet distributions, instead of merely (4.1), and the $k-1$ associated independent increment processes $H_{m,j,l}$ (with j fixed and $l \neq j$) exhibit dependence among themselves. In the time-continuous limit these dependences vanish, however, leading, as indicated, to independent beta processes.

Under some assumptions individuals will move among the states according to a nonhomogeneous Markov process, characterized by the $H_{j,l}$ cumulative rate functions, where it is also understood that (4.4) depends only upon the process being in j at time s, and not upon other aspects of the past. There is again a conjugacy theorem (see Hjort, 1990, Section 5) that the $H_{j,l}$ remain independent beta processes but with updated characteristics. In particular, the posterior means are

$$\widehat{H}_{j,k}(t) = \int_0^t \frac{c_j(s)h_{0,j,k}(s)\,\mathrm{d}s + \mathrm{d}N_{j,k}(s)}{c_j(s) + Y_j(s)},$$

with the Aalen estimators emerging as the prior strength parameters c_j tend to zero. There are similarly nonparametric Bayesian estimators and credibility bands for the full matrix of all transition probabilities $P_{j,l}(s,t)$; this properly generalizes the Aalen–Johansen estimator and associated frequentist tools (see Aalen and Johansen, 1978; Aalen, Borgan and Gjessing, 2008, Chapter 3).

There are various results in the event history literature to the effect that the Aalen estimators $\int_0^t \mathrm{d}N_{j,k}/Y_j$ and Aalen–Johansen estimators of transition probabilities retain validity and a clear interpretation also when the underlying processes dictating transitions between states are not quite Markovian, i.e. when the forces of transition (4.4) have a more complicated dependence on the past. For discussion of such issues see Datta and Satten (2001, 2002) and Aalen, Borgan and Gjessing (2008, Section 3.3). We envisage that similar conclusions hold true also for the Bayesian counterparts, i.e. that estimators derived with beta processes and Markovian assumptions retain validity also under broader assumptions. Calculations involving posterior variances, credibility bands, etc., will however need nontrivial modifications. See also Phelan (1990), who reaches results in a framework of Markov renewal processes, using beta processes.

4.1.3 Hazard regression models

Survival and event history inference are both more important and challenging when individuals are not considered as coming from the same homogeneous population, but rather have recorded characteristics in the form of covariate information. Suppose individual i has covariate vector x_i thought to influence his hazard rate function h_i. The classic semiparametric proportional hazards model of Cox takes $h_i(s) = h_0(s)r(x_i^{\rm t}\beta)$, most typically with $r(u) = \exp(u)$ as relative risk function, where h_0 is not modeled parametrically. For nonparametric Bayesian inference we would often have to work with a slight variation of this model assumption, namely that the cumulative hazard rate functions H_i satisfy

$$1 - {\rm d}H_i(s) = \{1 - {\rm d}H(s)\}^{r(x_i^{\rm t}\beta)} \quad \text{for } i = 1, \ldots, n, \tag{4.5}$$

for a common cumulative hazard rate H; this is consistent with

$$S_i(t) = \prod_{[0,t]}\{1 - {\rm d}H_i(s)\} = S(t)^{\exp(x_i^{\rm t}\beta)},$$

for the survival functions, with S associated with cumulative hazard rate H. The reason for needing (4.5) rather than $h_i(s) = h_0(s)r(x_i^{\rm t}\beta)$ is that various natural prior specifications entail discrete sample paths.

The arguably canonical Bayesian extension of Cox regression analysis is to let H in (4.5) be a beta process (c, H_0), independently of the finite-dimensional β, which is given some prior $\pi(\beta)$. Incidentally, the Jeffreys prior for this problem depends on the relative risk function r, and is for example constant over the parameter range when $r(u) = \exp(u)$, i.e. for the Cox type model; see De Blasi and Hjort (2007). Calculations are now more complicated than those associated with (4.2)–(4.3), as the likelihood contribution for the time interval $[s, s + {\rm d}s]$ takes the form

$$\prod_{i=1}^{n}\{1 - {\rm d}H(s)\}^{r(x_i^{\rm t}\beta)\{Y_i(s) - {\rm d}N_i(s)\}}[1 - \{1 - {\rm d}H(s)\}^{r(x_i^{\rm t}\beta)}]^{{\rm d}N_i(s)},$$

where $Y_i(s)$ is the risk indicator for individual i at s and ${\rm d}N_i(s)$ is 1 if that individual dies inside $[s, s + {\rm d}s]$ and 0 otherwise. The precise posterior density of β and the posterior distribution of H given β are worked out in Hjort (1990, Section 6), and various MCMC schemes have been developed for handling the calculations, see in particular Damien, Laud and Smith (1995, 1996) and Laud, Damien and Smith (1998). In particular, one may generate (H, β) samples from the posterior distribution, making Bayesian inference amenable for any quantity of interest. An individual carrying covariate characteristics x and having survived up to time t_0 might for example find the precise distribution of his median remaining lifetime, given all information; this is the random time point

$$\mathrm{med}(x, t_0) - t_0 = \mathrm{med}(x, t_0, H, \beta) - t_0 = \inf\Big\{t \geq t_0 \colon \prod_{[t_0,t]} \{1 - \mathrm{d}H(s)\}^{r(x^{\mathrm{t}}\beta)} \leq \tfrac{1}{2}\Big\} - t_0.$$

This would be an instance of tools of use for "personalized medicine," cf. the introduction to Chapter 7, where there is a shift of emphasis from traditional analysis of β_j population parameters to what matters for a given individual. The difference between the population median and personal median is well captured in Gould (1995).

There are issues with the Cox proportional hazard model that spell trouble under sets of plausible conditions. As explained in Aalen and Hjort (2002), Hjort (2003) and De Blasi and Hjort (2007), various natural assumptions about the biological mechanisms underlying survival imply that the relative risk function $r(u)$ needs to be bounded. Sticking to the traditional choice $r(u) = \exp(u)$ then leads to imbalance and biased hazard predictions, particularly in the extreme parts of the covariate space where $x^{\mathrm{t}}\beta$ is rather small or rather big. These considerations suggest that it is fruitful to work with the logistic form $r(u) = \exp(u)/\{1 + \exp(u)\}$ instead, and motivate using

$$\mathrm{d}H_i(s) = r(x_i^{\mathrm{t}}\gamma)\,\mathrm{d}H(s) \quad \text{for } i = 1, \ldots, n \tag{4.6}$$

instead of (4.5). De Blasi and Hjort (2007) develop a semiparametric Bayesian framework for this model, again involving a beta process for H and a prior for γ. Figure 4.1 provides an application of such analysis, pertaining to data from a Danish study of 205 melanoma patients, discussed extensively in Andersen, Borgan, Gill and Keiding (1993) and elsewhere. The covariate x for this particular illustration is the tumor thickness, expressed in mm, after subtracting the sample mean 2.92 mm. In Figure 4.1, the posterior distribution of the random remaining life median $\mathrm{med}(x, t_0) - t_0$ is displayed, for an individual with $x = 1$ (i.e. tumor thickness 3.92 mm), given that he has survived t_0 years after the operation, for respectively $t_0 = 0, 1, 2$. Calculations involve model (4.6) with the Jeffreys prior for γ and a certain beta process prior for H, and a MCMC regime for simulating (H, γ) from the appropriate posterior distribution, from which we then compute $\mathrm{med}(x, t_0) - t_0$.

We note that frequentist tools may also be developed for approaching the problem of predicting median remaining survival time, but these would involve normal large-sample approximations that by the nature of Figure 4.1 would not be expected to work well here.

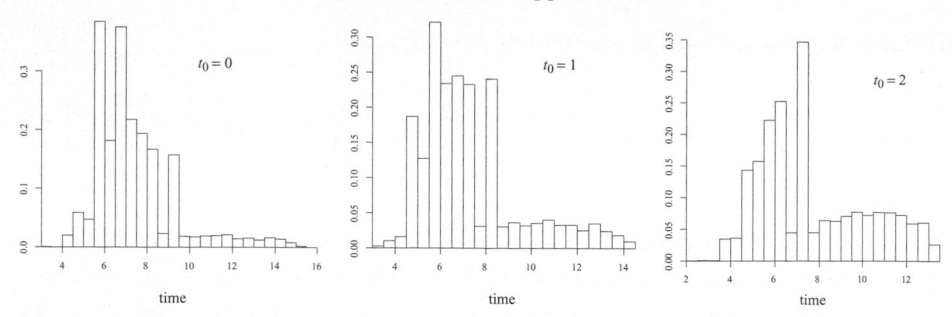

Figure 4.1 How long can one expect to survive further, when one already has survived a certain period after operation? For a patient with $x = 1$, corresponding to tumor thickness 3.92 mm, the figure displays the posterior distribution of the median remaining lifetime just after operation (*left panel*), after one year of survival (*central panel*) and after two years of survival (*right panel*). The time unit is years.

4.1.4 Semiparametric competing risks models

A special case of the Markov transitions model above is that of the competing risk model, where there are say k types of failure, or potential transitions, from a given initial state. Thus, with T the time to transition and D denoting the type of failure, the cause-specific hazard functions are

$$\mathrm{d}H_l(s) = \Pr\{T \in [s, s + \mathrm{d}s], D = l \mid T \geq s\} \quad \text{for } l = 1, \ldots, k.$$

There is a literature on cumulative incidence functions, here given as

$$P_l(t) = \Pr\{T \leq t, D = l\} = \int_0^t \prod_{[0,s]} \{1 - \mathrm{d}H_.(u)\}\, \mathrm{d}H_l(s) \quad \text{for } l = 1, \ldots, k,$$

writing $H_. = \sum_{l=1}^k H_l$ for the combined hazard rate function of T. Sometimes the emphasis is rather on the cause-specific conditional probabilities

$$q_l(t) = \Pr\{D = l \mid T = t\} = \lim_{\epsilon \to 0} \Pr\{D = l \mid T \in [t, t + \epsilon]\},$$

and how these vary with time; these may also be recovered from the incidence functions via $\mathrm{d}H_l(t)/\mathrm{d}H_.(t)$. Nonparametric Bayesian inference may proceed after placing independent beta process priors on $H_1, \ldots, H_k$.

Again the scope for applications becomes significantly wider (and more challenging) when covariate information is taken on board. Suppose data of the form (t_i, d_i, x_i) are available for individuals $i = 1, \ldots, n$, where t_i is failure time, d_i the type of failure, and x_i a vector of covariates of fixed dimension, say p; censoring is allowed via the notation $d_i = 0$. A class of semiparametric regression models for such data takes the form

$$\mathrm{d}H_{i,l}(s) = q_l(s, \theta) r(x_i^{\mathrm{t}} \beta_l)\, \mathrm{d}H(s) \quad \text{for } l = 1, \ldots, k \text{ and } i = 1, \ldots, n. \tag{4.7}$$

Here $q_l(s, \theta)$ represent parametrically modeled cause-specific conditional probabilities, satisfying $\sum_{l=1}^k q_l(s, \theta) = 1$; there are k regression vectors β_l, one for each of type of failure; and H is an unknown and nonparametrically modeled cumulative hazard rate function. A convenient class favored in De Blasi and Hjort (2009a, 2009b) is the normalized Gompertz type

$$q_l(s, \theta) = \frac{\exp(a_l + b_l s)}{\sum_{l'=1}^k \exp(a_{l'} + b_{l'} s)} \quad \text{for } l = 1, \ldots, k,$$

where we let $(a_k, b_k) = (0, 0)$, to ensure identifiability; thus the q_l functions have $2k - 2$ parameters. They further employ the logistic form $\exp(u)/\{1 + \exp(u)\}$ for the relative risk function $r(u)$. Issues of identifiability and interpretation are alleviated by pre-centering covariates; an "average individual" with $x_i = 0$ then has $q_l(s, \theta) r(0) \, \mathrm{d}H(s)$ in (4.7). Finally a beta process is used for this H, along with (typically Jeffreys type) priors for θ and the β_l; note that the jumps associated with $\mathrm{d}H_{i,l}(s)$ of (4.7) are really inside $[0, 1]$, as required. Prior knowledge that some of the x_i components have no or low impact on some of the k forces of transition $\mathrm{d}H_{i,l}(s)$ may be taken into account when setting up the corresponding β_l priors. They show how this leads to a full Bayesian analysis, with appropriate characterizations of the posterior density of $\theta, \beta_1, \ldots, \beta_k$ (of dimension $2k - 2 + kp$) and the posterior distribution of H given these parameters. Again, suitable MCMC schemes yield full posterior simulation of any quantity of interest, i.e. any function of θ, the β_l and H.

As an illustration we consider the mice data set described and analyzed in Andersen, Borgan, Gill and Keiding (1993). There are two groups of mice; 95 mice lived under ordinary laboratory conditions ($x = -1$) while 82 were kept in a germ-free environment ($x = 1$). There were three types of death: thymic lymphoma ($D = 1$), reticulum cell sarcoma ($D = 2$), and other causes ($D = 3$). With Jeffreys type priors for the four-dimensional θ involved in modeling $q_l(s, \theta)$ and the β_l parameters for $l = 1, 2, 3$ in (4.7), and a particular beta prior process for H, analysis of the type reported on in more detail in De Blasi and Hjort (2009b) leads to Figure 4.2, with posterior mean curves and 90% credibility bands for the cause-specific conditional probabilities $q_l(s)$ as a function of time. The figure indicates in particular that the chances that a death is caused by reticulum cell sarcoma behave rather differently in the two environments; this is a likely cause of death, after say 400 days, in a conventional environment, but rather less likely in the germ-free environment, for mice aged 400 days or more. This is an important and dynamic component of the competing risks story, though only a part thereof; we refer to De Blasi and Hjort (2009b) for further analysis.

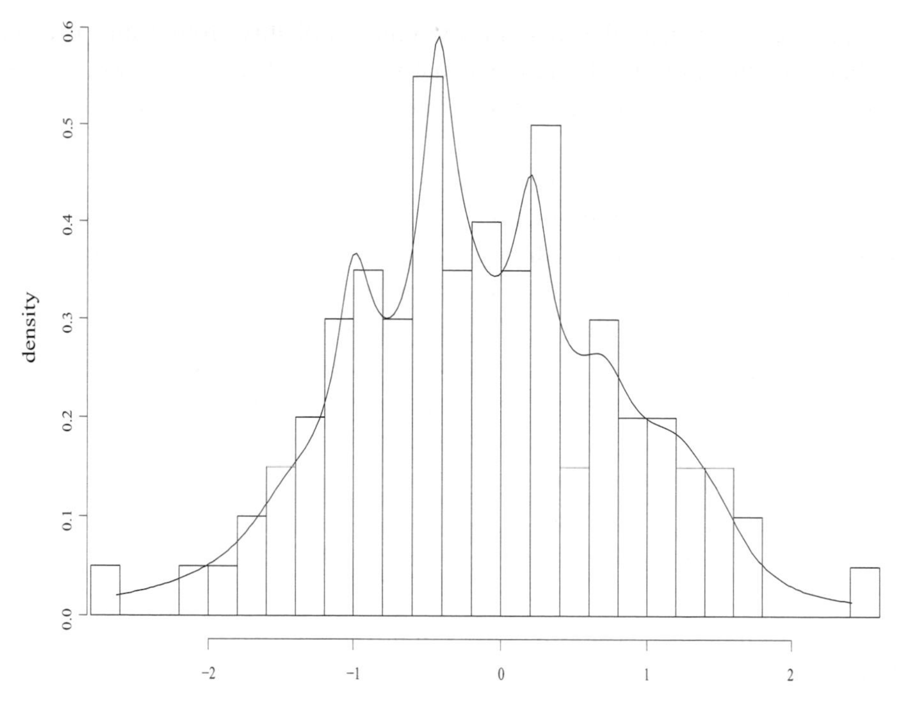

Figure 4.3 A histogram (with more cells than usual) over $n = 100$ data points from the standard normal, along with the automatic density estimator of (4.9).

with earlier in the frequentist literature, and is here given independent motivation from the noninformative Bayesian perspective. Also noteworthy is the fact that it is smooth in y, unlike what emerges from the corresponding arguments for the distribution function – the result there is the empirical distribution function F_n, which is piecewise constant with derivative zero almost everywhere. This encourages one to invert the $\widehat{Q}_0(y)$ quantile function to its density. The result is

$$\widehat{f}_0(x) = \Big[\sum_{i=1}^{n-1}(x_{(i+1)} - x_{(i)})\,\mathrm{be}(\widehat{F}_0(x); i, n-1)\Big]^{-1}, \tag{4.9}$$

with $\mathrm{be}(u; i, n-i)$ denoting the beta density with the indicated parameters, and $\widehat{F}_0$ being the distribution function associated with $\widehat{Q}_0$; here one needs to solve the equation $\widehat{Q}_0(y) = x$ with respect to y for each fixed x. The formula (4.9) may be viewed as a fully automatic Bayesian density estimator, being associated with the Dirichlet process for the noninformative limiting case of $a \to 0$. Its support is precisely the data range $[x_{(1)}, x_{(n)}]$, and there are no smoothing parameters to decide on. Figure 4.3 gives an illustration with 100 data points drawn from the standard normal; see Hjort and Petrone (2007) for further discussion.

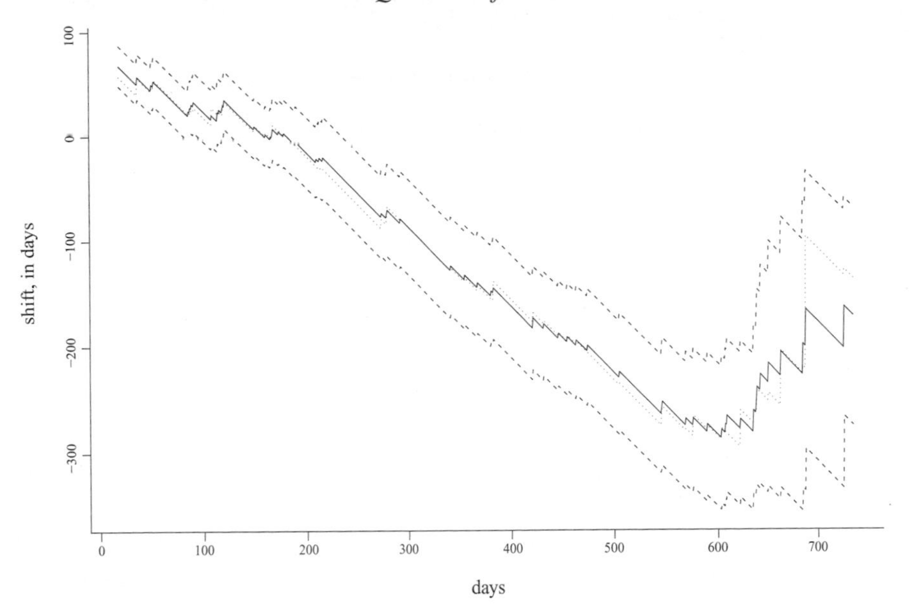

Figure 4.4 For the 65 guinea pigs in the control group and the 60 in the treatment group, we display the Bayes estimator (*solid line*) of the shift function associated with the two survival distributions, alongside Doksum's sample estimator (*dotted line*). Also given is the approximate pointwise 90% credibility band.

Quantile inference is rather more than caring for $Q = F^{-1}$ for a given sample; see for example Parzen (1979) for a wider discussion. Hjort and Petrone (2007) provide Bayesian methods for dealing with the Lorenz curve and the Gini index, for quantile regression, and finally for Parzen's (1979) comparison function $\pi(y) = G(F^{-1}(y))$ and Doksum's (1974) shift function $D(x) = G^{-1}(F(x)) - x$, in situations with independent data $x'_1, \ldots, x'_m$ from G and $x_1, \ldots, x_n$ from F. Figure 4.4 gives an illustration of Bayesian output for Doksum's shift function, using Doksum's 1974 data. These involve the effect of virulent tubercle bacilli, with 65 guinea pigs in the control group and 60 in the treatment group, the latter having received a dose of such bacilli. The figure displays the Bayes estimate $\widehat{D}_0(x)$, seen to be quite close to Doksum's direct frequentist estimate, along with a pointwise 90% credibility band. Here noninformative priors are used, i.e. Dirichlet processes with parameters say aF_0 and bG_0, but where a and b are sent to zero in the calculations. The interpretation of the shift function is that $X + D(X)$ has the same distribution as X'. The figure illustrates that the weaker pigs (those tending to die early) will tend to have longer lives with the treatment, whereas the stronger pigs (those tending to have long lives) are made drastically weaker, i.e. their lives will be shortened.

4.3 Shape analysis

We shall work with shapes of random two-dimensional objects. A general class of such objects emerges by considering the representation

$$R(s)\,\big(\cos(2\pi s), \sin(2\pi s)\big) \quad \text{for } 0 \le s \le 1$$

for the perimeter of the object, with $R(s)$ denoting the radius, in direction $2\pi s$, as measured from origo. For this to make sense in the context of shapes and objects we would typically need R to be continuous and to satisfy $R(0) = R(1)$. It is now clear that any random process R on $[0, 1]$, with these two properties, gives a random shape. There have been suggestions in the literature of such approaches, involving Gaussian processes for the random radius, see for example Grenander and Miller (1994) and Kent, Dryden and Anderson (2000). These constructions do not take the a priori positivity constraint into account, however; this somewhat negative aspect affects both simulation and the likelihood-based methods developed in these papers. Here we attempt instead to construct genuine random radius processes.

One such construction starts by defining $R(s) = \int K_b(s-u)\,\mathrm{d}G(u)$, where G is a gamma process (with independent and gamma distributed increments) and $K_b(u) = b^{-1}K(b^{-1}u)$ a scaled version of a basis kernel function K, taken to be continuous and symmetric on its support $[-\frac{1}{2}, \frac{1}{2}]$. We define this radius integral as being modulo the circle around which it lives, i.e. clockwise modulo its parameter interval $[0, 1]$. In other words, $u_1 = 0.99$ is as close a neighbor to $s = 0.07$ as is $u_2 = 0.15$ (namely 0.08 away), and so on. In order to make the random curve scale-free we normalize the G process, reaching $\bar{G} = G/G[0, 1]$. But such a normalized gamma process is precisely a Dirichlet process.

For illustration we focus here on shapes without prior landmarks, so their distributions are meant to be rotation invariant. This corresponds to gamma or Dirichlet processes with uniform base measures. Thus an effective model for random shapes, invariant with respect to scaling and to rotation, is

$$R(s)\big(\cos(2\pi s), \sin(2\pi s)\big) \quad \text{with} \quad R(s) = \int K_b(s-u)\,\mathrm{d}\bar{G}_a(u),$$

where $\bar{G}_a$ is Dirichlet with parameter aU, with U the uniform density on $[0, 1]$. Figure 4.5 displays four random shapes from this model, with $(a, b) = (100, 0.40)$; here we are using the smooth biweight kernel $K(u) = (15/8)(1-4u^2)^2$ on $[-\frac{1}{2}, \frac{1}{2}]$, so the K_b in question has support $[-0.20, 0.20]$. Properties of this random shape

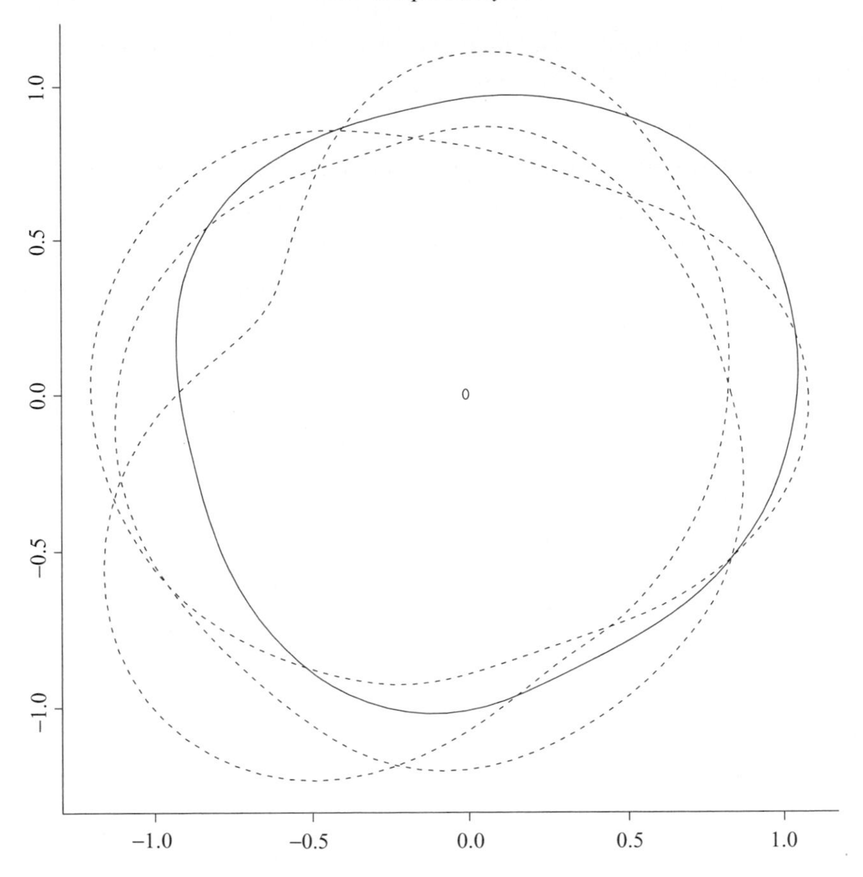

Figure 4.5 Four realizations of the two-parameter smoothed Dirichlet shape model, with $(a, b) = (100, 0.40)$.

mechanism include

$$\mathrm{E}\, R(s) = 1, \quad \mathrm{Var}\, R(s) = S(K)/(ab), \quad \mathrm{E}\, A = \pi\{1 + S(K)/ab\},$$

where $A = \int \pi R(s)^2 \, \mathrm{d}s$ is the random area of the object, and $S(K) = \int K^2 \, \mathrm{d}u$; for the biweight kernel, $S(K) = (15/8)^2(384/945) = 1.4286$. We note that higher values of ab imply more regularity and shapes closer to a circle.

Statistical issues might involve, for example, estimation and assessment of (a, b) from a sample of shapes from a population, perhaps to discriminate one population from another. Methods for handling such problems have been worked out by the author and will be published elsewhere. There are also extensions of the two-parameter model above for situations of template matching, where there might be landmarks, etc. See Dryden and Mardia (1998) for a broad introduction to the field of statistical shape analysis, where one now may envisage alternative methods based on smoothed Dirichlet processes.

4.4 Time series with nonparametric correlation function

As the chapters in this book in various ways demonstrate, there is a multitude of contributions in the literature to nonparametric modeling and analysis of mean functions in regression models, distribution functions, probability densities and hazard rate functions, etc. The Bayesian nonparametrics route may however be taken also for other types of functions in statistical modeling, such as link functions, variance heterogeneity and overdispersion functions in regression models, interaction functions for point process models, etc. The following briefly reports on a successful attempt at analyzing stationary time series data with nonparametric modeling of the correlation function. An earlier article in the same direction, but employing a different type of prior, is Choudhuri, Ghosal and Roy (2004).

Let $Y_1, Y_2, \ldots$ be a stationary time series with finite variance, with covariances $c(u) = \mathrm{cov}(Y_t, Y_{t+u})$ for $u = 0, \pm 1, \pm 2, \ldots$. A necessary and sufficient condition for the sequence of covariances to be well defined and positive definite is that

$$c(u) = 2\int_0^{\pi} \cos(u\lambda)\,\mathrm{d}F(\lambda) \quad \text{for all } u = 0, \pm 1, \pm 2, \ldots, \tag{4.10}$$

for some finite measure F on $[0, \pi]$, called the power spectrum measure of the time series. The simple case of a first-order autoregressive model, for example, corresponds to $c(u) = \sigma^2 \rho^u$, and to power spectrum density

$$f_0(\lambda) = \sigma^2 \frac{1}{2\pi} \frac{1 - \rho^2}{1 - 2\rho\cos(\lambda) + \rho^2} \quad \text{for } \lambda \in [0, \pi]. \tag{4.11}$$

A way of constructing a nonparametric prior for the covariances is now to view F of (4.10) as a random measure. It could for example be a gamma process, with independent increments of the form

$$\mathrm{d}F(\lambda) \sim \mathrm{Gamma}(a f_0(\lambda)\,\mathrm{d}\lambda, a),$$

centered at the autocorrelation model (4.11), with large a indicating the strength of belief in such a prior guess. Note that the variance of Y_t is $c(0) = 2F(\pi)$ and that the random correlation function may be expressed as

$$c(u)/c(0) = \int_0^{\pi} \cos(u\lambda)\,\mathrm{d}D(\lambda),$$

with $D = F/F(\pi)$ a Dirichlet process.

To perform Bayesian inference we need the posterior distribution of F (and hence of any $c(u)$ of relevance). Suppose data $y_1, \ldots, y_n$ stem from such a time series, with mean zero. Assuming normality, the log-likelihood function takes the form

$$\ell_n = -\tfrac{1}{2}\log|\Sigma| - \tfrac{1}{2}y^{\mathrm{t}}\Sigma^{-1}y - \tfrac{1}{2}n\log(2\pi),$$

with Σ the $n \times n$ matrix of $c(j-k)$. Several approximations have been used in the literature (for purposes different from those in focus now), of the Whittle variety, see Brillinger (2001) and Dzhaparidze (1986). One such is

$$\widetilde{\ell}_n = -\tfrac{1}{2}n\frac{1}{\pi}\int_0^{\pi}\Big\{\log f(\lambda) + \frac{I_n(\lambda)}{f(\lambda)}\Big\}\,\mathrm{d}\lambda + \text{constant},$$

in terms of the periodogramme function

$$I_n(\lambda) = \frac{1}{n}\frac{1}{2\pi}\Big|\sum_{k=1}^{n}\exp(-\mathrm{i}k\lambda)y_k\Big|^2.$$

This assumes F of (4.10) to have a density f. A further approximation that avoids this assumption is as follows: split $[0,\pi]$ into small intervals $(c_{j-1}, c_j]$ for $j = 1,\ldots,m$ and define increments

$$v_j = F(c_j) - F(c_{j-1}) \quad\text{and}\quad \widehat{v}_j = \widehat{F}_n(c_j) - \widehat{F}_n(c_{j-1}) = \int_{c_{j-1}}^{c_j} I_n(\lambda)\,\mathrm{d}\lambda,$$

where $\widehat{F}_n$ is the cumulative periodogramme. Then ℓ_n (and $\widetilde{\ell}_n$) are close to $\ell_n^* = -\frac{1}{2}n\frac{1}{\pi}\sum_{j=1}^{m}\Delta_j(\log v_j + \widehat{v}_j/v_j) + \text{constant}$, with $\Delta_j = c_j - c_{j-1}$ the length of the jth window.

It follows from this that if F is given a prior with independent increments, and $v_j \sim g_j(v_j)$, say, then these increments are approximately independent also given data, with updated densities close to

$$g_j(v_j \mid \text{data}) \propto g_j(v_j)\exp\Big\{-\tfrac{1}{2}n\frac{1}{\pi}\Delta_j(\log v_j + \widehat{v}_j/v_j)\Big\}.$$

If in particular the F increment v_j has an inverse gamma (a_j, b_j) prior, then

$$v_j \mid \text{data} \approx_d \text{inverse gamma}\,(a_j + \tfrac{1}{2}n\Delta_j/\pi,\ b_j + \tfrac{1}{2}n\Delta_j\widehat{v}_j/\pi),$$

with posterior mean

$$\mathrm{E}(v_j \mid \text{data}) = \frac{a_j - 1}{a_j - 1 + \frac{1}{2}n\Delta_j/\pi}\,\frac{b_j}{a_j - 1} + \frac{\frac{1}{2}n\Delta_j/\pi}{a_j - 1 + \frac{1}{2}n\Delta_j/\pi}\,\widehat{v}_j$$

for $j = 1,\ldots,m$. These are convex combinations of the prior means and the frequentist estimates. Via (4.10) this statement translates to the (approximate) posterior means of every covariance $c(u)$.

We may also work out what happens to the normalized increments $w_j = \sqrt{n}(v_j - \widehat{v}_j)$ as n increases. These are asymptotically independent, with

$$g_j(w_j\,|\,\text{data}) \approx g_j(\widehat{v}_j + w_j/\sqrt{n}) \exp\Big[-\tfrac{1}{2} n \frac{1}{\pi} \Delta_j \Big\{\log(\widehat{v}_j + w_j/\sqrt{n}) + \frac{\widehat{v}_j}{\widehat{v}_j + w_j/\sqrt{n}}\Big\}\Big].$$

A Taylor approximation indicates that the influence of the prior density g_j vanishes for large n and that

$$w_j \mid \text{data} \approx_d \text{N}(0, 2\pi\widehat{v}_j^2/\Delta_j).$$

If the data stem from a time series with underlying true power spectrum density $f_{\text{tr}}(\lambda)$, then the variance here is for large n also close to $2\pi f_{\text{tr}}(\lambda_j)^2 \Delta_j$, with λ_j inside the jth window.

There is a "fine limit" version of this story, in which the maximum width of the small intervals tends to zero, and with the number of intervals balanced against a growing sample size. For each random F in a rather large class of priors of the indicated type, there is a.s. process convergence of the normalized posterior process:

$$\sqrt{n}\{F(\lambda) - \widehat{F}_n(\lambda)\} \mid \text{data} \to_d A(\lambda) = W\Big(2\pi \int_0^\lambda f_{\text{tr}}(u)^2 \,\text{d}u\Big),$$

i.e. a time-transformed Brownian motion process. This is a perfect mirror result, in the Bernshteĭn–von Mises theorem tradition (cf. Chapter 2, Section 2.7), of the frequentist result that $\sqrt{n}(\widehat{F}_n - F_{\text{tr}})$ converges to precisely the same limit A in distribution. For details, see Hermansen and Hjort (2009).

The use of this machinery is to reach inference statements more nonparametric in nature than those usually associated with time series analyses. One typical task is to find the predictive distribution, that of the next datum Y_{n+1}. With given covariance function terms $c(u)$, this is a normal with mean $\xi_n = \sum_{k=1}^n b_k y_k$ and variance τ_n^2, say, with the b_k coefficients and the τ_n being explicit functions of the variance matrix Σ of $c(j - k)$. Parametric Bayes analysis might proceed for example by using a prior on (σ, ρ) of (4.11), and drawing posterior samples of the Σ matrix through the posterior distribution of (σ, ρ). The nonparametric apparatus instead draws such samples of the full Σ matrix via $c(j - k) = 2\int_0^\pi \cos((j - k)\lambda)\,\text{d}F(\lambda)$, and reaches conclusions free of any particular assumed parametric structure of the covariance function. We may also easily enough make joint inference for say $c(1), c(2), c(3)$.

4.5 Concluding remarks

Below we offer some concluding comments, some pointing to further research questions of relevance.

4.5.1 Bernshteĭn–von Mises theorems

The relevance and importance of Bernshteĭn–von Mises (BvM) theorems are discussed in Chapter 2; in particular, establishing such a theorem, for a certain class of priors, is also a stamp of approval and a guarantee that nothing may go very wrong when data accumulate. It is good, therefore, to report that BvM theorems indeed have been or may be proven for several of the classes of priors that have been discussed in this chapter.

That the BvM holds for the beta processes in the one-sample case and for nonhomogeneous Markov processes with censoring was made clear in Hjort (1990, Sections 4 and 5); the explicit structure of the posterior distributions make it easy to see there that the Bayes methods become asymptotically equivalent to those based on Aalen estimators and Kaplan–Meier methods; see also Kim and Lee (2004). The situation is more complicated for the Bayesian beta process approach to Cox proportional hazards regression, but again large-sample equivalence with the frequentist methods was essentially covered in Hjort (1990, Section 6), with later extensions in Kim (2006). That BvM holds in the more complicated proportional hazard model with logistic or bounded relative risk function, cf. Section 4.1.3, was shown in De Blasi and Hjort (2007). Similarly De Blasi and Hjort (2009a) have shown the BvM in the semiparametric competing risks model of Section 4.1.4. Finally, BvM results are demonstrated for both the quantile pyramids class of Hjort and Walker (2009, Section 8) and for the Dirichlet quantile process in Hjort and Petrone (2007, Section 7), and Hermansen and Hjort (2009) reach partial BvM conclusions for a class of nonparametrically modeled covariance functions for stationary time series.

4.5.2 Mixtures of beta processes

There are several ways of using beta processes to build nonparametric envelopes around given parametric models. Here we briefly consider two approaches, related respectively to mixtures of beta processes and to frailty mechanisms. We learn that even promising constructions may lead to methods with unfortunate behavior, if the details of the construction are not well set up; specifically, the BvM theorem might not hold.

Let $h_\theta(s)$ be a parametric hazard rate with cumulative $H_\theta(s)$. Giving θ some prior $\pi(\theta)$, let $H \mid \theta$ be a beta process (c_θ, H_θ), for some $c_\theta(\cdot)$. Thus H is a mixture of beta processes, and $\mathrm{E}\,\mathrm{d}H(s) = h^*(s)\,\mathrm{d}s$, with $h^*(s) = \int h_\theta(s)\pi(\theta)\,\mathrm{d}\theta$. Note that H has independent increments given θ, but not marginally. The posterior of θ, given a set of possibly censored data, is proportional to $\pi(\theta)L_n^*(\theta)$ for a certain L_n^*, see Hjort and Kim (2009). Examination of this function reveals that the behavior of Bayes estimators depends in partly unexpected ways on details of the c_θ functions, and there may in particular be inconsistency, even though there is full nonparametric support.

The second approach starts from a fixed beta process $\bar{H}$ with parameters (c, H_0), where $H_0(t) = t$ is the cumulative hazard rate of the unit exponential. One then defines $1 - \mathrm{d}H(s) = \{1 - \mathrm{d}\bar{H}(s)\}^{h_\theta(s)}$, and again equips θ with a prior $\pi(\theta)$. This is a frailty-motivated random cumulative hazard function, again centered at the parametric family. The posterior is proportional to $\pi(\theta)\widetilde{L}_n(\theta)$, for a certain $\widetilde{L}_n$ function, see again Hjort and Kim (2009) for details. This second approach appears more robust than the first, and a BvM result has been derived under some conditions on the c function.

4.5.3 Bayesian kriging

We saw in Section 4.4 how nonparametric modeling of the covariance function leads to fruitful Bayesian methods. Similarly, one may attempt to construct nonparametric models for the covariance function in spatial models, say $K(\|u\|) = \mathrm{cov}\{Z(x), Z(x+u)\}$ in an isotropic situation. This may be done using spectral representations, and under normality assumptions the posterior distribution of K may be characterized; see Hermansen (2008) for some of the details. This leads to new methods for spatial interpolation, so-called kriging. To indicate how, suppose the stationary random normal field $Z(\cdot)$ has been observed in locations $x_1, \ldots, x_n$, and let x be a new position. Then $Z(x)$ given data z_d, the vector of $Z(x_i)$, is a normal, with mean of the form

$$\widehat{m}(x, K) = \mu + c(K)^{\mathrm{t}}\Sigma(K)^{-1}(z_d - \mu\mathbf{1}).$$

Here μ is the mean of $Z(x)$, $\Sigma(K)$ is the $n \times n$ variance matrix of $K(\|x_i - x_j\|)$, and $c(K)$ the vector of $K(\|x - x_i\|)$. When K is known, $\widehat{m}(x, K)$ is the best spatial interpolator, called the kriging algorithm (typically with an estimate of μ inserted). The nonparametric Bayes version is now to include an additional layer of variation in K, leading to the more complicated interpolator

$$\widehat{m}(x) = \mathrm{E}\{\widehat{m}(x, K) \mid \mathrm{data}\}.$$

This conditional expectation would be computed by simulating say 1000 functions K_j from the appropriate posterior distribution, from which one computes $\Sigma(K_j)$, $c(K_j)$ and hence $\widehat{m}(x, K_j)$; the final result is the average of these.

4.5.4 From nonparametric Bayes to parametric survival models

It is important to realize that tools and concepts from Bayesian nonparametrics may lead to fruitful models themselves. An example is that of the two-parameter random shape model of Section 4.3, where the role of the Dirichlet process is to be smoothed and moulded into a radius function. Similarly, classes of models for survival and event history data emerge via nonparametric viewpoints, such as cumulative damage processes, the time to hit a threshold, etc.; see Gjessing, Aalen and Hjort (2003) and Hjort (2003) for general perspectives. Here we indicate two such examples.

First imagine that individuals $i = 1, \ldots, n$ with covariate vectors $x_1, \ldots, x_n$ carry latent gamma processes $Z_1, \ldots, Z_n$, of the form $Z_i(t) \sim \text{Gamma}(aM_i(t), 1)$, and that failure or transition occurs as soon as $Z_i(t)$ crosses the threshold ac_i:

$$T_i = \min\{t\colon Z_i(t) \geq ac_i\} \quad \text{for } i = 1, \ldots, n.$$

The survival functions are then $S_i(t) = \Pr\{T_i \geq t\} = G(ac_i, aM_i(t))$, writing $G(u, b)$ for the probability that $\text{Gamma}(b, 1) \leq u$. The covariates may be taken to influence the threshold, say with $c_i = \exp(x_i^{\mathrm{t}}\beta)$, and perhaps also with the individual clock M_i, say as in $M_i(t) = M_0(\exp(x_i^{\mathrm{t}}\gamma)t)$ for some base clock M_0. Such models are explored in Claeskens and Hjort (2008, Chapter 3), where they are shown often to fit survival data better than competing models. These models also have the possibility of "crossing hazards" and "crossing survival curves", unlike most of the conventional survival models such as Cox regression; the point is to include the biologically plausible possibility that some individuals have a faster running damage process clock than others.

Our second class of examples is associated with the jumps of a beta process. If individuals as above have latent beta processes $H_1, \ldots, H_n$ then we may imagine that a failure occurs for individual i as soon as the first beta process jump exceeds a threshold, or perhaps as soon as there have been say k such sharp enough shocks to the system. Working from such assumptions a variety of survival models emerge; see Hjort and Kim (2009). The Cox proportional hazard model may be seen to correspond to a special case.

Acknowledgements Parts of what I have reported on in this chapter stem from collaborative work and always stimulating discussions with Pierpaolo De Blasi, Gudmund Hermansen, Yongdai Kim, Sonia Petrone and Stephen Walker.

References

Aalen, O. O., Borgan, Ø. and Gjessing, H. K. (2008). *Event History Analysis: A Process Point of View*. New York: Springer-Verlag.

Aalen, O. O. and Hjort, N. L. (2002). Frailty models that yield proportional hazards. *Statistics and Probability Letters*, **58**, 335–42.

Aalen, O. O. and Johansen, S. (1978). An empirical transition matrix for nonhomogeneous Markov chains based on censored observations. *Scandinavian Journal of Statistics*, **5**, 141–50.

5

Hierarchical Bayesian nonparametric models with applications

Yee Whye Teh and Michael I. Jordan

Hierarchical modeling is a fundamental concept in Bayesian statistics. The basic idea is that parameters are endowed with distributions which may themselves introduce new parameters, and this construction recurses. In this review we discuss the role of hierarchical modeling in Bayesian nonparametrics, focusing on models in which the infinite-dimensional parameters are treated hierarchically. For example, we consider a model in which the base measure for a Dirichlet process is itself treated as a draw from another Dirichlet process. This yields a natural recursion that we refer to as a hierarchical Dirichlet process. We also discuss hierarchies based on the Pitman–Yor process and on completely random processes. We demonstrate the value of these hierarchical constructions in a wide range of practical applications, in problems in computational biology, computer vision and natural language processing.

5.1 Introduction

Hierarchical modeling is a fundamental concept in Bayesian statistics. The basic idea is that parameters are endowed with distributions which may themselves introduce new parameters, and this construction recurses. A common motif in hierarchical modeling is that of the conditionally independent hierarchy, in which a set of parameters are coupled by making their distributions depend on a shared underlying parameter. These distributions are often taken to be identical, based on an assertion of exchangeability and an appeal to de Finetti's theorem.

Hierarchies help to unify statistics, providing a Bayesian interpretation of frequentist concepts such as shrinkage and random effects. Hierarchies also provide ways to specify nonstandard distributional forms, obtained as integrals over underlying parameters. They play a role in computational practice in the guise of variable augmentation. These advantages are well appreciated in the world of parametric modeling, and few Bayesian parametric modelers fail to make use of some aspect of hierarchical modeling in their work.

Nonparametric Bayesian models also typically include many classical finite-dimensional parameters, including scale and location parameters, and hierarchical modeling concepts are often invoked in specifying distributions for these parameters. For example, the Dirichlet process DP(α, G_0) involves a concentration parameter α, which is generally given a prior distribution in nonparametric (and semiparametric) models that make use of the Dirichlet process. Moreover, the base measure, G_0, is often taken to be a parametric distribution and its parameters are endowed with prior distributions as well.

In this chapter we discuss a more thoroughgoing exploitation of hierarchical modeling ideas in Bayesian nonparametric statistics. The basic idea is that rather than treating distributional parameters such as G_0 parametrically, we treat them nonparametrically. In particular, the base measure G_0 in the Dirichlet process can itself be viewed as a random draw from some distribution on measures – specifically it can be viewed as a draw from the Dirichlet process. This yields a natural recursion that we refer to as a *hierarchical Dirichlet process*. Our focus in this chapter is on nonparametric hierarchies of this kind, where the tools of Bayesian nonparametric modeling are used recursively.

The motivations for the use of hierarchical modeling ideas in the nonparametric setting are at least as strong as they are in the parametric setting. In particular, nonparametric models involve large numbers of degrees of freedom, and hierarchical modeling ideas provide essential control over these degrees of freedom. Moreover, hierarchical modeling makes it possible to take the building blocks provided by simple stochastic processes such as the Dirichlet process and construct models that exhibit richer kinds of probabilistic structure. This breathes life into the nonparametric framework.

The chapter is organized as follows. In Section 5.2, we discuss the hierarchical Dirichlet process, showing how it can be used to link multiple Dirichlet processes. We present several examples of real-world applications in which such models are natural. Section 5.3 shows how the hierarchical Dirichlet process can be used to build nonparametric hidden Markov models; these are hidden Markov models in which the cardinality of the state space is unbounded. We also discuss extensions to nonparametric hidden Markov trees and nonparametric probabilistic context free grammars. In Section 5.4 we consider a different nonparametric hierarchy based on the Pitman–Yor model, showing that it is natural in domains such as natural language processing in which data often exhibit power-law behavior. Section 5.5 discusses the beta process, an alternative to the Dirichlet process which yields sparse featural representations. We show that the counterpart of the Chinese restaurant process is a distribution on sparse binary matrices known as the Indian buffet process. We also consider hierarchical models based on the beta process. In Section 5.6, we consider some semiparametric models that are based on nonparametric hierarchies. Finally, in Section 5.7 we present an overview of some of the algorithms that have been developed for posterior inference in hierarchical Bayesian nonparametric models.

In all of these cases, we use practical applications to motivate these constructions and to make our presentation concrete. Our applications range from problems in biology to computational vision to natural language processing. Several of the models that we present provide state-of-the-art performance in these application domains. This wide range of successful applications serves notice as to the growing purview of Bayesian nonparametric methods.

5.2 Hierarchical Dirichlet processes

The Dirichlet process (DP) is useful in models for which a component of the model is a discrete random variable of unknown cardinality. The canonical example of such a model is the DP mixture model, where the discrete variable is a cluster indicator. The *hierarchical Dirichlet process* (HDP) is useful in problems in which there are multiple groups of data, where the model for each group of data incorporates a discrete variable of unknown cardinality, and where we wish to tie these variables across groups (Teh, Jordan, Beal and Blei, 2006). For example, the HDP mixture model allows us to share clusters across multiple clustering problems.

The basic building block of a hierarchical Dirichlet process is a recursion in which the base measure G_0 for a Dirichlet process $G \sim \mathrm{DP}(\alpha, G_0)$ is itself a draw from a Dirichlet process: $G_0 \sim \mathrm{DP}(\gamma, H)$. This recursive construction has the effect of constraining the random measure G to place its atoms at the discrete locations determined by G_0. The major application of such a construction is to the setting of conditionally independent hierarchical models of grouped data.

More formally, consider an indexed collection of DPs, $\{G_j\}$, one for each of a countable set of groups and defined on a common probability space (Θ, Ω). The hierarchical Dirichlet process ties these random measures probabilistically by letting them share their base measure and letting this base measure be random:

$$\begin{aligned} G_0 \mid \gamma, H &\sim \mathrm{DP}(\gamma, H), \\ G_j \mid \alpha, G_0 &\sim \mathrm{DP}(\alpha, G_0), \qquad \text{for } j \in \mathcal{J}, \end{aligned} \tag{5.1}$$

where $\mathcal{J}$ is the index set. This conditionally independent hierarchical model induces sharing of atoms among the random measures G_j since each inherits its set of atoms from the same G_0. To understand the precise nature of the sharing induced by the HDP it is helpful to consider representations akin to the stick-breaking and Chinese restaurant representations of the DP. We consider these representations in the next three subsections before turning to a discussion of applications of the HDP.

Note that the recursive construction of the HDP can be generalized to arbitrary hierarchies in the obvious way. Each G_j is given a DP prior with base measure $G_{\text{pa}(j)}$, where pa(j) is the parent index of j in the hierarchy. As in the two-level hierarchy in equation (5.1), the set of atoms at the top level is shared throughout the hierarchy, while the multilevel hierarchy allows for a richer dependence structure on the weights of the atoms. Section 5.4 presents an instance of such a hierarchy in the setting of Pitman–Yor processes.

Other ways to couple multiple Dirichlet processes have been proposed in the literature; in particular the *dependent Dirichlet process* of MacEachern, Kottas and Gelfand (2001) provides a general formalism. Ho, James and Lau (2006) gives a complementary view of the HDP and its Pitman–Yor generalizations in terms of coagulation operators. See Teh, Jordan, Beal and Blei (2006) and Chapter 7 for overviews.

5.2.1 Stick-breaking construction

In this section we develop a stick-breaking construction for the HDP. This representation provides a concrete representation of draws from an HDP and it provides insight into the sharing of atoms across multiple DPs.

We begin with the stick-breaking representation for the random base measure G_0, where $G_0 \sim \text{DP}(\gamma, H)$. Given that this base measure is distributed according to a DP, we have (Sethuraman, 1994; Ishwaran and James, 2001; see also Section 2.2.3);

$$G_0 = \sum_{k=1}^{\infty} \beta_k \delta_{\theta_k^{**}}, \tag{5.2}$$

where

$$\begin{aligned} v_k \mid \gamma &\sim \text{Beta}(1, \gamma), \\ \beta_k &= v_k \prod_{l=1}^{k-1}(1 - v_l), \qquad \text{for } k = 1, \ldots, \infty \\ \theta_k^{**} \mid H &\sim H. \end{aligned} \tag{5.3}$$

We refer to the joint distribution on the infinite sequence $(\beta_1, \beta_2, \ldots)$ as the GEM(γ) distribution (Pitman, 2002) ("GEM" stands for Griffiths, Engen and McCloskey).

The random measures G_j are also distributed (conditionally) according to a DP. Moreover, the support of each G_j is contained within the support of G_0. Thus the stick-breaking representation for G_j is a reweighted sum of the atoms in G_0:

$$G_j = \sum_{k=1}^{\infty} \pi_{jk} \delta_{\theta_k^{**}}. \tag{5.4}$$

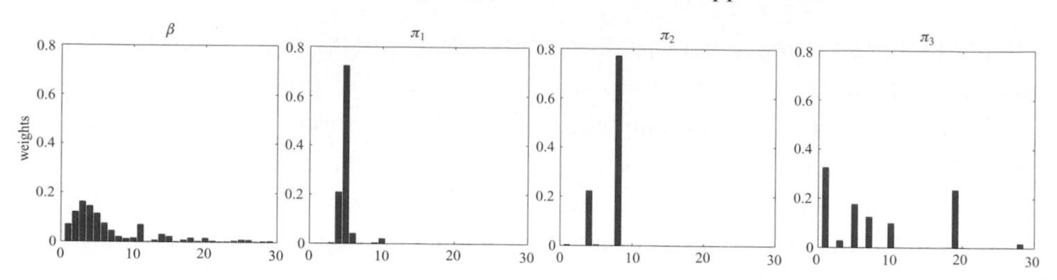

Figure 5.1 The HDP stick-breaking construction. The left panel depicts a draw of $\boldsymbol{\beta}$, and the remaining panels depict draws of $\boldsymbol{\pi}_1$, $\boldsymbol{\pi}_2$ and $\boldsymbol{\pi}_3$ conditioned on $\boldsymbol{\beta}$.

The problem reduces to finding a relationship between the weights $\boldsymbol{\beta} = (\beta_1, \beta_2, \ldots)$ and $\boldsymbol{\pi}_j = (\pi_{j1}, \pi_{j2}, \ldots)$. Let us interpret these weight vectors as probability measures on the discrete space $\{1, \ldots, \infty\}$. Taking partitions over integers induced by partitions on Θ, the defining property of the DP (Ferguson, 1973) implies:

$$\boldsymbol{\pi}_j \mid \alpha, \boldsymbol{\beta} \sim \mathrm{DP}(\alpha, \boldsymbol{\beta}). \tag{5.5}$$

Some algebra then readily yields the following explicit construction for $\boldsymbol{\pi}_j$ conditioned on $\boldsymbol{\beta}$:

$$\begin{aligned} v_{jk} \mid \alpha, \beta_1, \ldots, \beta_k &\sim \mathrm{Beta}\left(\alpha\beta_k, \alpha\left(1 - \sum_{l=1}^{k} \beta_l\right)\right), \\ \pi_{jk} &= v_{jk} \prod_{l=1}^{k-1} (1 - v_{jl}), \qquad \text{for } k = 1, \ldots, \infty. \end{aligned} \tag{5.6}$$

Figure 5.1 shows a sample draw of $\boldsymbol{\beta}$ along with draws from $\boldsymbol{\pi}_1$, $\boldsymbol{\pi}_2$ and $\boldsymbol{\pi}_3$ given $\boldsymbol{\beta}$.

From equation (5.3) we see that the mean of β_k is $\mathrm{E}[\beta_k] = \gamma^{k-1}(1+\gamma)^{-k}$ which decreases exponentially in k. The mean for $\boldsymbol{\pi}_j$ is simply its base measure $\boldsymbol{\beta}$; thus $\mathrm{E}[\pi_{jk}] = \mathrm{E}[\beta_k] = \gamma^{k-1}(1+\gamma)^{-k}$ as well. However the law of total variance shows that π_{jk} has higher variance than β_k: $\mathrm{Var}[\pi_{jk}] = \mathrm{E}[\frac{\beta_k(1-\beta_k)}{1+\alpha}] + \mathrm{Var}[\beta_k] > \mathrm{Var}[\beta_k]$. The higher variance is reflected in Figure 5.1 by the sparser nature of $\boldsymbol{\pi}_j$ relative to $\boldsymbol{\beta}$.

5.2.2 *Chinese restaurant franchise*

The Chinese restaurant process (CRP) describes the marginal probabilities of the DP in terms of a random partition obtained from a sequence of customers sitting at tables in a restaurant. There is an analogous representation for the HDP which we refer to as a *Chinese restaurant franchise* (CRF). In a CRF the metaphor of a Chinese restaurant is extended to a set of restaurants, one for each index in

$\mathcal{J}$. The customers in the jth restaurant sit at tables in the same manner as for the CRP, and this is done independently in the restaurants. The coupling among restaurants is achieved via a franchise-wide menu. The first customer to sit at a table in a restaurant chooses a dish from the menu and all subsequent customers who sit at that table inherit that dish. Dishes are chosen with probability proportional to the number of tables (franchise-wide) which have previously served that dish.

More formally, label the ith customer in the jth restaurant with a random variable θ_{ji} that is distributed according to G_j. Similarly, let θ^*_{jt} denote a random variable corresponding to the tth table in the jth restaurant; these variables are drawn independently and identically distributed (i.i.d.) according to G_0. Finally, the dishes are i.i.d. variables θ^{**}_k distributed according to the base measure H. We couple these variables as follows. Each customer sits at one table and each table serves one dish; let customer i in restaurant j sit at table t_{ji}, and let table t serve dish k_{jt}. Then let $\theta_{ji} = \theta^*_{jt_{ji}} = \theta^{**}_{k_{jt_{ji}}}$.

Let n_{jtk} be the number of customers in restaurant j seated around table t and being served dish k, let m_{jk} be the number of tables in restaurant j serving dish k, and let K be the number of unique dishes served in the entire franchise. We denote marginal counts with dots; e.g., $n_{j\cdot k}$ is the number of customers in restaurant j served dish k.

To show that the CRF captures the marginal probabilities of the HDP, we integrate out the random measures G_j and G_0 in turn from the HDP. We start by integrating out the random measure G_j; this yields a set of conditional distributions for the θ_{ji} described by a Pólya urn scheme:

$$\theta_{ji} \mid \theta_{j1}, \ldots, \theta_{j,i-1}, \alpha, G_0 \sim \sum_{t=1}^{m_{j\cdot}} \frac{n_{jt\cdot}}{\alpha + n_{j\cdot\cdot}} \delta_{\theta^*_{jt}} + \frac{\alpha}{\alpha + n_{j\cdot\cdot}} G_0. \tag{5.7}$$

A draw from this mixture can be obtained by drawing from the terms on the right-hand side with probabilities given by the corresponding mixing proportions. If a term in the first summation is chosen then the customer sits at an already occupied table: we increment $n_{jt\cdot}$, set $\theta_{ji} = \theta^*_{jt}$ and let $t_{ji} = t$ for the chosen t. If the second term is chosen then the customer sits at a new table: we increment $m_{j\cdot}$ by one, set $n_{jm_{j\cdot}\cdot} = 1$, draw $\theta^*_{jm_{j\cdot}} \sim G_0$, set $\theta_{ji} = \theta^*_{jm_{j\cdot}}$ and $t_{ji} = m_{j\cdot}$.

Notice that each θ^*_{jt} is drawn i.i.d. from G_0 in the Pólya urn scheme in equation (5.7), and this is the only reference to G_0 in that equation. Thus we can readily integrate out G_0 as well, obtaining a Pólya urn scheme for the θ^*_{jt}:

$$\theta^*_{jt} \mid \theta^*_{11}, \ldots, \theta^*_{1m_{1\cdot}}, \ldots, \theta^*_{j,t-1}, \gamma, H \sim \sum_{k=1}^{K} \frac{m_{\cdot k}}{\gamma + m_{\cdot\cdot}} \delta_{\theta^{**}_k} + \frac{\gamma}{\gamma + m_{\cdot\cdot}} H, \tag{5.8}$$

where we have presumed for ease of notation that $\mathcal{J} = \{1, \ldots, |\mathcal{J}|\}$. As promised, we see that the kth dish is chosen with probability proportional to the number of tables franchise-wide that previously served that dish ($m_{\cdot k}$).

The CRF is useful in understanding scaling properties of the clustering induced by an HDP. In a DP the number of clusters scales logarithmically (Antoniak, 1974). Thus $m_{j\cdot} = \mathrm{O}(\alpha \log \frac{n_{j\cdot\cdot}}{\alpha})$ where $m_{j\cdot}$ and $n_{j\cdot\cdot}$ are respectively the total number of tables and customers in restaurant j. Since G_0 is itself a draw from a DP, we have that $K = \mathrm{O}(\gamma \log \sum_j \frac{m_{j\cdot}}{\gamma}) = \mathrm{O}(\gamma \log(\frac{\alpha}{\gamma} \sum_j \log \frac{n_{j\cdot\cdot}}{\alpha}))$. If we assume that there are J groups and that the groups (the customers in the different restaurants) have roughly the same size N, $n_{j\cdot\cdot} = \mathrm{O}(N)$, we see that $K = \mathrm{O}(\gamma \log \frac{\alpha}{\gamma} J \log \frac{N}{\alpha}) = \mathrm{O}(\gamma \log \frac{\alpha}{\gamma} + \gamma \log J + \gamma \log \log \frac{N}{\alpha})$. Thus the number of clusters scales doubly logarithmically in the size of each group, and logarithmically in the number of groups. The HDP thus expresses a prior belief that the number of clusters grows very slowly in N. If this prior belief is inappropriate for a given problem, there are alternatives; in particular, in the language modeling example of Section 5.4.3 we discuss a hierarchical model that yields power-law scaling.

5.2.3 Posterior structure of the HDP

The Chinese restaurant franchise is obtained by integrating out the random measures G_j and then integrating out G_0. Integrating out the random measures G_j yields a Chinese restaurant for each group as well as a sequence of i.i.d. draws from the base measure G_0, which are used recursively in integrating out G_0. Having obtained the CRF, it is of interest to derive conditional distributions that condition on the CRF; this not only illuminates the combinatorial structure of the HDP but it also prepares the ground for a discussion of inference algorithms (see Section 5.7), where it can be useful to instantiate the CRF explicitly.

The state of the CRF consists of the dish labels $\boldsymbol{\theta}^{**} = \{\theta_k^{**}\}_{k=1,\ldots,K}$, the table t_{ji} at which the ith customer sits, and the dish k_{jt} served at the tth table. As functions of the state of the CRF, we also have the numbers of customers $\mathbf{n} = \{n_{jtk}\}$, the numbers of tables $\mathbf{m} = \{m_{jk}\}$, the customer labels $\boldsymbol{\theta} = \{\theta_{ji}\}$ and the table labels $\boldsymbol{\theta}^* = \{\theta_{jt}^*\}$. The relationship between the customer labels and the table labels is given as follows: $\theta_{jt}^* = \theta_{jk_{jt}}^{**}$ and $\theta_{ji} = \theta_{jt_{ji}}^*$.

Consider the distribution of G_0 conditioned on the state of the CRF. G_0 is independent from the rest of the CRF when we condition on the i.i.d. draws $\boldsymbol{\theta}^*$, because the restaurants interact with G_0 only via the i.i.d. draws. The posterior thus follows from the usual posterior for a DP given i.i.d. draws:

$$G_0 \mid \gamma, H, \boldsymbol{\theta}^* \sim \mathrm{DP}\left(\gamma + m_{\cdot\cdot}, \frac{\gamma H + \sum_{k=1}^K m_{\cdot k} \delta_{\theta_k^{**}}}{\gamma + m_{\cdot\cdot}}\right). \tag{5.9}$$

Note that values for $\mathbf{m}$ and $\boldsymbol{\theta}^{**}$ are determined given $\boldsymbol{\theta}^{*}$, since they are simply the unique values and their counts among $\boldsymbol{\theta}^{*}$.† A draw from equation (5.9) can be constructed as follows (using the defining property of a DP):

$$\begin{aligned} \beta_0, \beta_1, \ldots, \beta_K \mid \gamma, G_0, \boldsymbol{\theta}^* &\sim \text{Dirichlet}(\gamma, m_{\cdot 1}, \ldots, m_{\cdot K}), \\ G_0' \mid \gamma, H &\sim \text{DP}(\gamma, H), \\ G_0 &= \beta_0 G_0' + \sum_{k=1}^{K} \beta_k \delta_{\theta_k^{**}}. \end{aligned} \tag{5.10}$$

We see that the posterior for G_0 is a mixture of atoms corresponding to the dishes and an independent draw from $\text{DP}(\gamma, H)$.

Conditioning on this draw of G_0 as well as the state of the CRF, the posteriors for the G_j are independent. In particular, the posterior for each G_j follows from the usual posterior for a DP, given its base measure G_0 and i.i.d. draws $\boldsymbol{\theta}_j$:

$$G_j \mid \alpha, G_0, \boldsymbol{\theta}_j \sim \text{DP}\left(\alpha + n_{j\cdot\cdot}, \frac{\alpha G_0 + \sum_{k=1}^{K} n_{j\cdot K} \delta_{\theta_k^{**}}}{\alpha + n_{j\cdot\cdot}}\right). \tag{5.11}$$

Note that $\mathbf{n}_j$ and $\boldsymbol{\theta}^{**}$ are simply the unique values and their counts among the $\boldsymbol{\theta}_j$. Making use of the decomposition of G_0 into G_0' and atoms located at the dishes $\boldsymbol{\theta}^{**}$, a draw from equation (5.11) can thus be constructed as follows:

$$\begin{aligned} \pi_{j0}, \pi_{j1}, \ldots, \pi_{jK} \mid \alpha, \boldsymbol{\theta}_j &\sim \text{Dirichlet}(\alpha\beta_0, \alpha\beta_1 + n_{j\cdot 1}, \ldots, \alpha\beta_K + n_{j\cdot K}), \\ G_j' \mid \alpha, G_0 &\sim \text{DP}(\alpha\beta_0, G_0'), \\ G_j &= \pi_{j0} G_j' + \sum_{k=1}^{K} \pi_{jk} \delta_{\theta_k^{**}}. \end{aligned} \tag{5.12}$$

We see that G_j is a mixture of atoms at θ_k^{**} and an independent draw from a DP, where the concentration parameter depends on β_0.

The posterior over the entire HDP is obtained by averaging the conditional distributions of G_0 and G_j over the posterior state of the Chinese restaurant franchise given $\boldsymbol{\theta}$.

This derivation shows that the posterior for the HDP can be split into a "discrete part" and a "continuous part." The discrete part consists of atoms at the unique values $\boldsymbol{\theta}^{**}$, with different weights on these atoms for each DP. The continuous part is a separate draw from an HDP with the same hierarchical structure as the original HDP and global base measure H, but with altered concentration parameters. The continuous part consists of an infinite series of atoms at locations drawn i.i.d.

† Here we make the simplifying assumption that H is a continuous distribution so that draws from H are unique. If H is not continuous then additional bookkeeping is required.

from H. Although we have presented this posterior representation for a two-level hierarchy, the representation extends immediately to general hierarchies.

5.2.4 Applications of the HDP

In this section we consider several applications of the HDP. These models use the HDP at different depths in an overall Bayesian hierarchy. In the first example the random measures obtained from the HDP are used to generate data directly, and in the second and third examples these random measures generate latent parameters.

Information retrieval

The growth of modern search engines on the World Wide Web has brought new attention to a classical problem in the field of information retrieval (IR) – how should a collection of documents be represented so that relevant documents can be returned in response to a query? IR researchers have studied a wide variety of representations and have found empirically that a representation known as *term frequency-inverse document frequency*, or "tf-idf," yields reasonably high-quality rankings of documents (Salton and McGill, 1983). The general intuition is that the relevance of a document to a query should be proportional to the frequency of query terms it contains ("term frequency"), but that query terms that appear in many documents should be downweighted since they are less informative ("inverse document frequency").

Cowans (2004, 2006) has shown that the HDP provides statistical justification for the intuition behind tf-idf. Let x_{ji} denote the ith word in the jth document in some corpus of documents, where the range of x_{ji} is a discrete *vocabulary* Θ. Consider the following simple model for documents:

$$\begin{aligned} G_0 \mid \gamma, H &\sim \mathrm{DP}(\gamma, H), && \\ G_j \mid \alpha, G_0 &\sim \mathrm{DP}(\alpha, G_0), && \text{for } j \in \mathcal{J} \\ x_{ji} \mid G_j &\sim G_j && \text{for } i = 1, \ldots, n_j, \end{aligned} \tag{5.13}$$

where H is the global probability measure over the vocabulary Θ and where n_j is the number of words in the jth document. (Note that $n_j = n_{j\cdot\cdot}$ where the latter refers to the general notation introduced in Section 5.2.2; here and elsewhere we use n_j as a convenient shorthand.) In this model, G_j is a discrete measure over the vocabulary associated with document j and G_0 is a discrete measure over the vocabulary that acts to tie together word usages across the corpus. The model is presented as a graphical model in the left panel of Figure 5.2.

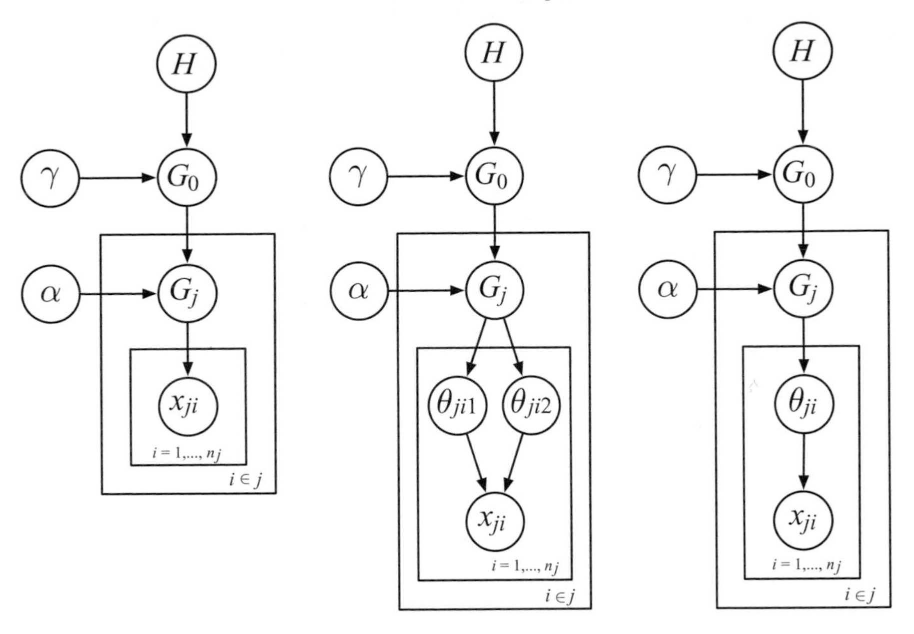

Figure 5.2 Graphical representations of HDP-based models. *Left:* An HDP model for information retrieval. *Center:* An HDP mixture model for haplotype phasing. *Right:* The HDP-LDA model for topic or admixture modeling.

Integrating out G_0 and the G_j as discussed in Section 5.2.2, we obtain the following marginal probabilities for words $\theta \in \Theta$ in the jth document:

$$\begin{aligned} p_j(\theta) &= \frac{n'_{j\theta} + \alpha p_0(\theta)}{n'_{j\cdot} + \alpha}, \\ p_0(\theta) &= \frac{m'_{\cdot\theta} + \gamma H(\theta)}{m'_{\cdot\cdot} + \gamma}, \end{aligned} \tag{5.14}$$

where $n'_{j\theta}$ is the term frequency (the number of occurrences of θ in document j) and $m'_{j\theta}$ is the number of tables serving dish θ in restaurant j in the CRF representation. (Note that the need for the specialized "prime" notation in this case is driven by the fact that Θ is a discrete space in this example. In particular, for each $\theta \in \Theta$ there may be multiple k such that $\theta_k^{**} = \theta$. The term frequency $n'_{j\theta} = \sum_{k:\theta_k^{**}=\theta} n_{j\cdot k}$ is the number of customers eating dish θ regardless of which menu entry they picked. Similarly, $m'_{j\theta} = \sum_{k:\theta_k^{**}=\theta} m_{jk}$.)

If we make the approximation that the number of tables serving a particular dish in a particular restaurant is at most one, then $m'_{\cdot\theta}$ is the document frequency – the number of documents containing word θ in the corpus. We now rank documents by computing a "relevance score" $R(j, Q)$, the log probability of a query Q under each document j:

$$R(j, Q) = \sum_{\theta \in Q} \log p_j(\theta)$$

$$= \sum_{\theta \in Q} \left(\log \left(1 + \frac{n'_{j\theta}}{\alpha \frac{m'_{\cdot\theta} + \gamma H(\theta)}{m'_{\cdot\cdot} + \gamma}} \right) - \log(n'_{j\cdot} + \alpha) + \log(\alpha p_0(\theta)) \right). \tag{5.15}$$

In this score the first term is akin to a tf-idf score, the second term is a normalization penalizing large documents, and the third term can be ignored as it does not depend on document identity j. Thus we see that a simple application of the HDP provides a principled justification for the use of inverse document frequency and document length normalization. Moreover, in small-scale experiments, Cowans (2004, 2006) found that this score improves upon state-of-the-art relevance scores (Robertson et al., 1992; Hiemstra and Kraaij, 1998).

Multipopulation haplotype phasing

We now consider a class of applications in which the HDP provides a distribution on latent parameters rather than on the observed data.

Haplotype phasing is an interesting problem in statistical genetics that can be formulated as a mixture model (Stephens, Smith and Donnelly, 2001). Consider a set of M binary markers along a chromosome. Chromosomes come in pairs for humans, so let θ_{i1} and θ_{i2} denote the binary-valued vectors of markers for a pair of chromosomes for the ith individual. These vectors are referred to as *haplotypes*, and the elements of these vectors are referred to as *alleles*. A *genotype* x_i is a vector which records the unordered pair of alleles for each marker; that is, the association of alleles to chromosome is lost. The *haplotype phasing* problem is to restore haplotypes (which are useful for predicting disease associations) from genotypes (which are readily assayed experimentally whereas haplotypes are not).

Under standard assumptions from population genetics, we can write the probability of the ith genotype as a mixture model:

$$p(x_i) = \sum_{\theta_{i1}, \theta_{i2} \in \mathcal{H}} p(\theta_{i1}) p(\theta_{i2}) p(x_i \mid \theta_{i1}, \theta_{i2}), \tag{5.16}$$

where $\mathcal{H}$ is the set of haplotypes in the population and where $p(x_i \mid \theta_{i1}, \theta_{i2})$ reflects the loss of order information as well as possible measurement error. Given that the cardinality of $\mathcal{H}$ is unknown, this problem is naturally formulated as a DP mixture modeling problem where a "cluster" is a haplotype (Xing, Jordan and Sharan, 2007).

Let us now consider a multipopulation version of the haplotype phasing problem in which the genotype data can be classified into (say) Asian, European and African subsets. Here it is natural to attempt to identify the haplotypes in each population and to share these haplotypes among populations. This can be achieved with the following HDP mixture model:

$$\begin{aligned} G_0 \mid \gamma, H &\sim \mathrm{DP}(\gamma, H), & & \\ G_j \mid \alpha, G_0 &\sim \mathrm{DP}(\alpha, G_0), & &\text{for each population } j \in \mathcal{J} \\ \theta_{ji1}, \theta_{ji2} \mid G_j &\overset{\text{i.i.d.}}{\sim} G_j, & &\text{for each individual } i = 1, \ldots, n_j \\ x_{ji} \mid \theta_{ji1}, \theta_{ji2} &\sim F_{\theta_{ji1}, \theta_{ji2}}, & & \end{aligned} \tag{5.17}$$

where $\theta_{ji1}, \theta_{ji2}$ denote the pair of haplotypes for the ith individual in the jth population. The model is presented as a graphical model in the center panel of Figure 5.2. Xing, Sohn, Jordan and Teh (2006) showed that this model performs effectively in multipopulation haplotype phasing, outperforming methods that lump together the multiple populations or treat them separately.

Topic modeling

A *topic model* or *mixed membership model* is a generalization of a finite mixture model in which each data point is associated with multiple draws from a mixture model, not a single draw (Blei, Ng and Jordan, 2003; Erosheva, 2003). As we will see, while the DP is the appropriate tool to extend finite mixture models to the nonparametric setting, the appropriate tool for nonparamametric topic models is the HDP.

To motivate the topic model formulation, consider the problem of modeling the word occurrences in a set of newspaper articles (for example, for the purposes of classifying future articles). A simple clustering methodology might attempt to place each article in a single cluster. But it would seem more useful to be able to cross-classify articles according to "topics"; for example, an article might be mainly about Italian food, but it might also refer to health, history and the weather. Moreover, as this example suggests, it would be useful to be able to assign numerical values to the degree to which an article treats each topic.

Topic models achieve this goal as follows. Define a *topic* to be a probability distribution across a set of *words* taken from some vocabulary W. A *document* is modeled as a probability distribution across topics. In particular, let us assume the following generative model for the words in a document. First choose a probability vector π from the K-dimensional simplex, and then repeatedly (1) select one of the K topics with probabilities given by the components of π and (2) choose a word from the distribution defined by the selected topic. The vector π thus encodes the expected fraction of words in a document that are allocated to each of the K topics. In general a document will be associated with multiple topics.

Another natural example of this kind of problem arises in statistical genetics. Assume that for each individual in a population we can assay the state of each of M *markers*, and recall that the collection of markers for a single individual is referred to as a *genotype*. Consider a situation in which K *subpopulations* which have hitherto remained separate are now thoroughly mixed (i.e., their mating patterns are that of a single population). Individual genotypes will now have portions that arise from the different subpopulations. This is referred to as "admixture." We can imagine generating a new admixed genotype by fixing a distribution π across subpopulations and then repeatedly (1) choosing a subpopulation according to π and (2) choosing the value of a marker (an "allele") from the subpopulation-specific distribution on the alleles for that marker. This formulation is essentially isomorphic to the document modeling formulation. (The difference is that in the document setting the observed words are generally assumed to be exchangeable, whereas in the genetics setting each marker has its own distribution over alleles.)

To specify a topic model fully, we require a distribution for π. Taking this distribution to be symmetric Dirichlet, we obtain the *latent Dirichlet allocation* (LDA) model, developed by Blei, Ng and Jordan (2003) and Pritchard, Stephens and Donnelly (2000) as a model for documents and admixture, respectively. This model has been widely used not only in the fields of information retrieval and statistical genetics, but also in computational vision, where a "topic" is a distribution across visual primitives, and an image is modeled as a distribution across topics (Fei-Fei and Perona, 2005).

Let us now turn to the problem of developing a Bayesian nonparametric version of LDA in which the number of topics is allowed to be open ended. As we have alluded to, this requires the HDP, not merely the DP. To see this, consider the generation of a single word in a given document. According to LDA, this is governed by a finite mixture model, in which one of K topics is drawn and then a word is drawn from the corresponding topic distribution. Generating all of the words in a single document requires multiple draws from this finite mixture. If we now consider a different document, we again have a finite mixture, with the same mixture components (the topics), but with a different set of mixing proportions (the document-specific vector π). Thus we have multiple finite mixture models. In the nonparametric setting they must be linked so that the same topics can appear in different documents.

We are thus led to the following model, which we refer to as HDP-LDA:

$$\begin{aligned} G_0 \mid \gamma, H &\sim \mathrm{DP}(\gamma, H), \\ G_j \mid \alpha, G_0 &\sim \mathrm{DP}(\alpha, G_0), && \text{for each document } j \in \mathcal{J} \\ \theta_{ji} \mid G_j &\sim G_j, && \text{for each word } i = 1, \ldots, n_j \\ x_{ji} \mid \theta_{ji} &\sim F_{\theta_{ji}}, \end{aligned} \tag{5.18}$$

where x_{ji} is the ith word in document j, H is the prior distribution over topics and $F_{\theta_{ji}}$ is the distribution over words. The model is presented as a graphical model in the right panel of Figure 5.2. Note that the atoms present in the random distribution G_0 are shared among the random distributions G_j. Thus, as desired, we have a collection of tied mixture models, one for each document.

Topic models can be generalized in a number of other directions. For example, in applications to document modeling it is natural to ask that topics occur at multiple levels of resolution. Thus, at a high level of resolution, we might wish to obtain topics that give high probability to words that occur throughout the documents in a corpus, while at a lower level we might wish to find topics that are focused on words that occur in specialized subsets of the documents. A Bayesian nonparametric approach to obtaining this kind of *abstraction hierarchy* has been presented by Blei, Griffiths, Jordan and Tenenbaum (2004). In the model presented by these authors, topics are arranged into a tree, and a document is modeled as a path down the tree. This is achieved by defining the tree procedurally in terms of a linked set of Chinese restaurants.

5.3 Hidden Markov models with infinite state spaces

Hidden Markov models (HMMs) are widely used to model sequential data and time series data (Rabiner, 1989). An HMM is a doubly stochastic Markov chain in which a state sequence, $\theta_1, \theta_2, \ldots, \theta_\tau$, is drawn according to a Markov chain on a discrete state space Θ with transition kernel $\pi(\theta_t, \theta_{t+1})$. A corresponding sequence of observations, $x_1, x_2, \ldots, x_\tau$, is drawn conditionally on the state sequence, where for all t the observation x_t is conditionally independent of the other observations given the state θ_t. We let $F_{\theta_t}(x_t)$ denote the distribution of x_t conditioned on the state θ_t; this is referred to as the "emission distribution."

In this section we show how to use Bayesian nonparametric ideas to obtain an "infinite HMM" – an HMM with a countably infinite state space (Beal, Ghahramani and Rasmussen, 2002; Teh, Jordan, Beal and Blei, 2006). The idea is similar in spirit to the passage from a finite mixture model to a DP mixture model. However, as we show, the appropriate nonparametric tool is the HDP, not the DP. The resulting model is thus referred to as the *hierarchical Dirichlet process hidden Markov model*

(HDP-HMM). We present both the HDP formulation and a stick-breaking formulation in this section; the latter is particularly helpful in understanding the relationship to finite HMMs. It is also worth noting that a Chinese restaurant franchise (CRF) representation of the HDP-HMM can be developed, and indeed Beal, Ghahramani and Rasmussen (2002) presented a precursor to the HDP-HMM that was based on an urn model akin to the CRF.

To understand the need for the HDP rather than the DP, note first that a classical HMM specifies a set of finite mixture distributions, one for each value of the current state θ_t. Indeed, given θ_t, the observation x_{t+1} is chosen by first picking a state θ_{t+1} and then choosing x_{t+1} conditional on that state. Thus the transition probability $\pi(\theta_t, \theta_{t+1})$ plays the role of a mixing proportion and the emission distribution F_{θ_t} plays the role of the mixture component. It is natural to consider replacing this finite mixture model by a DP mixture model. In so doing, however, we must take into account the fact that we obtain a *set* of DP mixture models, one for each value of the current state. If these DP mixture models are not tied in some way, then the set of states accessible in a given value of the current state will be disjoint from those accessible for some other value of the current state. We would obtain a branching structure rather than a chain structure. The solution to this problem is straightforward – we use the HDP to tie the DPs.

More formally, let us consider a collection of random transition kernels, $\{G_\theta : \theta \in \Theta\}$, drawn from an HDP:

$$\begin{aligned} G_0 \mid \gamma, H &\sim \mathrm{DP}(\gamma, H), \\ G_\theta \mid \alpha, G_0 &\sim \mathrm{DP}(\alpha, G_0) \qquad \text{for } \theta \in \Theta, \end{aligned} \tag{5.19}$$

where H is a base measure on the probability space $(\Theta, \mathcal{T})$. As we shall see, the random base measure G_0 allows the transitions out of each state to share the same set of next states. Let $\theta_0 = \theta_0^{**} \in \Theta$ be a predefined initial state. The conditional distributions of the sequence of latent state variables $\theta_1, \ldots, \theta_\tau$ and observed variables $x_1, \ldots, x_\tau$ are:

$$\begin{aligned} \theta_t \mid \theta_{t-1}, G_{\theta_{t-1}} &\sim G_{\theta_{t-1}}, \\ x_t \mid \theta_t &\sim F_{\theta_t} \qquad \text{for } t = 1, \ldots, \tau. \end{aligned} \tag{5.20}$$

A graphical model representation for the HDP-HMM is shown in Figure 5.3.

We have defined a probability model consisting of an uncountable number of DPs, which may raise measure-theoretic concerns. These concerns can be dealt with, however, essentially due to the fact that the sample paths of the HDP-HMM only ever encounter a finite number of states. To see this more clearly, and to understand the relationship of the HDP-HMM to the parametric HMM, it is helpful to consider a stick-breaking representation of the HDP-HMM. This

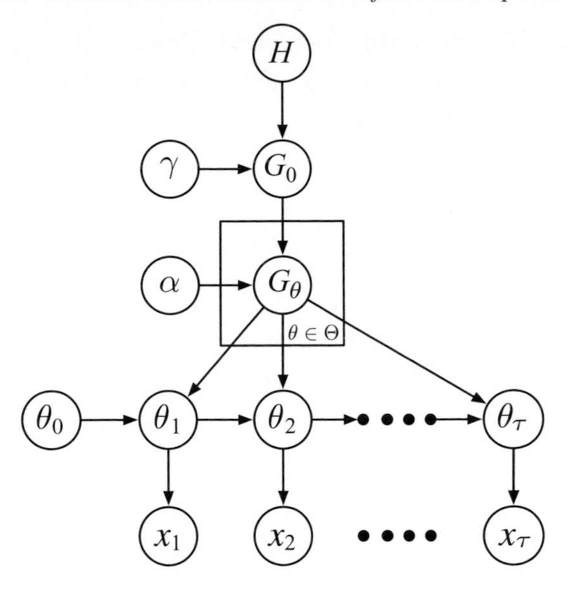

Figure 5.3 HDP hidden Markov model.

representation is obtained directly from the stick-breaking representation of the underlying HDP:

$$G_0 = \sum_{k=1}^{\infty} \beta_k \delta_{\theta_k^{**}},$$
$$G_{\theta_l^{**}} = \sum_{k=1}^{\infty} \pi_{\theta_l^{**}k} \delta_{\theta_k^{**}} \qquad \text{for } l = 0, 1, \ldots, \infty, \tag{5.21}$$

where

$$\theta_k^{**} \mid H \sim H,$$
$$\boldsymbol{\beta} \mid \gamma \sim \mathrm{GEM}(\gamma), \tag{5.22}$$
$$\boldsymbol{\pi}_{\theta_k^{**}} \mid \alpha, \boldsymbol{\beta} \sim \mathrm{DP}(\alpha, \boldsymbol{\beta}) \qquad \text{for } k = 1, \ldots, \infty.$$

The atoms θ_k^{**} are shared across G_0 and the transition distributions $G_{\theta_l^{**}}$. Since all states visited by the HMM are drawn from the transition distributions, the states possibly visited by the HMM with positive probability (given G_0) will consist only of the initial state θ_0^{**} and the atoms $\theta_1^{**}, \theta_2^{**}, \ldots$. Relating to the parametric HMM, we see that the transition probability from state θ_l^{**} to state θ_k^{**} is given by $\pi_{\theta_l^{**}k}$ and the distribution on the observations is given by $F_{\theta_k^{**}}$.

This relationship to the parametric HMM can be seen even more clearly if we identify the state θ_k^{**} with the integer k, for $k = 0, 1, \ldots, \infty$, and if we introduce integer-valued variables z_t to denote the state at time t. In particular, if $\theta_t = \theta_k^{**}$ is the state at time t, we let z_t take on value k, and write $\boldsymbol{\pi}_k$ instead of $\boldsymbol{\pi}_{\theta_k^{**}}$. The HDP-HMM can now be expressed as:

$$\begin{aligned} z_t \mid z_{t-1}, \boldsymbol{\pi}_{z_{t-1}} &\sim \boldsymbol{\pi}_{z_{t-1}}, \\ x_t \mid z_t, \theta_{z_t}^{**} &\sim F_{\theta_{z_t}^{**}}, \end{aligned} \tag{5.23}$$

with priors on the parameters and transition probabilities given by equation (5.23). This construction shows explicitly that the HDP-HMM can be interpreted as an HMM with a countably infinite state space.

A difficulty with the HDP-HMM as discussed thus far is that it tends to be poor at capturing state persistence; it has a tendency to create redundant states and rapidly switch among them. This may not be problematic for applications in which the states are nuisance variables and it is overall predictive likelihood that matters, but it can be problematic for segmentation or parsing applications in which the states are the object of inference and when state persistence is expected. This problem can be solved by giving special treatment to self-transitions. In particular, let G_θ denote the transition kernel associated with state θ. Fox, Sudderth, Jordan and Willsky (2008) proposed the following altered definition of G_θ (compare to equation (5.19)):

$$G_\theta \mid \alpha, \kappa, G_0, \theta \sim \mathrm{DP}\left(\alpha + \kappa, \frac{\alpha G_0 + \kappa \delta_\theta}{\alpha + \kappa}\right), \tag{5.24}$$

where δ_θ is a point mass at θ and where κ is a parameter that determines the extra mass placed on a self-transition. To see in more detail how this affects state persistence, consider the stick-breaking weights $\boldsymbol{\pi}_{\theta_k^{**}}$ associated with one of the countably many states θ_k^{**} that can be visited by the HMM. The stick-breaking representation of $G_{\theta_k^{**}}$ is altered as follows (compare to equation (5.23)):

$$\boldsymbol{\pi}_{\theta_k^{**}} \mid \alpha, \boldsymbol{\beta}, \kappa \sim \mathrm{DP}\left(\alpha + \kappa, \frac{\alpha \boldsymbol{\beta} + \kappa \delta_{\theta_k^{**}}}{\alpha + \kappa}\right). \tag{5.25}$$

Fox, Sudderth, Jordan and Willsky (2008) further place a vague gamma prior on $\alpha + \kappa$ and a beta prior on $\kappa/(\alpha + \kappa)$. The hyperparameters of these distributions allow prior control of state persistence. See also Beal, Ghahramani and Rasmussen (2002), who develop a related prior within the framework of their hierarchical urn scheme.

5.3.1 Applications of the HDP-HMM

In the following sections we describe a number of applications and extensions of the HDP-HMM. An application that we will not discuss, but is worth mentioning, is the application of HDP-HMMs to the problem of modeling recombination hotspots and ancestral haplotypes for short segments of single nucleotide polymorphisms (Xing and Sohn, 2007).

Speaker diarization

Speech recognition has been a major application area for classical parametric HMMs (Huang, Acero and Hon, 2001). In a typical application, several dozen states are used, roughly corresponding to the number of phoneme-like segments in speech. The observations x_t are spectral representations of speech over short time slices.

In many applications, however, the number of states is more fundamentally part of the inferential problem and it does not suffice simply to fix an arbitrary value. Consider an audio recording of a meeting in which the number of people participating in the meeting is unknown a priori. The problem of *speaker diarization* is that of segmenting the audio recording into time intervals associated with individual speakers (Wooters and Huijbregts, 2007). Here it is natural to consider an HDP-HMM model, where a state corresponds to an individual speaker and the observations are again short-term spectral representations. Posterior inference in the HDP-HMM yields estimates of the spectral content of each speaker's voice, an estimate of the number of speakers participating in the meeting, and a diarization of the audio stream.

Such an application of the HDP-HMM has been presented by Fox, Sudderth, Jordan and Willsky (2008), who showed that the HDP-HMM approach yielded a state-of-the-art diarization method. A noteworthy aspect of their work is that they found that the special treatment of self-transitions discussed in the previous section was essential; without this special treatment the tendency of the HDP-HMM to switch rapidly among redundant states led to poor speaker diarization performance.

Word segmentation

As another application of the HDP-HMM to speech, consider the problem of segmenting an audio stream into a sequence of words. Speech is surprisingly continuous with few obvious breaks between words and the problem of *word segmentation* – that of identifying coherent segments of "words" and their boundaries in continuous speech – is nontrivial. Goldwater, Griffiths and Johnson (2006b) proposed a statistical approach to word segmentation based upon the HDP-HMM. The latent states of the HMM correspond to words. An HDP-HMM rather than a parametric HMM is required for this problem, since there are an unbounded number of potential words.

In the model, an utterance is viewed as a sequence of phonemes, $\rho_1, \rho_2, \ldots, \rho_\tau$. The sequence is modeled by an HDP-HMM in which words are the latent states. A word is itself a sequence of phonemes. The model specification is as follows. First, the number of words n is drawn from a geometric distribution. Then a sequence of n words, $\theta_1, \theta_2, \ldots, \theta_n$, is drawn from an HDP-HMM:

$$\begin{aligned} G_0 \mid \gamma, H &\sim \mathrm{DP}(\gamma, H), \\ G_\theta \mid \alpha, G_0 &\sim \mathrm{DP}(\alpha, G_0), && \text{for } \theta \in \Theta \\ \theta_i \mid \theta_{i-1}, G_{\theta_{i-1}} &\sim G_{\theta_{i-1}}, && \text{for } i = 1, \ldots, n \end{aligned} \tag{5.26}$$

where $\theta_0 \sim G_\emptyset$ is a draw from an initial state distribution. Each G_θ is the transition distribution over next words, given the previous word θ. This is defined for every possible word θ, with Θ the set of all possible words (including the empty word θ_0 which serves as an initial state for the Markov chain). The base measure H over words is a simple independent phonemes model: the length of the word, $l \geq 1$, is first drawn from another geometric distribution, then each phoneme r_i is drawn independently from a prior over phonemes:

$$H(\theta = (r_1, r_2, \ldots, r_l)) = \eta_0 (1 - \eta_0)^{l-1} \prod_{t=1}^{l} H_0(r_t), \tag{5.27}$$

where H_0 is a probability measure over individual phonemes. The probability of the observed utterance is then a sum over probabilities of sequences of words such that their concatenation is $\rho_1, \rho_2, \ldots, \rho_\tau$.

Goldwater, Griffiths and Johnson (2006b) have shown that this HDP-HMM approach leads to significant improvements in segmentation accuracy.

Trees and grammars

A number of other structured probabilistic objects are amenable to a nonparametric treatment based on the HDP. In this section we briefly discuss some recent developments which go beyond the chain-structured HMM to consider objects such as trees and grammars.

A *hidden Markov tree* (HMT) is a directed tree in which the nodes correspond to states, and in which the probability of a state depends (solely) on its unique parent in the tree. To each state there is optionally associated an observation, where the probability of the observation is conditionally independent of the other observations given the state (Chou, Willsky and Benveniste, 1994).

We can generalize the HDP-HMM to a *hierarchical Dirichlet process hidden Markov tree* (HDP-HMT) model in which the number of states is unbounded. This is achieved by a generalization of the HDP-HMM model in which the transition matrix along each edge of the HMT is replaced with sets of draws from a DP (one draw for each row of the transition matrix) and these DPs are tied with the HDP. This model has been applied to problems in image processing (denoising, scene recognition) in which the HDP-HMT is used to model correlations among wavelet coefficients in multiresolution models of images (Kivinen, Sudderth and Jordan, 2007a, 2007b).

As a further generalization of the HDP-HMM, several groups have considered nonparametric versions of probabilistic grammars (Finkel, Grenager and Manning, 2007; Johnson, Griffiths and Goldwater, 2007; Liang, Petrov, Jordan and Klein, 2007). These grammars consist of collections of rules, of the form $A \to BC$, where this transition from a symbol A to a pair of symbols BC is modeled probabilistically. When the number of grammar symbols is unknown a priori, it is natural to use the HDP to generate symbols and to tie together the multiple occurrences of these symbols in a parse tree.

5.4 Hierarchical Pitman–Yor processes

As discussed in Chapters 3 and 4, a variety of alternatives to the DP have been explored in the Bayesian nonparametrics literature. These alternatives can provide a better fit to prior beliefs than the DP. It is therefore natural to consider hierarchical models based on these alternatives. In this section we shall describe one such hierarchical model, the hierarchical Pitman–Yor (HPY) process, which is based on the Pitman–Yor process (also known as the two-parameter Poisson–Dirichlet process). We briefly describe the Pitman–Yor process here; Example 3.27 in Chapter 3 as well as Perman, Pitman, and Yor (1992), Pitman and Yor (1997) and Ishwaran and James (2001) present further material on the Pitman–Yor process. In Section 5.4.3 we describe an application of the HPY process to language modeling and present a spatial extension of the HPY process and an application to image segmentation.

5.4.1 Pitman–Yor processes

The Pitman–Yor process is a two-parameter generalization of the DP, with a discount parameter $0 \leq d < 1$ and a concentration parameter $\alpha > -d$. When $d = 0$ the Pitman–Yor process reduces to a DP with concentration parameter α. We write $G \sim \text{PY}(d, \alpha, H)$ if G is a Pitman–Yor process with the given parameters and base measure H. The stick-breaking construction and the Chinese restaurant process have natural generalizations in the Pitman–Yor process. A draw G from

that power laws characterize many of the statistics within the domain. In particular, using a database of images that were manually segmented and labeled by humans (Oliva and Torralba, 2001), Sudderth and Jordan (2009) have shown that both the segment sizes and the label occurrences (e.g., "sky," "grass") follow long-tailed distributions that are well captured by the Pitman–Yor process. This suggests considering models in which the marginal distributions at each site in an image are governed by Pitman–Yor processes. Moreover, to share information across a collection of images it is natural to consider HPY priors. In this section we describe a model based on such an HPY prior (Sudderth and Jordan, 2009). Our focus is the problem of image segmentation, where the observed data are a collection of images (an image is a collection of gray-scale or color values at each point in a two-dimensional grid) and the problem is to output a partition of each image into segments (a segment is a coherent region of the image, as defined by human labelings).

Let us consider a generative model for image texture and color, simplifying at first in two ways: (1) we focus on a single image and (2) we neglect the issue of spatial dependence within the image. Thus, for now we focus simply on obtaining Pitman–Yor marginal statistics for segment sizes and segment labels within a single image. Let us suppose that the image is represented as a large collection of *sites*, where a site is a local region in the image (often referred to as a *pixel* or a *super-pixel*). Let $\boldsymbol{\pi} \sim \text{GEM}(d, \alpha)$ be a draw from the two-parameter GEM distribution. For each site i, let t_i denote the segment assignment of site i, where $t_i \sim \text{Discrete}(\boldsymbol{\pi})$ are independent draws from $\boldsymbol{\pi}$. Given a large number of sites of equal size, the total area assigned to segment t will be roughly π_t, and segment sizes will follow Pitman–Yor statistics.

We also assign a label to each segment, again using a two-parameter GEM distribution. In particular, let $\boldsymbol{\beta} \sim \text{GEM}(\eta, \gamma)$ be a distribution across labels. For each segment t we label the segment by drawing $k_t \sim \text{Discrete}(\boldsymbol{\beta})$ independently. We also let θ_k^{**} denote an "appearance model"† for label type k, where the θ_k^{**} are drawn from some prior distribution H. Putting this together, the label assigned to site i is denoted k_{t_i}. The visual texture and color at site i are then generated by a draw from the distribution $\theta_{k_{t_i}}^{**}$.

To obtain a spatially dependent Pitman–Yor process, Sudderth and Jordan (2009) adapt an idea of Duan, Guindani and Gelfand (2007), who used a latent collection of Gaussian processes to define a spatially dependent set of draws from a Dirichlet process. In particular, to each index t we associate a zero-mean Gaussian process, u_t. At a given site i, we thus have an infinite collection of Gaussian random variables,

† This parameter is generally a multinomial parameter encoding the probabilities of various discrete-valued texture and color descriptors.

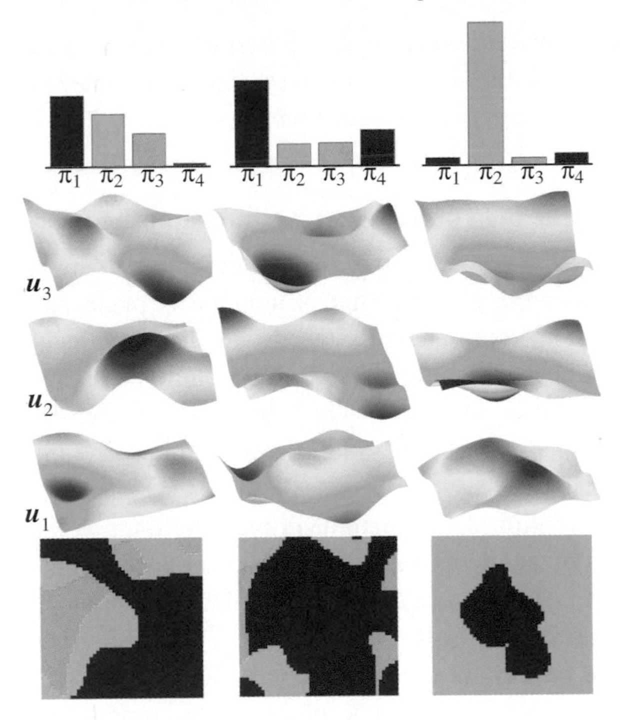

Figure 5.5 Draws from dependent Pitman–Yor processes. *Top:* the random proportions π_j. *Middle:* draws from Gaussian processes, one for each entry in π_j. *Bottom:* resulting segmentation.

$\{u_{ti}\}_{t=1,\ldots,\infty}$. By an appropriate choice of thresholds for this infinite sequence of Gaussian variables, it is possible to mimic a draw from the distribution π (by basing the selection on the first Gaussian variable in the sequence that is less than its threshold). Indeed, for a single site, this is simply a change-of-variables problem from a collection of beta random variables to a collection of Gaussian random variables. The Gaussian process framework couples the choice of segments at nearby sites via the covariance function. Figure 5.5 gives an example of three draws from this model, showing the underlying random distribution π (truncated to four values), the corresponding collection of draws from Gaussian processes (again truncated), and the resulting segmented image.

This framework applies readily to multiple images by coupling the label distribution β and appearance models θ_k^{**} across multiple images. Letting $j \in \mathcal{J}$ index the images in the collection, we associate a segment distribution π_j with each image and associate a set of Gaussian processes with each image to describe the segmentation of that image.

The image segmentation problem can be cast as posterior inference in this HPY-based model. Given an image represented as a collection of texture and color descriptors, we compute the maximum a posteriori set of segments for the sites. Sudderth and Jordan (2009) have shown that this procedure yields a state-of-the-art unsupervised image segmentation algorithm.

5.5 The beta process and the Indian buffet process

The DP mixture model embodies the assumption that the data can be partitioned or clustered into discrete classes. This assumption is made particularly clear in the Chinese restaurant representation, where the table at which a data point sits indexes the class (the mixture component) to which it is assigned. If we represent the restaurant as a binary matrix in which the rows are the data points and the columns are the tables, we obtain a matrix with a single one in each row and all other elements equal to zero.

A different assumption that is natural in many settings is that objects can be described in terms of a collection of binary *features* or *attributes*. For example, we might describe a set of animals with features such as diurnal/nocturnal, avian/non-avian, cold-blooded/warm-blooded, etc. Forming a binary matrix in which the rows are the objects and the columns are the features, we obtain a matrix in which there are multiple ones in each row. We will refer to such a representation as a *featural representation*.

A featural representation can of course be converted into a set of clusters if desired: if there are K binary features, we can place each object into one of 2^K clusters. In so doing, however, we lose the ability to distinguish between classes that have many features in common and classes that have no features in common. Also, if K is large, it may be infeasible to consider models with 2^K parameters. Using the featural representation, we might hope to construct models that use on the order of K parameters to describe 2^K classes.

In this section we discuss a Bayesian nonparametric approach to featural representations. In essence, we replace the Dirichlet/multinomial probabilities that underlie the Dirichlet process with a collection of beta/Bernoulli draws. This is achieved via the *beta process*, a stochastic process whose realizations provide a countably infinite collection of coin-tossing probabilities. We also discuss some other representations of the beta process that parallel those for the DP. In particular we describe a stick-breaking construction as well as an analog of the Chinese restaurant process known as the *Indian buffet process*.

5.5.1 The beta process and the Bernoulli process

The beta process is an instance of a general class of stochastic processes known as *completely random measures* (Kingman, 1967; see also Chapters 3). The key property of completely random measures is that the random variables obtained by evaluating a random measure on disjoint subsets of the probability space are mutually independent. Moreover, draws from a completely random measure are discrete (up to a fixed deterministic component). Thus we can represent such a draw as a weighted collection of atoms on some probability space, as we do for the DP. (Note, however, that the DP is not a completely random measure because the weights are constrained to sum to one for the DP; thus, the independence assertion does not hold for the DP. The DP can be obtained by *normalizing* a completely random measure (specifically the gamma process; see Section 3.3.1).)

Applications of the beta process in Bayesian nonparametric statistics have mainly focused on its use as a model for random hazard functions (Hjort, 1990; see also Chapters 3 and 4). In this case, the probability space is the real line and it is the cumulative integral of the sample paths that is of interest (yielding a random, nondecreasing step function). In the application of the beta process to featural representations, on the other hand, it is the realization itself that is of interest and the underlying space is no longer restricted to be the real line.

Following Thibaux and Jordan (2007), let us thus consider a general probability space (Θ, Ω) endowed with a finite *base measure* B_0 (note that B_0 is not a probability measure; it does not necessarily integrate to one). Intuitively we wish to partition Θ into small regions, placing atoms into these regions according to B_0 and assigning a weight to each atom, where the weight is a draw from a beta distribution. A similar partitioning occurs in the definition of the DP, but in that case the aggregation property of Dirichlet random variables immediately yields a consistent set of marginals and thus an easy appeal to Kolmogorov's theorem. Because the sum of two beta random variables is not a beta random variable, the construction is somewhat less straightforward in the beta process case.

The general machinery of completely random processes deals with this issue in an elegant way. Consider first the case in which B_0 is absolutely continuous and define the *Lévy measure* on the product space $[0, 1] \otimes \Theta$ in the following way:

$$\nu(\mathrm{d}\omega, \mathrm{d}\theta) = c\omega^{-1}(1-\omega)^{c-1}\mathrm{d}\omega B_0(\mathrm{d}\theta), \tag{5.37}$$

where $c > 0$ is a *concentration parameter*. Now sample from a nonhomogeneous Poisson process with the Lévy measure ν as its rate measure. This yields a set of atoms at locations $(\omega_1, \theta_1), (\omega_2, \theta_2) \dots$. Define a realization of the beta process as

$$B = \sum_{k=1}^{\infty} \omega_k \delta_{\theta_k}, \tag{5.38}$$

where δ_{θ_k} is an atom at θ_k with ω_k its mass in B. We denote this stochastic process as $B \sim \mathrm{BP}(c, B_0)$. Figure 5.6(a) provides an example of a draw from $\mathrm{BP}(1, U[0, 1])$, where $U[0, 1]$ is the uniform distribution on $[0, 1]$.

customers who have previously sampled dish k; that is, $Z^*_{nk} \sim \text{Bernoulli}(\frac{m_k}{c+n-1})$. Having sampled from the dishes previously sampled by other customers, customer n then goes on to sample an additional number of new dishes determined by a draw from a Poisson$(\frac{c}{c+n-1}\alpha)$ distribution.

To derive the IBP from the beta process, consider first the distribution equation (5.40) for $n = 0$; in this case the base measure is simply B_0. Drawing from $B \sim \text{BP}(B_0)$ and then drawing $Z_1 \sim \text{BeP}(B)$ yields atoms whose locations are distributed according to a Poisson process with rate B_0; the number of such atoms is Poisson(α). Now consider the posterior distribution after $Z_1, \ldots, Z_{n-1}$ have been observed. The updated base measure is $\frac{c}{c+n-1}B_0 + \frac{1}{c+n-1}\sum_{i=1}^{n-1} Z_i$. Treat the discrete component and the continuous component separately. The discrete component, $\frac{1}{c+n-1}\sum_{i=1}^{n-1} Z_i$, can be reorganized as a sum over the unique values of the atoms; let m_k denote the number of times the kth atom appears in one of the previous Z_i. We thus obtain draws $\omega_k \sim \text{Beta}((c+n-1)q_k, (c+n-1)(1-q_k))$, where $q_k = \frac{m_k}{c+n-1}$. The expected value of ω_k is $\frac{m_k}{c+n-1}$ and thus (under Bernoulli sampling) this atom appears in Z_n with probability $\frac{m_k}{c+n-1}$. From the continuous component, $\frac{c}{c+n-1}B_0$, we generate Poisson$(\frac{c}{c+n-1}\alpha)$ new atoms. Equating "atoms" with "dishes," and rows of Z^* with draws Z_n, we have obtained exactly the probabilistic specification of the IBP.

5.5.3 *Stick-breaking constructions*

The stick-breaking representation of the DP is an elegant constructive characterization of the DP as a discrete random measure (Chapter 2). This construction can be viewed in terms of a metaphor of breaking off lengths of a stick, and it can also be interpreted in terms of a size-biased ordering of the atoms. In this section, we consider analogous representations for the beta process. Draws $B \sim \text{BP}(c, B_0)$ from the beta process are discrete with probability one, which gives hope that such representations exist. Indeed, we will show that there are two stick-breaking constructions of B, one based on a size-biased ordering of the atoms (Thibaux and Jordan, 2007), and one based on a stick-breaking representation known as the inverse Lévy measure (Wolpert and Ickstadt, 1998).

The size-biased ordering of Thibaux and Jordan (2007) follows straightforwardly from the discussion in Section 5.5.2. Recall that the Indian buffet process is defined via a sequence of draws from Bernoulli processes. For each draw, a Poisson number of new atoms are generated, and the corresponding weights in the base measure B have a beta distribution. This yields the following truncated representation:

$$B_N = \sum_{n=1}^{N} \sum_{k=1}^{K_n} \omega_{nk} \delta_{\theta_{nk}}, \tag{5.41}$$

where

$$\begin{aligned} K_n \mid c, B_0 &\sim \text{Poisson}(\tfrac{c}{c+n-1}\alpha), \\ \omega_{nk} \mid c &\sim \text{Beta}(1, c+n-1), && \text{for } n = 1, \ldots, \infty \\ \theta_{nk} \mid B_0 &\sim B_0/\alpha && \text{for } k = 1, \ldots, K_n. \end{aligned} \tag{5.42}$$

It can be shown that this size-biased construction B_N converges to B with probability one. The expected total weight contributed at step N is $c\alpha/\{(c+N)(c+N-1)\}$, while the expected total weight remaining, in $B - B_N$, is $\frac{c\alpha}{c+N}$. The expected total weight remaining decreases to zero as $N \to \infty$, but at a relatively slow rate. Note also that we are not guaranteed that atoms contributed at later stages of the construction will have small weight – the sizes of the weights need not be in decreasing order.

The stick-breaking construction of Teh, Görür and Ghahramani (2007) can be derived from the inverse Lévy measure algorithm of Wolpert and Ickstadt (1998). This algorithm starts from the Lévy measure of the beta process, and generates a sequence of weights of decreasing size using a nonlinear transformation of a one-dimensional Poisson process to one with uniform rate. In general this approach does not lead to closed forms for the weights; inverses of the incomplete beta function need to be computed numerically. However for the one-parameter beta process (where $c = 1$) we do obtain a simple closed form:

$$B_K = \sum_{k=1}^{K} \omega_k \delta_{\theta_k}, \tag{5.43}$$

where

$$\begin{aligned} v_k \mid \alpha &\sim \text{Beta}(1, \alpha), \\ \omega_k &= \prod_{l=1}^{k} (1 - v_l), \\ \theta_k \mid B_0 &\sim B_0/\alpha && \text{for } k = 1, \ldots, \infty. \end{aligned} \tag{5.44}$$

Again $B_K \to B$ as $K \to \infty$, but in this case the expected weights decrease exponentially to zero. Further, the weights are generated in strictly decreasing order, so we are guaranteed to generate the larger weights first.

The stick-breaking construction for the one-parameter beta process has an intriguing connection to the stick-breaking construction for the DP. In particular, both constructions use the same beta-distributed breakpoints v_k; the difference is that for the DP we use the lengths of the sticks just broken off as the weights while for the beta process we use the remaining lengths of the sticks. This is depicted graphically in Figure 5.7.

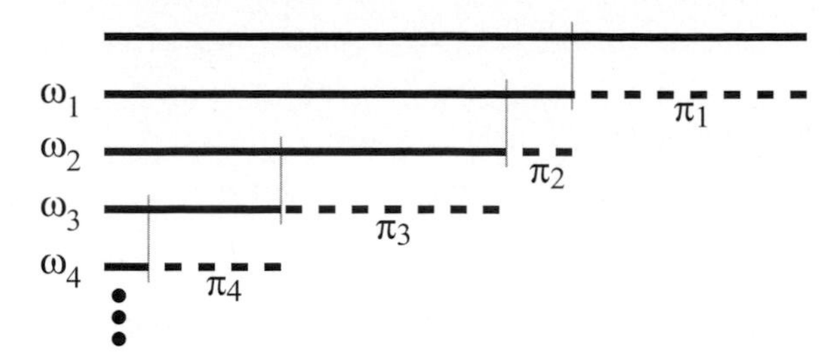

Figure 5.7 Stick-breaking construction for the DP and the one-parameter BP. The lengths π_i are the weights for the DP and the lengths ω_i are the weights for the BP.

5.5.4 Hierarchical beta processes

Recall the construction of the hierarchical Dirichlet process: a set of Dirichlet processes are coupled via a random base measure. A similar construction can be carried out in the case of the beta process: let the common base measure for a set of beta processes be drawn from an underlying beta process (Thibaux and Jordan, 2007). Under this hierarchical Bayesian nonparametric model, the featural representations that are chosen for one group will be related to the featural representations that are used for other groups.

We accordingly define a *hierarchical beta process* (HBP) as follows:

$$\begin{aligned} B_0 \mid \kappa, A &\sim \mathrm{BP}(\kappa, A), & \\ B_j \mid c, B_0 &\sim \mathrm{BP}(c, B_0), & \text{for } j \in \mathcal{J} \\ Z_{ji} \mid B_j &\sim \mathrm{BeP}(B_j) & \text{for } i = 1, \ldots, n_j, \end{aligned} \tag{5.45}$$

where $\mathcal{J}$ is the set of groups and there are n_j individuals in group j. The hyperparameter c controls the degree of coupling among the groups: larger values of c yield realizations B_j that are closer to B_0 and thus a greater degree of overlap among the atoms chosen in the different groups.

As an example of the application of the HBP, Thibaux and Jordan (2007) considered the problem of document classification, where there are $|\mathcal{J}|$ groups of documents and where the goal is to classify a new document into one of these groups. In this case, Z_{ji} is a binary vector that represents the presence or absence in the ith document of each of the words in the vocabulary Θ. The HBP yields a form of regularization in which the group-specific word probabilities are shrunk towards each other. This can be compared to standard Laplace smoothing, in which word probabilities are shrunk towards a fixed reference point. Such a reference point can be difficult to calibrate when there are rare words in a corpus, and Thibaux and Jordan (2007) showed empirically that the HBP yielded better predictive performance than Laplace smoothing.

5.5.5 Applications of the beta process

In the following sections we describe a number of applications of the beta process to hierarchical Bayesian featural models. Note that this is a rather different class of applications than the traditional class of applications of the beta process to random hazard functions.

Sparse latent variable models

Latent variable models play an essential role in many forms of statistical analysis. Many latent variable models take the form of a regression on a latent vector; examples include principal component analysis, factor analysis and independent components analysis. Paralleling the interest in the regression literature in sparse regression models, one can also consider sparse latent variable models, where each observable is a function of a relatively small number of latent variables. The beta process provides a natural way of constructing such models. Indeed, under the beta process we can work with models that define a countably infinite number of latent variables, with a small, finite number of variables being *active* (i.e., non-zero) in any realization.

Consider a set of n observed data vectors, $x_1, \ldots, x_n$. We use a beta process to model a set of latent features, $Z_1, \ldots, Z_n$, where we capture interactions among the components of these vectors as follows:

$$\begin{aligned} B \mid c, B_0 &\sim \mathrm{BP}(c, B_0), \\ Z_i \mid B &\sim \mathrm{BeP}(B) \qquad \text{for } i = 1, \ldots, n. \end{aligned} \tag{5.46}$$

As we have seen, realizations of beta and Bernoulli processes can be expressed as weighted sums of atoms:

$$\begin{aligned} B &= \sum_{k-1}^{\infty} \omega_k \delta_{\theta_k}, \\ Z_i &= \sum_{k=1}^{\infty} Z_{ik}^* \delta_{\theta_k}. \end{aligned} \tag{5.47}$$

We view θ_k as parametrizing feature k, while Z_i denotes the features that are active for item i. In particular, $Z_{ik}^* = 1$ if feature k is active for item i. The data point x_i is modeled as follows:

$$\begin{aligned} y_{ik} \mid H &\sim H, \\ x_i \mid Z_i, \boldsymbol{\theta}, \mathbf{y}_i &\sim F_{\{\theta_k, y_{ik}\}_{k:Z_{ik}^*=1}} \qquad \text{for } k = 1, \ldots, \infty, \end{aligned} \tag{5.48}$$

where y_{ik} is the value of feature k if it is active for item i, and the distribution $F_{\{\theta_k, y_{ik}\}_{k:Z_{ik}^*=1}}$ depends only on the active features, their values, and their parameters.

Note that this approach defines a latent variable model with an infinite number of sparse latent variables, but for each data item only a finite number of latent variables are active. The approach would often be used in a predictive setting in which the latent variables are integrated out, but if the sparseness pattern is of interest per se, it is also possible to compute a posterior distribution over the latent variables.

There are several specific examples of this sparse latent variable model in the literature. One example is an independent components analysis model with an infinite number of sparse latent components (Knowles and Ghahramani, 2007; Teh, Görür and Ghahramani, 2007), where the latent variables are real valued and x_i is a noisy observation of the linear combination $\sum_k Z_{ik}^* y_{ik}\theta_k$. Another example is the "noisy-or" model of Wood, Griffiths and Ghahramani (2006), where the latent variables are binary and are interpreted as presence or absence of diseases, while the observations x_i are binary vectors indicating presence or absence of symptoms.

Relational models

The beta process has also been applied to the modeling of relational data (also known as dyadic data). In the relational setting, data are relations among pairs of objects (Getoor and Taskar, 2007); examples include similarity judgments between two objects, protein–protein interactions, user choices among a set of options, and ratings of products by customers.

We first consider the case in which there is a single set of objects and relations are defined among pairs of objects in that set. Formally, define an observation as a relation x_{ij} between objects i and j in a collection of n objects. Each object is modeled using a set of latent features as in equation (5.46) and equation (5.47). The observed relation x_{ij} between objects i and j then has a conditional distribution that is dependent only on the features active in objects i and j. For example, Navarro and Griffiths (2007) modeled subjective similarity judgments between objects i and j as normally distributed with mean $\sum_{k=1}^{\infty} \theta_k Z_{ik}^* Z_{jk}^*$; note that this is a weighted sum of features active in both objects. Chu, Ghahramani, Krause and Wild (2006) modeled high-throughput protein–protein interaction screens where the observed binding affinity of proteins i and j is related to the number of overlapping features $\sum_{k=1}^{\infty} Z_{ik}^* Z_{jk}^*$, with each feature interpreted as a potential protein complex consisting of proteins containing the feature. Görür, Jäkel and Rasmussen (2006) proposed a nonparametric *elimination by aspects* choice model where the probability of a user choosing object i over object j is modeled as proportional to a weighted sum, $\sum_{k=1}^{\infty} \theta_k Z_{ik}^* (1 - Z_{jk}^*)$, across features active for object i that are not active for object j. Note that in these examples, the parameters of the model, θ_k, are the atoms of the beta process.

Relational data involving separate collections of objects can be modeled with the beta process as well. Meeds, Ghahramani, Neal and Roweis (2007) modeled movie ratings, where the collections of objects are movies and users, and the relational data consist of ratings of movies by users. The task is to predict the ratings of movies not yet rated by users, using these predictions to recommend new movies to users. These tasks are called *recommender systems* or *collaborative filtering*. Meeds, Ghahramani, Neal and Roweis (2007) proposed a featural model where movies and users are modeled using separate IBPs. Let Z^* be the binary matrix of movie features and Y^* the matrix of user features. The rating of movie i by user j is modeled as normally distributed with mean $\sum_{k=1}^{\infty} \sum_{l=1}^{\infty} \theta_{kl} Z^*_{ik} Y^*_{jl}$. Note that this dyadic model cannot be represented using two independent beta processes, since there is a parameter θ_{kl} for each combination of features in the two IBPs. The question of what random measure underlies this model is an interesting one.

5.6 Semiparametric models

The nonparametric priors introduced in previous sections can be combined with more traditional finite-dimensional priors, as well as hierarchies of such priors. In the resulting *semiparametric* models, the object of inference may be the finite-dimensional parameter, with the nonparametric component treated as a nuisance parameter to be integrated out. In other cases, the finite-dimensional parameters are to be integrated out and aspects of the nonparametric component are the inferential focus. In this section we describe two such semiparametric models based on the HDP. The first model couples the stick-breaking representation of the HDP with a Gaussian hierarchy, while the other is based on the Chinese restaurant franchise representation of the HDP.

5.6.1 Hierarchical DPs with random effects

An important characteristic of the HDP is that the same atoms appear in different DPs, allowing clusters to be shared across the different groups. The *hierarchical DP with random effects* (HDP+RE) model of Kim and Smyth (2007) generalizes the HDP by allowing atoms in different DPs to differ from each other to better capture group-specificity of cluster parameters. This model is based on the stick-breaking representation for HDPs. We begin with the standard representation for the common random base measure $G_0 \sim \mathrm{DP}(\gamma, H)$:

$$\begin{aligned} \boldsymbol{\beta} \mid \gamma &\sim \mathrm{GEM}(\gamma), \\ \theta_k^{**} \mid H &\sim H, \\ G_0 &= \sum_{k=1}^{\infty} \beta_k \delta_{\theta_k^{**}} \qquad \text{for } k = 1, \ldots, \infty. \end{aligned} \tag{5.49}$$

For each group $j \in \mathcal{J}$, the weights and atoms for the group-specific G_j differ from G_0 in the following way:

$$
\begin{aligned}
\boldsymbol{\pi}_j \mid \boldsymbol{\beta} &\sim \mathrm{DP}(\alpha, \boldsymbol{\beta}), && \text{for } j \in \mathcal{J} \\
\theta^*_{jk} \mid \theta^{**}_k &\sim T_{\theta^{**}_k}, && \\
G_j &= \sum_{k=1}^{\infty} \pi_{jk} \delta_{\theta^*_{jk}} && \text{for } k = 1, \ldots, \infty,
\end{aligned}
\tag{5.50}
$$

where T_θ is a distribution centered at θ; for example, T_θ might be a normal distribution with mean θ.

Kim and Smyth (2007) used the HDP+RE model to model bumps in functional magnetic resonance imaging (fMRI) data. fMRI analyses report areas of high metabolic activity in the brain that are correlated with external stimuli in an attempt to discover the function of local brain regions. Such areas of activity often show up as *bumps* in fMRI images, and each bump can be modeled well using a normal density. An fMRI image then consists of multiple bumps and can be modeled with a DP mixture. Each individual brain might have slightly different structure and might react differently to the same stimuli, while each fMRI machine has different characteristics. The HDP+RE model naturally captures such variations while sharing statistical strength across individuals and machines.

5.6.2 Analysis of densities and transformed DPs

In this section we describe another approach to introducing group-specific parameters within the DP framework. The common base measure G_0 is still given a DP prior as in equation (5.49), while the group-specific random measures are defined differently:

$$
\begin{aligned}
H_0 &= \sum_{k=1}^{\infty} \beta_k T_{\theta^{**}_k}, && \\
G_j \mid H_0 &\sim \mathrm{DP}(\alpha, H_0) && \text{for } j \in \mathcal{J}.
\end{aligned}
\tag{5.51}
$$

In the particular case in which H and T_θ are normal distributions with fixed variances, this model has been termed the *analysis of densities* (AnDe) model by Tomlinson and Escobar (2003), who used it for sharing statistical strength among multiple density estimation problems.

Sudderth, Torralba, Freeman and Willsky (2008) called the model given by equations (5.49) and (5.51) a *transformed DP*. The transformed DP is very similar to an HDP, the difference being that the atoms in G_0 are replaced by distributions parametrized by the atoms. If these distributions are smooth the measures G_j will not share atoms as in the HDP. Instead each atom in G_j is drawn from $T_{\theta_k^{**}}$ with probability β_k. Identifying an atom of G_j with θ_k^{**}, the Chinese restaurant franchise representation for the HDP can be generalized to the transformed DP. We have customers (draws from G_j) going into restaurants (G_j) and sitting around tables (draws from H_0), while tables are served dishes (atoms in G_0) from a franchise-wide menu (G_0). In the HDP the actual dish served at the different tables that order the same dish are identical. For the transformed DP the dishes that are served at different tables ordering the same dish on the menu can take on distinct values.

Sudderth, Torralba, Freeman and Willsky (2008) used the transformed DP as a model for visual scene analysis that can simultaneously segment, detect and recognize objects within the scenes. Each image is first preprocessed into a set of low-level descriptors of local image appearances, and equations (5.49) and (5.51) are completed with a mixture model for these descriptors:

$$\begin{aligned} \theta_{ji} \mid G_j &\sim G_j, \\ x_{ji} \mid \theta_{ji} &\sim F_{\theta_{ji}} \qquad \text{for } j \in \mathcal{J} \text{ and } i = 1, \ldots, n_j, \end{aligned} \tag{5.52}$$

where x_{ji} is one of n_j image descriptors in image j and $F_{\theta_{ji}}$ is a distribution over image descriptors parameterized by θ_{ji}.

The Chinese restaurant franchise representation of the transformed DP translates to a hierarchical representation of visual scenes, with scenes consisting of multiple objects and objects consisting of descriptors of local image appearances. To see this, note that customers (x_{ji}) are clustered into tables (object instances), and tables are served dishes from a global menu (each object instance belongs to an object category). There could be multiple tables in the same restaurant serving variations (different "seasonings") on a given dish from the global menu. This corresponds to the fact that there could be multiple instances of the same object category in a single visual scene, with each instance being in a different location or having different poses or lighting conditions (thus yielding *transformed* versions of an object category template).

5.7 Inference for hierarchical Bayesian nonparametric models

In this section we discuss algorithmic aspects of inference for the hierarchical Bayesian nonparametric models that we have discussed in earlier sections. Our treatment will be brief and selective; in particular, we focus on relatively simple algorithms that help to convey basic methodology and provide a sense of some of the options that are available. An underlying theme of this section is that the various mathematical representations available for nonparametric models – including

A drawback of the CRF sampler is that it couples sampling in the various restaurants (since all DPs are integrated out). This coupling makes deriving a CRF sampler for certain models (e.g. the HDP-HMM) difficult. An alternative is to construct samplers that use a mixed representation – some DPs in stick-breaking representation and some in CRP representation – and thereby decouple the restaurants (Teh, Jordan, Beal and Blei, 2006).

The CRF-based sampler can be easily extended to arbitrary hierarchies. It can also be extended to the hierarchical Pitman–Yor process discussed in Section 5.4.

Posterior representation sampler

In Section 5.2.3 we showed that the posterior of the HDP consists of a discrete part corresponding to mixture components associated with data and a continuous part corresponding to components not associated with data. This representation can be used to develop a sampler which represents only the discrete part explicitly. In particular, referring to equations (5.10) and (5.12), the posterior representation sampler maintains only the weights $\boldsymbol{\beta}$ and $\{\boldsymbol{\pi}_j\}_{j\in\mathcal{J}}$. (The atoms $\{\theta_k^{**}\}_{k=1,\ldots,K}$ can be integrated out in the conjugate setting.) We also make use of cluster index variables z_{ji}, defined so that $\theta_{ji} = \theta_{z_{ji}}^{**}$ (i.e., $z_{ji} = k_{jt_{ji}}$ in the CRF representation).

The sampler iterates between two phases: the sampling of the cluster indices $\{z_{ji}\}$, and the sampling of the weights $\boldsymbol{\beta}$ and $\{\boldsymbol{\pi}_j\}$. The sampling of the cluster indices is a simple variation on the Gibbs updates in the CRF sampler described above. In particular, we define the following Gibbs conditionals:

$$z_{ji} = \begin{cases} k & \text{with probability} \propto \pi_{jk} f_k(\{x_{ji}\}), \\ k^{\text{new}} & \text{with probability} \propto \pi_{j0} f_{k^{\text{new}}}(\{x_{ji}\}). \end{cases} \quad (5.57)$$

If a new component k^{new} is chosen, the corresponding atom is instantiated in the sampler. Specifically, the weights corresponding to this new atom can be generated as follows:

$$\begin{aligned} v_0 \mid \gamma &\sim \text{Beta}(\gamma, 1), \\ (\beta_0^{\text{new}}, \beta_{K+1}^{\text{new}}) &= (\beta_0 v_0, \beta_0(1 - v_0)), \\ v_j \mid \alpha, \beta_0, v_0 &\sim \text{Beta}(\alpha\beta_0 v_0, \alpha\beta_0(1 - v_0)), \\ (\pi_{j0}^{\text{new}}, \pi_{jK+1}^{\text{new}}) &= (\pi_{j0} v_j, \pi_{j0}(1 - v_j)) \qquad \text{for } j \in \mathcal{J}. \end{aligned} \quad (5.58)$$

Finally we set $z_{ji} = K + 1$ and increment K.

The second phase resamples the weights $\{\boldsymbol{\pi}_j\}_{j\in\mathcal{J}}$ and $\boldsymbol{\beta}$ conditioned on the cluster indices $\{z_{ji}\}$. The approach is to first integrate out the random measures, leaving a CRP representation as in Section 5.2.2, then the weights $\{\boldsymbol{\pi}_j\}_{j\in\mathcal{J}}$ and $\boldsymbol{\beta}$ can be sampled conditionally on the state of the CRF using equations (5.10) and (5.12). Because we are conditioning on $\{z_{ji}\}$, and customers with different values of z_{ji} cannot be assigned to the same table, each restaurant effectively gets split into independent "sub-restaurants," one for each value of k. (See also the related direct assignment sampler in Teh, Jordan, Beal and Blei (2006).) Let $n_{j\cdot k}$ be the number of observations in group j assigned to component k, and let m_{jk} be the random number of tables in a sub-restaurant with $n_{j\cdot k}$ customers and concentration parameter $\alpha\beta_k$. The $\{m_{jk}\}$ are mutually independent and thus a draw for each of them can be simulated using the CRP. We can now sample the $\boldsymbol{\beta}$ and $\{\boldsymbol{\pi}_j\}$ using equations (5.10) and (5.12).

5.7.2 Inference for HDP hidden Markov models

The posterior representation sampler of Section 5.7.1 can also be used to derive a Gibbs sampler for the HDP-HMM. Consider the formulation of the HDP-HMM given in equations (5.23) and (5.24) where we make use of a sequence of latent indicator variables $z_1, \ldots, z_\tau$. We again assume that H is conjugate to F_θ. Note that the posterior of the HDP prior for the model (given $z_1, \ldots, z_\tau$) can be decomposed into a discrete part consisting of K atoms (corresponding to the K states currently visited by $z_1, \ldots, z_\tau$), as well as a continuous part consisting of unused atoms. The weights on the K atoms (equivalently the transition probabilities among the K states currently used by the HDP-HMM) can be constructed from a CRF representation of the HDP:

$$\begin{aligned}(\beta_0, \beta_1, \ldots, \beta_K) &\sim \text{Dirichlet}(\gamma, m_{\cdot 1}, \ldots, m_{\cdot K}),\\ (\pi_{j0}, \pi_{j1}, \ldots, \pi_{jK}) &\sim \text{Dirichlet}(\alpha\beta_0, \alpha\beta_1 + n_{j\cdot 1}, \ldots, \alpha\beta_K + n_{j\cdot K})\\ &\quad \text{for } j = 1, \ldots, K,\end{aligned} \tag{5.59}$$

where $n_{j\cdot k}$ is the number of transitions from state j to state k (equivalently the number of customers eating dish k in restaurant j), while $m_{\cdot k}$ is the number of tables serving dish k in the CRF representation of the HDP. The conditional probabilities for the Gibbs update of z_t are as follows:

$$z_t = \begin{cases} k & \text{with probability} \propto \pi_{z_{t-1}k}\pi_{kz_{t+1}} f_k(\{x_t\}),\\ k^{\text{new}} & \text{with probability} \propto \pi_{z_{t-1}0}\beta_{z_{t+1}} f_{k^{\text{new}}}(\{x_t\}).\end{cases} \tag{5.60}$$

The three factors on the right-hand side are the probability of transitioning into the current state, the probability of transitioning out of the current state, and the conditional probability of the current observation x_t respectively. The $\beta_{z_{t+1}}$ factor arises because transitions from the new state k^{new} have not been observed before so we need to use the conditional prior mean $\boldsymbol{\beta}$. The weights $\boldsymbol{\beta}$ and transition probabilities $\boldsymbol{\pi}_j$ can be updated as for the posterior representation sampler for plain HDPs.

This simple Gibbs sampler can converge very slowly due to strong dependences among the latent states (Scott, 2002). To obtain a faster algorithm we would like to update the latent states in a block via the forward-backward algorithm for HMMs; the traditional form of this algorithm cannot, however, be applied directly to the HDP-HMM since there are an infinite number of possible states. The solution is to limit the number of states to a finite number so that the forward-backward algorithm becomes feasible. Fox, Sudderth, Jordan and Willsky (2008) proposed doing this via a truncation of the stick-breaking process (cf. Ishwaran and James, 2001), while Van Gael, Saatci, Teh, and Ghahramani (2008) proposed a slice sampling approach which adaptively limits the number of states to a finite number (Neal, 2003; Walker, 2007).

5.7.3 Inference for beta processes

In this section we describe a Gibbs sampler for the beta process latent variable model described in Section 5.5.5. This sampler is based on the stick-breaking representation of the beta process.

Recall that the model is defined in terms of a set of feature weights $\{\omega_k\}_{k=1,\ldots,\infty}$ and the atoms (feature parameters) $\{\theta_k\}_{k=1,\ldots,\infty}$. Moreover, corresponding to each data item x_i, we have a set of binary feature "activities" $\{Z^*_{ik}\}_{k=1,\ldots,\infty}$ and latent feature values $\{y_{ik}\}_{k=1,\ldots,\infty}$. The observed data item x_i depends on $\{\theta_k, y_{ik}\}_{k:Z^*_{ik}=1}$.

The conditional distributions defining the model are given in equations (5.44) and (5.48), where $p(Z^*_{ik} = 1 \mid \omega_k) = \omega_k$. Gibbs sampling in this model is straightforward except for a few difficulties which we describe below along with their resolution.

The main difficulty with a Gibbs sampler is that there are an infinite number of random variables that need to be sampled. To circumvent this problem, Teh, Görür and Ghahramani (2007) propose to use slice sampling (Neal, 2003; Walker, 2007) to truncate the representation adaptively to a finite number of features. Consider an auxiliary variable s with conditional distribution:

$$s \mid Z^*, \{\omega_k\}_{k=1,\ldots,\infty} \sim \text{Uniform}\left[0, \min_{k:\exists i, Z^*_{ik}=1} \omega_k\right], \tag{5.61}$$

where the supremum in the range of s is the smallest feature weight ω_k among the currently active features. Conditioned on the current state of the other variables a new value for s can easily be sampled. Conditioned on s, features for which $\omega_k < s$ are forced to be inactive since making them active would make s lie outside its range. This means that we only need to update the finite number of features for which $\omega_k > s$. This typically includes all the active features, along with a small number of inactive features (needed for the sampler to explore the use of new features).

A related issue concerns the representation of the model within the finite memory of the computer. Using the auxiliary variable s it is clear that we need only represent features $1, \ldots, K$, where K is such that $\omega_{K+1} < s$; that is, the model is truncated after feature K. As the values of s and the feature weights change over the course of Gibbs sampling this value of K changes as well. If K is decreased we simply delete the last few features, while if K is increased we sample the variables ω_k, θ_k and y_{ik} corresponding to these new features from their conditional distributions given the current state of the represented features.

The final issue is the problem of sampling the feature weights $\omega_1, \ldots, \omega_K$. Unlike the case of DPs, it is easier in this case to work with the weights directly instead of the stick-breaking variables v_k. In particular, Teh, Görür and Ghahramani (2007) showed that the joint probability for the weights is:

$$p(\omega_1, \ldots, \omega_K) = \mathbb{I}(0 \leq \omega_K \leq \cdots \leq \omega_1 \leq 1)\alpha^K \omega_K^\alpha \prod_{k=1}^{K} \omega_k^{-1}, \tag{5.62}$$

where $\mathbb{I}(\cdot) = 1$ if the predicate is true and 0 otherwise. For $k = 1, \ldots, K - 1$ the conditional probability of ω_k given the other variables can be computed from equation (5.62) and the conditional probability of $Z^*_{1k}, \ldots, Z^*_{nk}$ given ω_k. For ω_K we also have to condition on $Z^*_{ik} = 0$ for all i and $k > K$; this probability can be computed using the Lévy–Khintchine representation for the beta process (Teh, Görür and Ghahramani, 2007).

5.7.4 Inference for hierarchical beta processes

In this section we present an inference algorithm for the hierarchical beta process given in equation (5.45). The observed data are the variables Z_{ji}; these binary vectors denote (in the language of document classification) the presence or absence of words in document i of group j. The underlying measure space Θ is interpreted as the vocabulary. (Each element in Θ is referred to as a "word.") Let $\theta_1, \ldots, \theta_K \in \Theta$ denote the words that are observed among the documents. That is, these are the $\theta \in \Theta$ such that $Z_{ji}(\theta) = 1$ for some i and j.

Because both the beta and Bernoulli processes are completely random measures, the posterior over B_0 and B_j for $j \in \mathcal{J}$ decomposes into a discrete part over the observed vocabulary $\{\theta_1, \ldots, \theta_K\}$ and a continuous part over $\Theta \backslash \{\theta_1, \ldots, \theta_K\}$. The discrete part further factorizes over each observed word θ_k. Thus it is sufficient to focus separately on inference for each observed word and for the continuous part corresponding to unobserved words.

For a fixed θ_k, let $a = A(\theta_k)$, $\omega_0 = B_0(\theta_k)$, $\omega_j = B_j(\theta_k)$ and $z_{ji} = Z_{ji}(\theta_k)$. The slice of the HBP corresponding to θ_k has the following joint distribution:

$$\begin{aligned} \omega_0 \mid c_0, a &\sim \text{Beta}(c_0 a, c_0(1-a)), & \\ \omega_j \mid c_j, \omega_0 &\sim \text{Beta}(c_j \omega_0, c_j(1-\omega_0)), & \text{for } j \in \mathcal{J} \\ z_{ji} \mid \omega_j &\sim \text{Bernoulli}(\omega_j) & \text{for } i = 1, \ldots, n_j. \end{aligned} \tag{5.63}$$

Note that the prior over ω_0 is improper if A is continuous and $a = 0$. This beta hierarchy is a special case of the finite Dirichlet hierarchy of equations (5.10) and (5.12) and it is straightforward to use the posterior representation sampler described in Section 5.7.1 to sample from the posterior given the observed z_{ji}. Thibaux and Jordan (2007) described an alternative where the ω_j are integrated out and rejection sampling is used to sample from ω_0.

Finally, we consider the continuous part of the posterior. This component is not simply the prior, since we have to condition on the fact that no words in $\Theta \backslash \{\theta_1, \ldots, \theta_K\}$ have been observed among the documents. Thibaux and Jordan (2007) solved this problem by noting that the posterior factors over the levels indexed by n in the size-biased ordering in equation (5.42). Focusing on each level separately, they derived a posterior distribution on the number of atoms in each level, combining this with the posterior over the level-specific weights to obtain the overall posterior.

5.8 Discussion

Our goal in this chapter has been to place hierarchical modeling in the same central role in Bayesian nonparametrics that it plays in other areas of Bayesian statistics. Indeed, one of the principal arguments for hierarchical modeling in parametric statistics is that it provides control over the large numbers of degrees of freedom that arise, for example, in random effects models. Such an argument holds a fortiori in the nonparametric setting.

Nonparametric priors generally involve hyperparameters, some of which are finite dimensional and some of which are infinite dimensional. Sharing the finite-dimensional parameter among multiple draws from such a prior is a natural modeling strategy that mimics classical hierarchical modeling concepts. It is our contention, however, that this form of control is far too limited, and that the infinite-dimensional parameters should generally also be shared. We have made this point principally by considering examples in applied problem domains. In domains such as computational vision, information retrieval and genetics, nonparametric models provide natural descriptions of the complex objects under study; in particular, it is natural to describe an image, a document or a genome as the realization of a stochastic process. Now, in considering collections of such objects it is natural to want to share details of the realization among the objects in the collection – we wish to share parts of objects, features, recurring phrases and motifs. This can be achieved by coupling multiple draws from a nonparametric prior via their infinite-dimensional parameters.

Another advantage of hierarchical modeling in the classical setting is that it expands the repertoire of distributional forms that can be considered. For example, heavy-tailed distributions can be obtained by placing a prior on the scale parameter of lighter tailed distributions. Although this point has been little explored to date in the nonparametric setting, we expect that it will be a fruitful direction for further research. In particular, there are stringent computational constraints that limit the nonparametric repertoire, and hierarchical constructions offer one way forward. Indeed, as we have seen, computationally oriented constructions such as urn models and stick-breaking representations often carry over naturally to hierarchical nonparametric models.

Finally, it is worth noting a difficulty that is raised by hierarchical modeling. Although Bayesian hierarchies help to control hyperparameters, they do not remove the need to specify distributions for hyperparameters. Indeed, when hyperparameters are placed high in a hierarchy it can be difficult to give operational meaning to such hyperparameters. One approach to coping with this issue involves considering the marginal probabilities that are induced by a nonparametric prior. For example, we argued that the marginals induced by a Pitman–Yor prior exhibit long tails that provide a good match to the power-law behavior found in textual data and image statistics. Further research is needed to develop this kind of understanding for a wider range of hierarchical Bayesian nonparametric models and problem domains.

Acknowledgements We would like to thank David Blei, Jan Gasthaus, Sam Gershman, Tom Griffiths, Kurt Miller, Vinayak Rao and Erik Sudderth for their helpful comments on the manuscript.

References

Antoniak, C. E. (1974). Mixtures of Dirichlet processes with applications to Bayesian nonparametric problems. *Annals of Statistics*, **2**, 1152–74.

Beal, M. J., Ghahramani, Z. and Rasmussen, C. E. (2002). The infinite hidden Markov model. In *Advances in Neural Information Processing Systems*, Volume 14, ed. T. G. Dietterich, S. Becker and Z. Ghahramani, 577–92. Cambridge, Mass.: MIT Press.

Blei, D. M., Griffiths, T. L., Jordan, M. I. and Tenenbaum, J. B. (2004). Hierarchical topic models and the nested Chinese restaurant process. In *Advances in Neural Information Processing Systems*, Volume 16, ed. S. Thrun, B. Schölkopt and L. K. Saul. Cambridge, Mass.: MIT Press.

Blei, D. M., Ng, A. Y. and Jordan, M. I. (2003). Latent Dirichlet allocation. *Journal of Machine Learning Research*, **3**, 993–1022.

Neal, R. M. (2000). Markov chain sampling methods for Dirichlet process mixture models. *Journal of Computational and Graphical Statistics*, **9**, 249–65.

Neal, R. M. (2003). Slice sampling. *Annals of Statistics*, **31**, 705–67.

Oliva, A. and Torralba, A. (2001). Modeling the shape of the scene: A holistic representation of the spatial envelope. *International Journal of Computer Vision*, **42**, 145–75.

Perman, M., Pitman, J. and Yor, M. (1992). Size-biased sampling of Poisson point processes and excursions. *Probability Theory and Related Fields*, **92**, 21–39.

Pitman, J. (2002). Combinatorial stochastic processes. *Technical Report 621*, Department of Statistics, University of California at Berkeley. Lecture notes for St. Flour Summer School.

Pitman, J. and Yor, M. (1997). The two-parameter Poisson–Dirichlet distribution derived from a stable subordinator. *Annals of Probability*, **25**, 855–900.

Pritchard, J., Stephens, M. and Donnelly, P. (2000). Inference of population structure using multilocus genotype data. *Genetics*, **155**, 945–59.

Rabiner, L. (1989). A tutorial on hidden Markov models and selected applications in speech recognition. *Proceedings of the IEEE*, **77**, 257–85.

Robertson, S. E., Walker, S., Hancock-Beaulieu, M., Gull, A. and Lau, M. (1992). Okapi at TREC. In *Text REtrieval Conference*, 21–30.

Salton, G. and McGill, M. (1983). *An Introduction to Modern Information Retrieval*. New York: McGraw-Hill.

Scott, S. L. (2002). Bayesian methods for hidden Markov models: Recursive computing in the 21st century. *Journal of the American Statistical Association*, **97**, 337–51.

Sethuraman, J. (1994). A constructive definition of Dirichlet priors. *Statistica Sinica*, **4**, 639–50.

Stephens, M., Smith, N. and Donnelly, P. (2001). A new statistical method for haplotype reconstruction from population data. *American Journal of Human Genetics*, **68**, 978–89.

Sudderth, E. and Jordan, M. I. (2009). Shared segmentation of natural scenes using dependent Pitman-Yor processes. In *Advances in Neural Information Processing Systems*, Volume 21.

Sudderth, E., Torralba, A., Freeman, W. and Willsky, A. (2008). Describing visual scenes using transformed objects and parts. *International Journal of Computer Vision*, **77**, 291–330.

Teh, Y. W. (2006a). A Bayesian interpretation of interpolated Kneser-Ney. *Technical Report TRA2/06*, School of Computing, National University of Singapore.

Teh, Y. W. (2006b). A hierarchical Bayesian language model based on Pitman–Yor processes. In *Proceedings of the 21st International Conference on Computational Linguistics and 44th Annual Meeting of the Association for Computational Linguistics*, 985–92. Morristown, NJ: Association for Computational Linguistics.

Teh, Y. W., Görür, D. and Ghahramani, Z. (2007). Stick-breaking construction for the Indian buffet process. In *Proceedings of the International Conference on Artificial Intelligence and Statistics*, Volume 11, ed. M. Meila and X. Shen, 556–63. Brookline, Mass.: Microtone.

Teh, Y. W., Jordan, M. I., Beal, M. J. and Blei, D. M. (2006). Hierarchical Dirichlet processes. *Journal of the American Statistical Association*, **101**, 1566–81.

Teh, Y. W., Kurihara, K. and Welling, M. (2008). Collapsed variational inference for HDP. In *Advances in Neural Information Processing Systems*, Volume 20.

Thibaux, R. and Jordan, M. I. (2007). Hierarchical beta processes and the Indian buffet process. In *Proceedings of the International Workshop on Artificial Intelligence and Statistics*, Volume 11, 564–71.

Tomlinson, G. and Escobar, M. (2003). Analysis of densities. Talk given at the Joint Statistical Meeting.

Van Gael, J., Saatci, Y., Teh, Y. W. and Ghahramani, Z. (2008). Beam sampling for the infinite hidden Markov model. In *Proceedings of the International Conference on Machine Learning*, Volume 25, 1088–95.

Walker, S. G. (2007). Sampling the Dirichlet mixture model with slices. *Communications in Statistics: Simulation and Computation*, **36**, 45.

Wolpert, R. L. and Ickstadt, K. (1998). Simulations of Lévy random fields. In *Practical Nonparametric and Semiparametric Bayesian Statistics*, 227–42. Springer-Verlag.

Wood, F., Griffiths, T. L. and Ghahramani, Z. (2006). A non-parametric Bayesian method for inferring hidden causes. In *Proceedings of the Conference on Uncertainty in Artificial Intelligence*, Volume 22, 536–43. AUAI Press.

Wooters, C. and Huijbregts, M. (2007). The ICSI RT07s speaker diarization system. In *Lecture Notes in Computer Science*. Springer.

Xing, E. P., Jordan, M. I. and Sharan, R. (2007). Bayesian haplotype inference via the Dirichlet process. *Journal of Computational Biology*, **14**, 267–84.

Xing, E. P. and Sohn, K. (2007). Hidden Markov Dirichlet process: Modeling genetic recombination in open ancestral space. *Bayesian Analysis 2*, ed. J. M. Bernado et al. Oxford: Oxford University Press.

Xing, E. P., Sohn, K., Jordan, M. I. and Teh, Y. W. (2006). Bayesian multi-population haplotype inference via a hierarchical Dirichlet process mixture. *Proceedings of the 23rd International Conference on Machine Learning*. ACM International Conference Proceeding Series, Volume 148, 1049–56. New York: ACM Press.

where $Q(F)$ can be thought about beliefs about the limiting empirical distribution function F_n

$$Q(F) = \lim_{n\to\infty} P(F_n).$$

Traditional parametric models can be seen as restricting the span of $Q(\cdot)$ to certain families of distributions of standard form such as "Gaussian" or "gamma." That is, they give probability zero to distributions F' outside of the parametric family, $Q(F') = 0$ for $F \notin \Omega_Q$ where Ω_Q denotes the space of parametric probability measures defined under $Q(\cdot)$. The rationale for nonparametric modeling is to induce a $Q(\cdot)$ which has wider coverage. This then allows for richer forms of dependence structures in models for observations $P(x_1, \ldots, x_n)$. However, it is worth repeating that whatever the span of $Q(\cdot)$ the dimension of $P(\cdot)$ is finite.

One of the great strengths in the hierarchical structures described by Teh and Jordan is when the observations are colored as $P(x_1, \ldots, x_{n_x}, y_1, \ldots, y_{n_y}, z_1, \ldots)$ for different classes of exchangeable observations $\{x, y, z, \ldots\}$. Here the hierarchical construction allows for information to be shared and information transfer through the hierarchy structure. For certain classes of inference problems there are clear benefits of endowing $Q(\cdot)$ with a nonparametric measure, as ably demonstrated in Chapter 5. As the field moves forward it will be interesting to see what additional classes of problems are amenable to benefit from nonparametric structures. As a modeler it is always worth asking what is the price to pay for additional freedom. The types of nonparametric priors are to date reasonably small with the Dirichlet process by far the most widespread. Yet jumping from the finite to the infinite is a big step. For example, the Dirichlet process prior only has a single parameter α, $F \sim \mathrm{DP}(\alpha, F_0)$ to control for departures of realizations of the infinite-dimensional random measure F from F_0. This can be seen to be quite restrictive. Often we will have strong prior beliefs about certain properties of $P(x_1, \ldots, x_n)$ in particular for the marginal densities $P(x_i)$ such as unimodality, symmetry, or "smoothness" in $P(x_i, x_j)$ as $|x_i - x_j| \to 0$. Such properties can be hard to control when $Q(\cdot)$ has such freedom. It is testament to the creativity of modelers that we have come so far with such currently restrictive nonparametric structures. Green and Richardson (2001) provide some interesting discussion on the implications of going nonparametric.

As a final remark in this section, and as Teh and Jordan note in the last line of Chapter 5, there is a danger in giving a model too much rope or to put it another way "nothing can come from nothing."

Having considered some of the operational motivation of Bayesian nonparametrics and the interpretation as beliefs about limiting distributions on observations we now turn to some of the computational challenges arising in inference.

6.3 Recent advances in computation for Dirichlet process mixture models

A powerful driving force behind the development and application of Bayesian nonparametric methods has been the availability of efficient computational methods using Markov chain Monte Carlo methods (Smith and Roberts, 1993; Tierney, 1994). The standard model to be fitted is the Dirichlet process mixture model (Lo, 1984). This model can be written hierarchically as

$$y_i \sim g(y_i|\phi_i) \quad i = 1, 2, \ldots, n$$

$$\phi_i \sim F \quad i = 1, 2, \ldots, n$$

$$F \sim \mathrm{DP}(M, H),$$

where $g(y_i|\phi)$ is a probability distribution with parameters ϕ, $M > 0$ is a positive mass parameter and H is a distribution. Standard results for the Dirichlet process show that the realized distribution F can be expressed as

$$F = \sum_{k=1}^{\infty} w_k \delta_{\theta_k} \tag{6.1}$$

where $w_1, w_2, w_3, \ldots$ are a sequence of positive random variables such that $\sum_{i=1}^{\infty} w_i = 1$ almost surely and $\theta_1, \theta_2, \theta_3, \ldots$ are an infinite sequence of independent and identically distributed random variables with distribution H, which has density h for a continuous distribution. This implies that the marginal distribution of y_i is a mixture model with an infinite number of components. The common computational approach to fitting this model introduces latent allocation variables $(s_1, s_2, \ldots, s_n)$ which yield the modified hierarchical model

$$y_i \sim g\left(y_i|\theta_{s_i}\right) \quad i = 1, 2, \ldots, n$$

$$p(s_i = k) = w_k \quad i = 1, 2 \ldots, n, \quad k = 1, 2, 3 \ldots$$

$$w_1, w_2, w_3, \cdots \sim \mathrm{GEM}(M)$$

$$\theta_1, \theta_2, \theta_3, \cdots \overset{\text{i.i.d.}}{\sim} H,$$

where GEM represents the distribution of the weights in a Dirichlet process (see Pitman, 2002). The collection of latent variables will be denoted $s = (s_1, s_2, \ldots, s_n)$. This form of model makes explicit the independence of $(\theta_1, \theta_2, \theta_3, \ldots)$ and $(w_1, w_2, w_3, \ldots)$ under the prior which is an important property for posterior simulation of Dirichlet process mixture models (and many other infinite mixture

5. If $s_i \neq s_j$, then we have the following.
   (a) Propose to merge the s_ith and s_jth clusters and propose the vector s^{merge} where
   - $s_i^{\text{merge}} = s_j$, $s_j^{\text{merge}} = s_j$,
   - if $k \in \mathcal{S}$, $s_k^{\text{merge}} = s_j$,
   - $s_k^{\text{merge}} = s_k$ if $k \notin \{i, j\} \cup \mathcal{S}$.

   (b) For $k \in \mathcal{S}$ calculate the probability under a Gibbs sampling step of allocating s_k^{launch} to s_k where s_k could only take the values s_i or s_j. Let this probability be q_k':

$$\alpha = \min\left\{1, \frac{p\left(y\,|s^{\text{merge}}\right)p\left(s^{\text{merge}}\right)\prod_{k\in\mathcal{S}} q_k'}{p\left(y|s\right)p\left(s\right)}\right\}.$$

The number of intermediate steps of Gibbs sampling, t, is a tuning parameter and can be chosen to encourage good mixing of the chain, see Jain and Neal (2004) for some guidelines. Extensions of this algorithm to the nonconjugate case are given in Jain and Neal (2007).

Posterior simulation of Dirichlet process mixture models using Pólya urn schemes are computationally efficient but the extension of these methods to other nonparametric priors depends on the availability of a Pólya urn scheme. This has led to interest in alternative methods which avoid integrating out the random measure F from the posterior. We describe three possible approaches: truncation of F, retrospective sampling methods and slice sampling. The truncation method replaces the infinite-dimensional random distribution function

$$F_\infty \stackrel{d}{=} \sum_{k=1}^{\infty} w_k \delta_{\theta_k}$$

with a finite version

$$F_N \stackrel{d}{=} \sum_{k=1}^{N} w_k \delta_{\theta_k}$$

where now $\sum_{k=1}^{N} w_k = 1$ almost surely. This leads to a finite-dimensional posterior distribution which can be simulated directly using standard methods. Truncated versions of the Dirichlet process were initially considered by several authors including Muliere and and Secchi (1996), Muliere and Tardella (1998), Ishwaran and Zarepour (2000) and Ishwaran and James (2000, 2001). The use of these methods raises two important concerns: (1) how to choose the number of atoms N to include in the truncation to avoid a large probability of a big difference between F and F_N; and (2) how to update efficiently a potentially large number of parameters. Ishwaran and James (2001) describe solutions to both problems for the more general class of stick-breaking priors. This class of process is parameterized by two infinite sequences of positive numbers, $a = (a_1, a_2, a_3, \ldots)$ and $b = (b_1, b_2, b_3, \ldots)$ and defines the weights in equation (6.1) by $w_k = v_k \prod_{j<k}(1 - v_j)$ for a sequence of independent random variables $v_1, v_2, v_3, \ldots$ for which $v_i \sim \text{Be}(a_i, b_i)$. The Dirichlet process arises if $a_i = 1$ and $b_i = M$ for all i. The process can be simply truncated by setting $v_N = 1$ since $w_k = 0$ for $k > N$.

Ishwaran and James (2001) argue that N should be chosen to control the difference between the probability of the observations under the truncated and infinite-dimensional priors. The probability under F_N is given by

$$\pi_N(y) = \int \left(\prod_{i=1}^{n} g(y_i|\theta_i) F_N(\mathrm{d}\theta_i) \right) \Pi_N(F_N)$$

where Π_N represents the probability law of F_N (this definition extends to the infinite-dimensional case by setting $N = \infty$). The distance between the probability of the sample under the finite approximation, $\pi_N(y)$, and the infinite-dimensional version, $\pi_\infty(y)$, can be measured using the L_1 distance which will be denoted $\| \cdot \|_1$. Ishwaran and James (2001) derive the following bound:

$$\| \pi_N(X) - \pi_\infty(X) \|_1 \leq 4 \left[1 - \mathbb{E} \left\{ \left(\sum_{k=1}^{N-1} w_k \right)^n \right\} \right].$$

The value of N can be chosen to make this bound small. In the case of a Dirichlet process prior, they derive a simpler, approximate expression:

$$4 \left[1 - \mathbb{E} \left\{ \left(\sum_{k=1}^{N-1} w_k \right)^n \right\} \right] \approx 4n \exp\{-(N-1)/M\}.$$

This allows the simple calculation of a value of N that gives a particular level of error.

The value of N derived for the approximation is potentially large. Ishwaran and James (2001) show that efficient methods for updating a large number of parameters are available for the stick-breaking processes. The posterior distribution is now defined on $(s_1, s_2, \ldots, s_n)$, $v = (v_1, v_2, \ldots, v_{N-1})$ and $\theta = (\theta_1, \theta_2, \ldots, \theta_N)$. Both θ and v can be updated in a block which leads to good mixing properties. The Gibbs sampler has the following steps.

1. The elements of θ are conditionally independent under the full conditional distribution. The full conditional of θ_j is proportional to $h(\theta_j) \prod_{\{i|s_i=j\}} g(y_i|\theta_j)$.
2. The elements of v are conditionally independent under the full conditional distribution. The full conditional of v_j is $\text{Be}(a_j + n_j, b_j + m_j)$ where $n_j = \sum_{i=1}^{n} \text{I}(s_i = j)$ and $m_j = \sum_{i=1}^{n} \text{I}(s_i > j)$.

3. Each element of $s_1, s_2, \ldots, s_n$ can be updated using the standard mixture model update. The parameter s_i is updated from the distribution $p(s_i = j) \propto w_j g(y_i|\theta_j)$, $j = 1, 2, \ldots, N$.

Replacing an infinite-dimensional object with a closely approximating finite-dimensional version can allow useful extensions to standard methods, such as the nested Dirichlet processes (Rodriguez, Dunson and Gelfand, 2008). Further examples are given in Chapter 7.

A drawback with truncation methods is the need to define a truncation point N. The result of Ishwaran and James (2001) gives a condition for stick-breaking processes but fitting results would need to be derived if other classes of nonparametric prior were to be fitted. The truncation point also depends on the parameters of the prior which will often be assumed unknown and given a prior themselves. Finding a value of the truncation point which works well for a range of values of these parameters which have good posterior support may be difficult. Therefore it is useful to develop methods which have random truncation points. Remarkably, it is possible to define methods that use a finite number of parameters in the sampler but which have the correct posterior under the infinite-dimensional prior. There are two classes of such methods: retrospective samplers and slice samplers.

The first proposed method was the retrospective sampler for Dirichlet process mixture models (Papaspiliopoulos and Roberts, 2008) which exploits the following observation for stick-breaking processes. We can simulate a draw, ϕ, from the unknown distribution F using standard inversion sampling for discrete distributions. The algorithm is as follows.

1. Simulate $u \sim \text{unif}(0, 1)$.
2. Find the value of k for which $\sum_{j=1}^{k} w_j \leq u < \sum_{j=1}^{k+1} w_j$.
3. Set $\phi = \theta_k$.

A finite value of i must exist since $\sum_{j=1}^{i+1} w_j$ is an increasing sequence bounded by 1. The draw is from a distribution with an infinite number of atoms but only involves a finite number of random variables $v_1, v_2, \ldots, v_k$ and $\theta_1, \theta_2, \ldots, \theta_k$. A sample of allocation variables $s_1, s_2, \ldots, s_n$ can be simulated in the same way and the maximum of these sampled values must also be finite. Let that value be $K_n = \max\{s_1, s_2, \ldots, s_n\}$. Papaspiliopoulos and Roberts (2008) show that posterior inference is possible with a Gibbs sampling scheme. Firstly, $v_1, v_2, \ldots, v_{K_n}$ and $\theta_1, \theta_2, \ldots, \theta_{K_n}$ can be updated using steps (1) and (2) of the algorithm of Ishwaran and James (2001) with K_n as the truncation point. Secondly, the allocation s_i is updated using Metropolis–Hastings methods in the following way.

1. Calculate $q_j = g(y_i|\theta_j)$ for $j = 1, 2, \ldots, K_n$ and let $q^\star_{K_n} = \max_{1\le j\le K_n}\{q_j\}$. Simulate $u \sim \text{unif}(0, 1)$ and let $c = \sum_{j=1}^{K_n} w_j q_j + q^\star_{K_n}\left(1 - \sum_{j=1}^{K_n} w_j\right)$. If

$$u < \frac{\sum_{j=1}^{K_n} w_j q_j}{c}$$

then find k for which

$$\frac{\sum_{j=1}^{k-1} w_j q_j}{c} < u < \frac{\sum_{j=1}^{k} w_j q_j}{c}$$

and set $s_i' = k, \theta' = \theta$ and $v' = v$. Otherwise, set $\theta' = \theta, v' = v$ and $k = K_n+1$ and use the following algorithm.

Step 1: simulate $v_k' \sim \text{Be}(a_k, b_k)$ and $\theta_k' \sim H$.
Step 2: if

$$u < \frac{\sum_{j=1}^{K_n} w_j q_j + q^\star_{K_n}\sum_{j=K_n+1}^{k} w_j}{c},$$

set $s_i' = k$. Otherwise set $k = k + 1$ and return to step 1.

Let $K_n' = \max\{s_1, \ldots, s_{i-1}, s_i', s_{i+1}, \ldots, s_n\}$ and if $K_n' > K_n$ define $q_j = g(y_i|\theta_j')$ for $K_n < j \le K_n'$. The proposed value is accepted with probability

$$\alpha = \begin{cases} \frac{q^\star_{K_n'}}{g(y_i|\theta_{s_i})} \frac{\sum_{j=1}^{K_n} w_j q_j + q^\star_{K_n}\left(1-\sum_{j=1}^{K_n} w_j\right)}{\sum_{j=1}^{K_n} w_j q_j + q^\star_{K_n'}\left(1-\sum_{j=1}^{K_n} w_j\right)} & \text{if } K_n' < K_n \\ 1 & \text{if } K_n' = K_n \\ \frac{g\left(y_i\middle|\theta_{s_i'}\right)}{q^\star_{K_n}} \frac{\sum_{j=1}^{K_n} w_j q_j + q^\star_{K_n}\left(1-\sum_{j=1}^{K_n} w_j\right)}{\sum_{j=1}^{K_n} w_j q_j + q^\star_{K_n'}\left(1-\sum_{j=1}^{K_n} w_j\right)} & \text{if } K_n' > K_n. \end{cases}$$

The method can be easily extended to posterior inference in more complicated nonparametric models. See, for example, Jackson, Dary, Doucet and Fitzgerald (2007), Dunson and Park (2008) and Griffin and Steel (2008).

A second method for posterior inference without truncation error is described by Walker (2007) who proposes a slice sampling scheme for the Dirichlet process mixture models. Slice sampling (Damien, Wakefield and Walker, 1999; Neal, 2003) has become a standard computational tool for simulating from nonstandard distribution of a finite number of variables, x, by introducing latent variables, u, which preserve the marginal distribution of x. Careful choice of the latent variables will define Gibbs sampling schemes for the joint distribution of x and u which have standard distributions. Sequences of draws of x from the Gibbs samplers will have the correct marginal distribution. In nonparametric models, slice sampling ideas can be used to define a random truncation point. Latent variables $(u_1, u_2, \ldots, u_n)$ are introduced such that

$$p(s_i = k, u_i) = \mathrm{I}(u_i < w_k)$$

which guarantees that the marginal distribution of s_i, integrating across u_i, is correct, i.e. $p(s_i = k) = w_k$. Let K be such that $\prod_{j=1}^{K}(1 - v_j) < \min\{u_i\}$. The posterior distribution is proportional to

$$p(w_1, w_2, \ldots, w_K) \prod_{j=1}^{n} \mathrm{I}(u_j < w_{s_j})\, g(y_j|\theta_{s_j}) \prod_{i=1}^{K} h(\theta_i).$$

Consider now updating s_i conditional on u_i. The probability that $s_i = k$ is proportional to $\mathrm{I}(u_i < w_k) g(y_i|\theta_k)$. There are only a finite number of w_k greater than u_i and so the full conditional of s_i given u_i will be a discrete distribution with only a finite number of elements. A sufficient condition to find all such atoms is to find the first K for which $\prod_{i=1}^{K}(1 - v_i) < u_i$. The Gibbs sampler is as follows.

1. Simulate s_i from the the following discrete distribution

$$p(s_i = j) \propto \sum_{j=1}^{K} \mathrm{I}(w_j > u_i) g(y_i|\theta_j)$$

   for $i = 1, 2, \ldots, n$.

2. The parameters $\theta_1, \theta_2, \ldots, \theta_K$ are independent under the full conditional distribution and θ_i is drawn from the distribution proportional to

$$h(\theta_i) \prod_{\{k|s_k = i\}} g(y_k|\theta_i).$$

3. The parameter v_i is updated from a truncated beta distribution which can be inversion sampled. Simulate a uniform random variable ξ and set

$$v_i = 1 - [\xi(1 - b)^M + (1 - \xi)(1 - a)^M]^{1/M},$$

   where

$$a = \max\left\{0, \max_{\{i|s_i = k\}} \frac{u_i}{\prod_{j<k}(1 - v_j)}\right\}$$

$$b = \min\left\{1, \min_{\{i|s_i > k\}} 1 - \frac{u_i}{v_{s_i} \prod_{j<s_i; j \neq k}(1 - v_j)}\right\}.$$

4. The parameter $u_i \sim \text{unif}(0, w_{s_i})$ where $\text{unif}(a, b)$ represents a uniform distribution on $(a, b]$. Find the smallest value of k for which $\prod_{j=1}^{k}(1 - V_j) < \min\{u_i\}$ which may involve simulating more breaks v_i from a $\text{Be}(1, M)$ distribution and atoms $\theta_i \sim H$.

A more efficent version of the scheme is introduced by Kalli, Griffin and Walker (2008) who notice that steps 3 and 4 can be replaced by the following step.

3*. Simulate v_i from $\text{Be}(1 + n_i, M + m_i)$ where $n_j = \sum_{i=1}^{n} \text{I}(s_i = j)$ and $m_j = \sum_{i=1}^{n} \text{I}(s_i > j)$. Simulate $u_i \sim \text{unif}(0, w_{s_i})$. Find the smallest value of k for which $\prod_{j=1}^{k}(1 - v_j) < \min\{u_i\}$ which may involve simulating more breaks v_i from a $\text{Be}(1, M)$ distribution and atoms $\theta_i \sim H$.

The sampling now has similar steps to the Gibbs sampler introduced by Ishwaran and James (2001). The differences are the full conditional of s and the introduction of a random truncation point. The sampler has been extended to more general stick-breaking processes by Kalli, Griffin and Walker (2008) and to the Indian buffet process by Teh, Görür and Ghahramani (2007).

The method is applied to the rather general class of normalized random measures (as discussed in Chapter 3) by Griffin and Walker (2008). These measures express the weights w_k in equation (6.1) as $w_k = J_k / J$ where $\{J_k\}_{k=1}^{\infty}$ are the jumps of a Lévy process with Lévy density $w(x)$ and $J = \sum_{l=1}^{\infty} J_l$. The weights will follow a GEM process if the jumps follow a gamma process, which has Lévy density $w(x) = Mx^{-1}\exp\{-x\}$. The algorithm exploits the fact that the jumps can be represented as the transformation of a unit intensity Poisson process $(\tau_1, \tau_2, \tau_3, \ldots)$. The jumps are $J_k = W^{-1}(\tau_k)$ where W^{-1} is the inverse of $W^+(x) = \int_x^{\infty} w(y)\, dy$. An advantage of this approach is that the jumps are ordered $J_1 > J_2 > J_3 > \ldots$ and it is straightforward to find the largest K for which $J_K > u$ for any u. They introduce latent variables $(u_1, u_2, \ldots, u_n)$ and ν to define the augmented likelihood

$$\nu^{n-1}\exp\{-\nu J\}\prod_{i=1}^{n} \text{I}(u_i < J_{s_i}) g(y_i|\theta_{s_i}).$$

This form is not suitable for MCMC since it involves the infinite random sum J. However, letting $L = \min_{1 \le i \le n} u_i$, all jumps J_k for which $J_k < L$ can be integrated out to define a workable augmented likelihood

$$\nu^{n-1}\exp\left\{-\nu\sum_{l=1}^{K} J_l\right\} \text{E}\left[\exp\left\{-\nu\sum_{l=K+1}^{\infty} J_l\right\}\right]\prod_{i=1}^{n} \text{I}(u_i < J_{s_i}) g(y_i|\theta_{s_i})$$

3. *updating the atoms* (random effects specific to each cluster) by independent sampling from normal posteriors

$$(\theta_h \mid -) \overset{\text{ind}}{\sim} \mathrm{N}\left(\frac{\sigma_0^{-2}\mu_0 + \sigma^{-2}\sum_{i:S_i=h}\sum_{j=1}^{n_i} y_{ij}}{\sigma_0^{-2} + \sigma^{-2}\sum_{i:S_i=h} n_i}, \frac{1}{\sigma_0^{-2} + \sigma^{-2}\sum_{i:S_i=h} n_i}\right),$$
$$h = 1, \ldots, N;$$

4. *updating the hyperparameters* (μ_0, σ_0^{-2}) by sampling from the conditionally conjugate normal-gamma posterior

$$(\mu_0, \sigma_0^{-2} \mid -) \sim \mathrm{N}(\mu_0; \widehat{\mu}_0, \widehat{\tau}\sigma_0^2)G(\sigma_0^{-2}; \widehat{a}_0, \widehat{b}_0),$$

   with $\mathrm{N}(\mu_0; \mu_{00}, \tau\sigma_0^2)G(\sigma_0^{-2}; a_0, b_0)$ the prior, $\widehat{\tau} = 1/(\tau^{-1} + N)$, $\widehat{\mu}_0 = \widehat{\tau}(\tau^{-1}\mu_{00} + \sum_{h=1}^{N}\theta_h)$, $\widehat{a}_0 = a_0 + N/2$, and $\widehat{b}_0 = b_0 + 1/2(\tau^{-1}\mu_{00}^2 + \sum_{h=1}^{N}\theta_h^2 - \widehat{\tau}^{-1}\widehat{\mu}_0^2)$; and

5. *updating the within-subject precision* σ^{-2} from its conditionally conjugate gamma posterior

$$(\sigma^{-2} \mid -) \sim G\left(a_1 + \frac{1}{2}\sum_{i=1}^{n} n_i, b_1 + \frac{1}{2}\sum_{i=1}^{n}\sum_{j=1}^{n_i}(y_{ij} - \theta_{S_i})^2\right),$$

   where $G(a_1, b_1)$ is the prior for σ^{-2}, with $G(a, b)$ corresponding to the gamma distribution parameterized to have mean a/b and variance a/b^2.

Each of these steps is quite easy to implement, so MCMC computation in the analysis of variance model with a Dirichlet process prior on the random effects distribution is essentially no more difficult than posterior computation in the parametric case in which a normal distribution is assumed for the random effects. One potential issue with the specification in which we let $P \sim \mathrm{DP}(\alpha P_0)$ is that it assumes a discrete distribution for the random effects, so that different subjects have exactly the same random effects value. This may be a useful approximation, but it may be more realistic to suppose that each subject has their own unique random effect value. Hence, we may instead want to characterize the random effects distribution using an unknown continuous density. This can be easily accomplished using a minor modification to the above specification to let $\mu_i \sim \mathrm{N}(\mu_{0i}, \sigma_{0i}^2)$, with $(\mu_{0i}, \sigma_{0i}^2) \sim Q$, and $Q \sim \mathrm{DP}(\alpha Q_0)$. In this case, the random effect distribution, P, is characterized as a DP mixture (DPM) of normals (Lo, 1984; Escobar and West, 1995).

The DPM of normals for the random effects distribution can be used for clustering of subjects into groups having similar, but not identical, random effects. The blocked Gibbs sampler is easily modified to accommodate this case. One common concern with the blocked Gibbs sampler, and other approaches that rely on truncation of the stick-breaking representation to a finite number of terms, is that in bypassing the infinite-dimensional representation, we are effectively fitting a finite (and hence parametric) mixture model. For example, if we let $N = 25$ as a truncation level, a natural question is how this is better or intrinsically different than fitting a finite mixture model with 25 components. One answer is that N is not the number of components occupied by the subjects in your sample, but is instead an upper bound on the number of clusters. In most cases, taking a conservative upper bound, such as $N = 25$ or $N = 50$, should be sufficient, since mixture models are most useful when there are relatively few components. In addition, because the weights in the infinite stick-breaking representation (7.2) decrease rapidly for typical choices of α, we also obtain an accurate approximation to the DP for modest N.

However, some recent approaches avoid the need for truncation. Walker (2007) proposed a slice sampling approach. Papaspiliopoulos and Roberts (2008) proposed an alternative retrospective MCMC algorithm, which is an easy to implement modification to the blocked Gibbs sampler that allows one to add adaptively, but not delete, components as needed as the MCMC algorithm progresses. This can actually result in substantially improved efficiency in some cases. To clarify, note that one would typically choose a conservative truncation level in implementing the blocked Gibbs sampler, which would then require updating of the stick-breaking weights and atoms for many components that are not needed in that they are assigned very low probabilities and are not occupied by any subjects in the sample. The retrospective sampling approach instead allows one to conduct computation for the number of components that are needed, though to take advantage of this efficiency gain it is typically necessary to run a short preliminary chain of 10–100 iterations to choose good starting values. Otherwise, the retrospective MCMC approach may add a large number of components in the first few sampling steps, and then one is unable to delete these components later in the sampling, resulting in a large computational burden.

7.2.3 General random effects models

Until this point, I have focused for illustration on the simple variance component model in (7.1). However, it is straightforward to extend the ideas to much richer classes of random effects models. For example, Kleinman and Ibrahim (1998a) placed a DP prior on the distribution of the random effects in a linear mixed effects model, while Kleinman and Ibrahim (1998b) extended this approach to the broader class of generalized linear mixed models. Mukhopadhyay and Gelfand (1997) proposed a wide class of DP mixtures of GLMs. Müller and Rosner (1997) used a DPM to obtain a flexible nonlinear hierarchical model for blood count data, while in more recent work, Müller, Quintana and Rosner (2007) proposed a semiparametric model for multilevel repeated measurement data. Walker and Mallick (1997) used Pólya tree (PT) priors for nonparametric modeling of random effect and frailty distributions. The PT prior is another popular and computationally attractive nonparametric Bayes prior. From a data analysis perspective, it could be used as an alternative to a DP prior on P in (7.2). For a recent article on mixtures of PT priors, refer to Hanson (2006).

The linear mixed effects model (Laird and Ware, 1982) is used routinely for the analysis of data from longitudinal studies and studies having multilevel designs (e.g., patients nested within study centers). Focusing on the longitudinal data case, let $\mathbf{y}_i = (y_{i1}, \ldots, y_{i,n_i})'$ denote the response data for subject i, with y_{ij} the observation at time t_{ij}, for $j = 1, \ldots, n_i$. Then, the linear mixed effects model has the form

$$\begin{aligned} y_{ij} &= \mathbf{x}_{ij}'\boldsymbol{\beta} + \mathbf{z}_{ij}'\mathbf{b}_i + \epsilon_{ij}, \quad \epsilon_{ij} \sim \mathrm{N}(0, \sigma^2), \\ \mathbf{b}_i &\sim P, \end{aligned} \tag{7.6}$$

where $\mathbf{x}_{ij} = (x_{ij1}, \ldots, x_{ijp})'$ are fixed effect predictors, $\boldsymbol{\beta} = (\beta_1, \ldots, \beta_p)'$, $\mathbf{z}_{ij} = (z_{ij1}, \ldots, z_{ijq})'$ are random effect predictors, and P is a random effect distribution on $\Re^q$.

It is straightforward to allow P to be unknown through the use of a DP prior to induce a discrete random effects distribution, or a DP mixture of Gaussians to induce an unknown continuous random effects density. In fact, such models have been increasingly used in applications. For example, van der Merwe and Pretorius (2003) applied a linear mixed effects model with a DP prior for the random effects distribution in an animal breeding application. Ohlssen, Sharples and Spiegelhalter (2007) provide a recent overview of the literature on Bayesian semiparametric random effects models, and provide a tutorial on routine implementation in WinBUGS. In addition, there is an R package, `DPpackage`, which provides R functions for efficiently fitting a broad variety of semiparametric Bayes hierarchical models, including not only DPMs but also Pólya tree models (Jara, 2007); see Section 8.6.

However, some subtle issues arise in semiparametric modeling of random effects distributions, which should be carefully considered in fitting such models. In particular, in expression (7.6), the posterior distribution of the random moments of P may impact inferences on $\boldsymbol{\beta}$. In parametric models, it is standard practice to use a multivariate normal distribution with mean zero for P, so that the coefficients $\boldsymbol{\beta}$ are then interpretable as fixed effects. In particular, we require that $\mathrm{E}(y_{ij} \mid \mathbf{x}_{ij}, \mathbf{z}_{ij}) = \mathbf{x}_{ij}'\boldsymbol{\beta}$, which is not true if the posterior expectation of the mean of P is non-zero. Constraining the base measure P_0 to have zero mean is not sufficient to ensure that P is centered on zero a posteriori.

One way to mitigate this problem is to use a centered parameterization. For example, one can let $y_{ij} = \mathbf{x}_{ij}'\boldsymbol{\beta}_i + \epsilon_{ij}$, with $\boldsymbol{\beta}_i \sim P$. In this case, the fixed effect regression coefficients correspond to the mean of the distribution P. One limitation of this is that one needs to assume that $\mathbf{x}_{ij} = \mathbf{z}_{ij}$, though the extension to allow $\mathbf{z}_{ij}$ to correspond to a subset of the predictors in $\mathbf{x}_{ij}$ is straightforward. Another limitation of using a centered parameterization is that the fixed effect coefficients $\boldsymbol{\beta}$ corresponding to the mean of P, which may not be directly available if one implements a computation approach, such as the collapsed Gibbs sampler, that marginalizes out P.

Two recent alternatives were proposed by Li, Lin and Müller (2009) and Dunson, Yang and Baird (2007). The Li, Lin and Müller (2009) approach uses post-processing to adjust for bias in using a DP prior for the random effects distribution. The Dunson, Yang and Baird (2007) approach instead induces a centered DP or centered DP mixture prior by considering the DP or DPM prior for the random effects distribution as a parameter-expanded version of a centered process with mean and/or variance constraints. The centered DPM is an alternative to previously proposed approaches, which constrain a random effects distribution to have median 0 (e.g., Burr and Doss, 2005).

7.2.4 Latent factor regression models

Linear mixed effects models and generalized linear mixed effects models are appropriate when the same type of response is measured repeatedly over time but not when data on a subject consist of a multivariate vector of different types of responses. In many biomedical studies, one may measure multiple surrogates of a latent predictor or health response of interest. For example, single-cell gel electrophoresis measures the frequency of DNA strand breaks on the individual cell level through different surrogates for the amount of DNA in the tail of an image that resembles a comet. As these different surrogates are on different scales, one can consider a latent factor regression model, such as

$$\begin{aligned} y_{ij} &= \mu_j + \lambda_j \eta_i + \epsilon_{ij}, \quad \epsilon_{ij} \sim \mathrm{N}(0, \sigma_j^2), \\ \eta_i &= \mathbf{x}_i'\boldsymbol{\beta} + \delta_i, \quad \delta_i \sim P, \end{aligned} \tag{7.7}$$

where $\mathbf{y}_i = (y_{i1}, \ldots, y_{ip})'$ are p different measures of the frequency of strand breaks in cell i, $\boldsymbol{\mu} = (\mu_1, \ldots, \mu_p)'$ are intercept parameters for the different surrogates, $\boldsymbol{\lambda} = (\lambda_1, \ldots, \lambda_p)'$ are factor loadings, with $\lambda_j > 0$ for $j = 1, \ldots, p$, η_i is a continuous frequency of DNA strand breaks latent variable, $\boldsymbol{\epsilon} = (\epsilon_{i1}, \ldots, \epsilon_{ip})'$ are idiosyncratic measurement errors, $\boldsymbol{\Sigma} = \mathrm{diag}(\sigma_1^2, \ldots, \sigma_p^2)$ is a diagonal covariance matrix, $\mathbf{x}_i = (x_{i1}, \ldots, x_{ip})'$ are predictors of frequency of strand breaks (e.g., dose of a possible genotoxic agent), and P is an unknown latent variable residual distribution.

The distribution of the frequency of strand breaks tends to be right skewed, often with a secondary mode in the right tail. Hence, in order to limit parametric assumptions, one can use a DPM of normals for the latent variable residual distribution, P. However, latent factor regression models require some restrictions for identifiability. In the parametric case of expression (7.7), one would typically let P correspond to the standard normal distribution, which is automatically restricted to have mean 0 and variance 1. Then, the coefficient β_j would have a simple interpretation as the number of standard deviations the latent trait is shifted for each unit change in the jth predictor. To induce mean 0 and variance 1 constraints on the latent variable distribution in the semiparametric case, one can use a centered DPM prior for P, as in Dunson, Yang and Baird (2007). Such an approach is very easy to implement, since a blocked Gibbs sampler can be implemented as if a DPM of normals were used for P, followed by a simple post-processing step. One can also apply related approaches in a much broader class of latent variable models that allow mixed categorical and continuous measurements and multiple latent variables.

7.3 Nonparametric Bayes functional data analysis

7.3.1 Background

In many applications, interest focuses on studying variability in random functions. Some examples of random functions include hormone trajectories over time and brain images collected using MRI technology. Functional data analysis (FDA) methods are used when data consist of error-prone observations on random functions that may differ for the different subjects under study (Ramsay and Silverman, 1997). In order to study heterogeneity among subjects and to borrow strength across the different subjects in estimating their functions, one may consider hierarchical models of the form

$$\begin{aligned} y_i(t) &= \eta_i(t) + \epsilon_i(t), \quad \epsilon_i(t) \sim \mathrm{N}(0, \sigma^2) \\ \eta_i &\sim P, \end{aligned} \tag{7.8}$$

where $y_i(t)$ is an error-prone observation of the function η_i for subject i at time t, $i = 1, \ldots, n$, $\epsilon_i(t)$ is a measurement error, and P is a distribution on Ω, the space of $\mathcal{T} \to \Re$ functions. In practice, it is not possible to observe η_i directly at any time, and we only have measurements of $y_i(t)$ for $t \in \mathbf{t}_i = (t_{i1}, \ldots, t_{i,n_i})'$.

In this section, we consider a variety of semiparametric Bayes approaches for functional data analysis. Section 7.3.2 describes methods based on basis function expansions of η_i. Section 7.3.3 reviews methods that avoid explicit basis function representations using functional Dirichlet processes. Section 7.3.4 provides an overview of recent kernel-based approaches, and Section 7.3.5 considers methods for joint modeling of related functions and of functional predictors with response variables.

7.3.2 Basis functions and clustering

In nonlinear regression and functional data analysis, it is common to simplify modeling by assuming that the unknown functions fall in the linear span of some pre-specified set of basis functions. For example, focusing on hierarchical model (7.8), suppose that

$$\eta_i(t) = \sum_{h=1}^{p} \beta_{ih} b_h(t), \quad \forall t \in \mathcal{T}, \tag{7.9}$$

where $\boldsymbol{\beta}_i = (\beta_{i1}, \ldots, \beta_{ip})'$ are basis coefficients specific to subject i, and $\mathbf{b} = \{b_h\}_{h=1}^{t}$ is a set of basis functions. For example, if η_i could be assumed to be a smooth function, cubic spline basis functions of the following form may be reasonable:

$$\mathbf{b}(t) = \{1, t, t^2, t^3, (t - \xi_1)_+^3, (t - \xi_2)_+^3, \ldots, (t - \xi_q)_+^3\},$$

where $\boldsymbol{\xi} = (\xi_1, \ldots, \xi_q)'$ are knot locations, and x_+ returns 0 for negative x and x for positive x.

Assuming the basis functions are pre-specified (e.g., by choosing a grid of a modest number of equally spaced knots), models (7.8) and (7.9) imply that

$$y_{ij} = \mathbf{x}_{ij}' \boldsymbol{\beta}_i + \epsilon_{ij}, \quad \epsilon_{ij} \sim \mathrm{N}(0, \sigma^2), \tag{7.10}$$

where $y_{ij} = y_i(t_{ij})$, $\mathbf{x}_{ij} = [b_1(t_{ij}), b_2(t_{ij}), \ldots, b_p(t_{ij})]'$, and $\epsilon_{ij} = \epsilon_i(t_{ij})$, for $i = 1, \ldots, n$, $j = 1, \ldots, n_i$. Hence, letting $\boldsymbol{\beta}_i \sim Q$, one can simply use a linear mixed effects model for functional data analysis. As a flexible semiparametric approach, one can place a DP prior on Q. As noted by Ray and Mallick (2006) in the setting of a wavelet model, such an approach induces functional clustering.

To clarify, note that the DP prior for Q implies that each subject is allocated into one of $k \leq n$ clusters, with subjects in a cluster having identical values for the basis coefficients. In particular, letting $S_i = h$ denote that subject i is allocated to cluster h, we would have $\boldsymbol{\beta}_i = \boldsymbol{\theta}_h^*$ for all subjects having $S_i = h$. Hence, all the subjects in a cluster would also have identical functional trajectories, with subjects in cluster h having $\eta_i(t) = \mathbf{b}(t)\boldsymbol{\theta}_h^*$, for all $t \in \mathcal{T}$. Note that this provides a semiparametric Bayes alternative to frequentist latent class trajectory models (Muthén and Shedden, 1999) and growth mixture models (Jones, Nagin and Roeder, 2001). Such approaches rely on finite mixture models, with the EM algorithm typically used to obtain maximum likelihood estimates.

Assuming a DP prior for Q implies that individuals in a functional cluster have exactly the same value of the measurement error-corrected function, η_i. This assumption may be overly restrictive and may result in estimation of a large number of functional clusters in some cases. Hence, it may be more realistic to suppose that every individual has a unique function, η_i, but that the functions for individuals in a cluster are similar to each other. This can be accomplished by using a DP mixture of multivariate Gaussians as the prior for Q.

Model (7.8) can be easily modified to include fixed and random effect covariates. Posterior computation is straightforward using a very similar approach to that described in Section 7.2.2 for the simple variance component model (7.1). However, some complications can arise in interpretation of the MCMC output. In particular, the Bayesian semiparametric approach has the appealing property of allowing uncertainty in the number of clusters and the allocation of subjects to these clusters. This has the side effect that the number of clusters and the meaning of the clusters will change across the MCMC iterations. Hence, it can be quite challenging to obtain meaningful posterior summaries of cluster-specific parameters. This problem is not unique to the functional clustering application, and is typically referred to in the literature as the label switching problem (Stephens, 2000; Jasra, Holmes and Stephens, 2005).

Frequentist analyses of mixture models that are fitted with the EM algorithm do not face this issue, because the EM algorithm converges to a point estimate corresponding to a local mode. This point estimate includes the cluster probabilities and each of the cluster-specific parameters (e.g., basis coefficients). In performing inferences on the cluster-specific parameters, one ignores the numeric labels for each of the clusters. However, EM algorithm-based analyses of mixture models also face problems in locating a global mode even when multiple starting points are used. In addition, such methods rely on pre-specification or selection of a fixed number of clusters, while the Bayesian semiparametric approach automatically allows uncertainty in the number of clusters.

Fortunately label switching is only a problem if one is interested in performing cluster-specific inferences instead of simply accounting for uncertainty in clustering when conducting predictions or performing inferences on global features. I use the term "global features" to denote any functional of interest that is not cluster specific. For example, global features may include fixed effect regression coefficients and values of $\eta_i(t)$, for specific subjects or averaged across subjects. Such features can be estimated easily from the MCMC output without worrying about the fact that there are latent cluster indices that are changing in meaning and dimension over the

MCMC iterates. For example, one can obtain a posterior mean and 95% credible interval by collecting $\eta_i(t)$ for each of a large number of MCMC iterates, and then averaging the samples and calculating the 2.5th and 97.5th percentiles.

The real problem occurs when one wants to estimate the functional trajectories specific to each cluster, and a variety of strategies have been proposed. One technique is to attempt to relabel the clusters at each MCMC iteration using a post-processing algorithm. For examples of post-processing approaches, refer to Stephens (2000) and Jasra, Holmes and Stephens (2005). Such approaches tend to be time consuming to implement and do not seem to address the problem fully. For example, in running an MCMC algorithm under models (7.8) and (7.9) with a DP prior on the distribution of the basis coefficients, the number of clusters may vary substantially over the MCMC iterations. However, in re-labeling to line up the clusters from the different iterations, one needs to assume that there is some coherence in the clusters after re-labeling. Typically, this would at least require fixing of the number of clusters.

Given the very high-dimensional set of possible partitions of subjects into clusters, it is not at all unlikely that one may visit dramatically different configurations over the MCMC iterations, particularly if an efficient MCMC is used. Hence, instead of attempting to align clusters that are in some sense unalignable, it is useful to view the partition of subjects as a model selection problem, with the MCMC approaches outlined in Section 7.2.2 providing an approach for model averaging. As is well known in the literature on Bayesian model selection, model averaging is most useful for prediction, and one may commonly encounter difficulties in interpretation when averaging over models in which the parameters have different meanings. In such settings, it is necessary to perform model selection to maintain interpretability.

Carrying over this idea to the setting of DP mixture models, one could attempt to identify an optimal partition of subjects into clusters, with this optimal clustering then used to obtain some insight into how the clusters differ. Before reviewing some of the approaches available to identify an optimal clustering in this setting, it is important to note that one should be very careful to avoid over-interpretation of the estimated partition. Even if one is able to identify the optimal partition from among the very high-dimensional set of possible partitions, this partition may have extremely low posterior probability, and there may be a very large number of partitions having very similar posterior probability to the optimal partition. This is essentially the same problem that is faced in model selection in high-dimensional settings. That said, it is sometimes impossible to bypass the need for selection given the scientific interests in a study. In such settings, the approaches of Medvedovic and Sivaganesan (2002), Dahl (2006) and Lau and Green (2007) are quite useful.

7.3.3 Functional Dirichlet process

The basis function expansion shown in (7.9) clearly requires an explicit choice of a set of basis functions. For well-behaved, smooth functions of time or a single predictor, it may be sufficient to choose splines, with knots specified at a modest-dimensional grid of equally spaced locations. However, when $\mathcal{T}$ is multidimensional (e.g., corresponding to a subset of $\Re^2$ in image or spatial applications or to $\Re^r$ in multivariate regression applications), it can be difficult to pre-specify an appropriate basis. Crandell and Dunson (2009) proposed a modification, which allows unknown numbers and locations of knots, while placing a DP prior on the distribution of the basis coefficients. An alternative is to avoid using an explicit basis function representation entirely by instead relying on a functional Dirichlet process (FDP).

The FDP provides a direct approach to specify a prior for P in (7.8) by letting $P \sim \text{DP}(\alpha P_0)$, where P_0 corresponds to a Gaussian process (GP). Hence, using the stick-breaking representation, we have

$$\eta_i \sim P = \sum_{h=1}^{\infty} \pi_h \delta_{\theta_h}, \quad \theta_h \sim \text{GP}(\mu, \mathcal{C}), \tag{7.11}$$

where $\{\pi_h\}_{h=1}^{\infty}$ are defined as in (7.2) and $\boldsymbol{\theta} = \{\theta\}_{h=1}^{\infty}$ are functional atoms sampled independently from a Gaussian process with mean function μ and covariance function $\mathcal{C}$, for $h = 1, \ldots, \infty$. The FDP has the same mathematical structure as the dependent DP proposed by MacEachern (1999), which will be discussed in Section 7.6.2. Gelfand, Kottas and MacEachern (2005) used the FDP for modeling of spatial data.

Under the FDP in (7.11) subjects will be allocated to functional clusters. Letting $S_i = h$ denote that subject i is allocated to the hth of the $k \leq n$ clusters represented in the data set, we have $\eta_i = \theta^*_{S_i}$, for $i = 1, \ldots, n$. In this case, instead of using a finite vector of basis coefficients to characterize the functions specific to each cluster, we take independent draws from a Gaussian process. This avoids the need to specify explicitly a set of basis functions. However, it is necessary to choose mean and covariance functions in the GP, and the number of functional clusters and the cluster-specific estimates can be quite sensitive to the particular choice made. For a recent book on Gaussian processes, including a discussion of the role of the covariance function, refer to Rasmussen and Williams (2006).

In implementing posterior computation, it is clearly not possible to estimate the functions across the infinitely many locations in $\mathcal{T}$. Hence, in practice one performs computation for finitely many points, typically corresponding to locations at which data are collected along with a tightly spaced grid of additional locations. The GP base measure then implies a multivariate normal base measure across this finite grid, and posterior computation can proceed as for DP mixtures of Gaussians. However, instead of updating the finite-dimensional mean and covariance in the base multivariate normal distribution, one estimates parameters characterizing the mean and covariance functions. For the covariance function, this would typically involve Metropolis–Hastings steps, after assuming a particular form, such as exponential, Gaussian or Matérn.

7.3.4 Kernel-based approaches

As there is often concern in practice about the impact of basis and covariance function specification on inferences, estimation and prediction, it is appealing to consider alternatives. In the frequentist literature on function estimation, it is common to consider kernel-based approaches. In particular, focusing initially on the mean regression function estimation problem in which $\mathrm{E}(Y \mid X = x) = \eta(x)$, there is a rich literature on estimation of η subject to the constraint that $\eta \in \mathcal{H}_K$, where $\mathcal{H}_K$ is a reproducing kernel Hilbert space (RKHS) defined by the uniformly bounded Mercer kernel K.

In their representer theorem, Kimeldorf and Wahba (1971) show that the solution to a least squares minimization problem subject to an RKHS norm penalty lies in a subspace of $\mathcal{H}_K$ represented as follows:

$$\eta(x) = \sum_{i=1}^{n} w_i K(x, x_i), \tag{7.12}$$

where $\mathbf{w} = (w_1, \ldots, w_n)'$ are unknown coefficients. Tipping (2001), Sollich (2002) and Chakraborty, Ghosh and Mallick (2005) consider Bayesian kernel-based methods based on choosing a prior for $\mathbf{w}$. Such approaches implicitly assume that the support of the prior lies in a subspace of $\mathcal{H}_K$ represented as in (7.12), which is somewhat unnatural in that (7.12) was derived in solving an optimization problem.

Pillai et al. (2007) and Liang et al. (2007) noted that a fully Bayesian solution would instead place a prior for η with large support in $\mathcal{H}_K$. Pillai et al. (2007) accomplished this through the integral representation

$$\eta(x) = \int_{\mathcal{T}} K(x, u) \mathrm{d}\gamma(u), \quad \forall x \in \mathcal{T}, \tag{7.13}$$

where $\gamma \in \Gamma$ and $\eta \in \mathcal{G}$. When Γ corresponds to the space of all signed Borel measures, then $\mathcal{G} = \mathcal{H}_K$. Pillai et al. considered a variety of specific possibilities for γ, focusing on Lévy process priors, while Liang et al. (2007) instead used a decomposition that expressed γ as a product of a GP and a DP. In the Liang et al. (2007) specification, the DP component essentially places a random probability measure

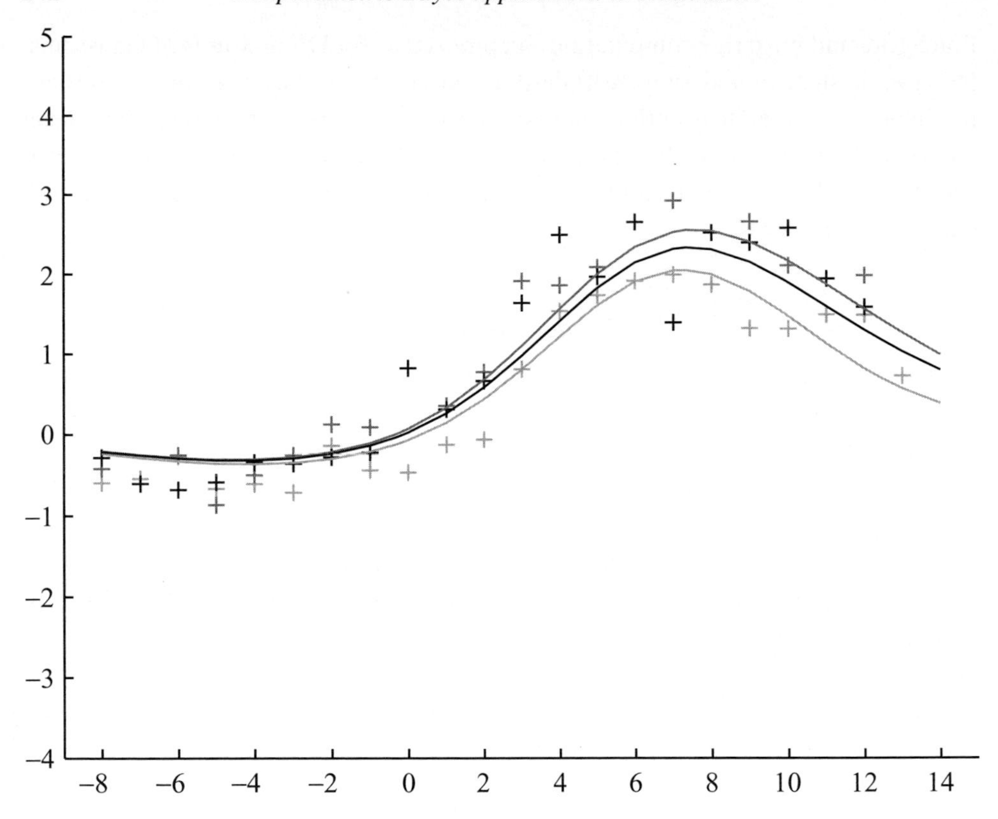

Figure 7.2 Posterior mean progesterone curves (*solid lines*) and observed progesterone levels for three selected nonconceptive cycles. Estimated progesterone curves and observed data are linked by color.

on the locations of the kernels, while the GP places a prior on the coefficients at the resulting countably infinite collection of locations.

MacLehose and Dunson (2009) generalized the Liang et al. (2007) formulation to the functional data analysis setting by letting $\eta_i(x) = \int K(x, u)\mathrm{d}\gamma_i\,(u)$, and then choosing a nonparametric Bayes hierarchical prior for $\boldsymbol{\gamma} = \{\gamma_1, \ldots, \gamma_n\}$. In particular, their specification relied on functional DP and hierarchical DP (HDP) (Tomlinson, 1998; Teh, Jordan, Beal and Blei, 2006) components. The HDP is a prior for modeling of related distributions through incorporating dependence by assuming a common base measure in DPs for each of the distributions, with this base measure allowed to be unknown through use of a DP prior. For further details on the HDP, refer to Chapter 5.

An appealing feature of the MacLehose and Dunson (2009) approach relative to the approaches described in Sections 7.3.2 and 7.3.3 is the allowance for local borrowing of information through local selection of kernels and locally dependent weights. To illustrate this, consider an application to progesterone trajectory data previously analyzed in Brumback and Rice (1998). Data were available for 51 women who provided samples over 91 cycles, of which 22 were conception cycles. Taking results from MacLehose and Dunson (2009), Figure 7.2 shows the raw data and estimated posterior mean progesterone curves for three women randomly selected from the nonconception group. This figure demonstrates the borrowing of information. In particular, during the baseline phase prior to ovulation (day 0 in the figure), the progesterone values are quite similar, so there is strong borrowing of information and the estimates are essentially equivalent. However, following ovulation the curves smoothly deviate, with the approach favoring similar shapes across the curves. Note that the methods of Sections 7.3.2 and 7.3.3 instead borrow information only through global clustering and through the parametric base model. Alternative methods for local borrowing of information will be discussed in detail in Section 7.4.

7.3.5 Joint modeling

In biomedical studies, there is very commonly interest in studying the relationship between functional predictors and response variables. For example, the functional predictor may correspond to the longitudinal trajectory in the level of an environmental exposure, such as air pollution, or to a diagnostic image, while the response corresponds to an indicator of an adverse health condition. In such cases, there is substantial interest in building flexible joint models for relating a subject's functional predictor to their health status, adjusting for possible confounding factors, such as age and demographic variables.

To focus on a motivating application, we consider an epidemiologic study of early pregnancy loss (EPL) in which daily urine samples were collected in order to measure levels of hormone metabolites over time prior to conception and through early pregnancy (Wilcox et al., 1988). Our interest is in studying how progesterone trajectories following ovulation predict EPL. EPLs are identified when hCG rises soon after implantation but then declines back to baseline levels instead of continuing to rise. Progesterone plays a critical role in maintaining the pregnancy, so some clinicians have even suggested treatment with exogenous progesterone as a possible intervention to reduce risk of EPL. Affordable home devices are available for measuring progesterone metabolites in urine, so an algorithm could potentially be programmed into such a device to alert the woman when she is at risk of impending loss.

Motivated by this application, Bigelow and Dunson (2009) develop a Bayesian nonparametric approach for joint modeling of functional predictors with a response variable. Their proposed approach relies on a simple extension of the method proposed in Section 7.3.2. In order to facilitate applications to other settings, I will initially present the approach in more generality than considered in Bigelow and Dunson (2009). In particular, suppose for subject i, we have data $\mathbf{y}_i = (\mathbf{y}'_{i1}, \ldots, \mathbf{y}'_{ip})'$, where $\mathbf{y}_{ij} = (y_{ij1}, \ldots, y_{ij,n_{ij}})'$ is a vector of observations of type j, for $j = 1, \ldots, p$. For example, $\mathbf{y}_{i1}$ may consist of error-prone measurements of a functional predictor, while y_{i2} is 0/1 indicator of a health response. More generally, the $\mathbf{y}_{ij}$ may correspond to several different types of information collected on a subject.

We define separate models for each of the components of the data vector as follows:

$$\begin{aligned} \mathbf{y}_{ij} &\sim f_j(\boldsymbol{\beta}_{ij}; \boldsymbol{\phi}_j), \quad j = 1, \ldots, p, \\ \boldsymbol{\beta}_i = (\boldsymbol{\beta}'_{i1}, \ldots, \boldsymbol{\beta}'_{ip})' &\sim P, \end{aligned} \tag{7.14}$$

where $f_j(\boldsymbol{\beta}_{ij}; \boldsymbol{\phi}_j)$ is the likelihood for component j, defined in terms of the subject-specific parameters $\boldsymbol{\beta}_{ij} = (\beta_{ij1}, \ldots, \beta_{ij,p_j})$ and population parameters $\boldsymbol{\phi}_j$, and the different component models are linked through P, the joint distribution for $\boldsymbol{\beta}_i$. In parametric joint modeling of multivariate data having a variety of measurement scales, it is common to use latent factor and structural equation models that incorporate shared latent variables in the different component models. By allowing P to be unknown through a nonparametric Bayes approach, we obtain a more flexible class of joint models.

Bigelow and Dunson (2009) propose to use a DP prior for P with the following structure:

$$P = \sum_{h=1}^{\infty} \pi_h \delta_{\Theta_h}, \quad \Theta_h \sim P_0 = \otimes_{j=1}^{p} P_{0j}, \tag{7.15}$$

where $\boldsymbol{\pi} = \{\pi_h\}_{h=1}^{\infty}$ are as defined in (7.2), $\Theta_h = \{\Theta_{hj}\}_{j=1}^{p}$ is a collection of atoms corresponding to the parameters in each of the p different components, and the base measure P_0 used in generating these atoms is characterized as a product measure of probability measures for each component. For example, in joint modeling of a longitudinal trajectory with a 0/1 response, P_{01} may correspond to a Gaussian process and P_{02} to a beta distribution, so that each DP cluster then contains a random function along with a corresponding probability of an adverse response. In this manner, a semiparametric joint model is defined for data having different scales, with dependence in the different types of data collected for a subject induced through allocating individuals to clusters having different parameters for each component model.

Instead of using a GP for the functional predictor component, Bigelow and Dunson (2009) use multivariate adaptive splines with unknown numbers and locations of knots. They applied this approach to progesterone metabolite and EPL

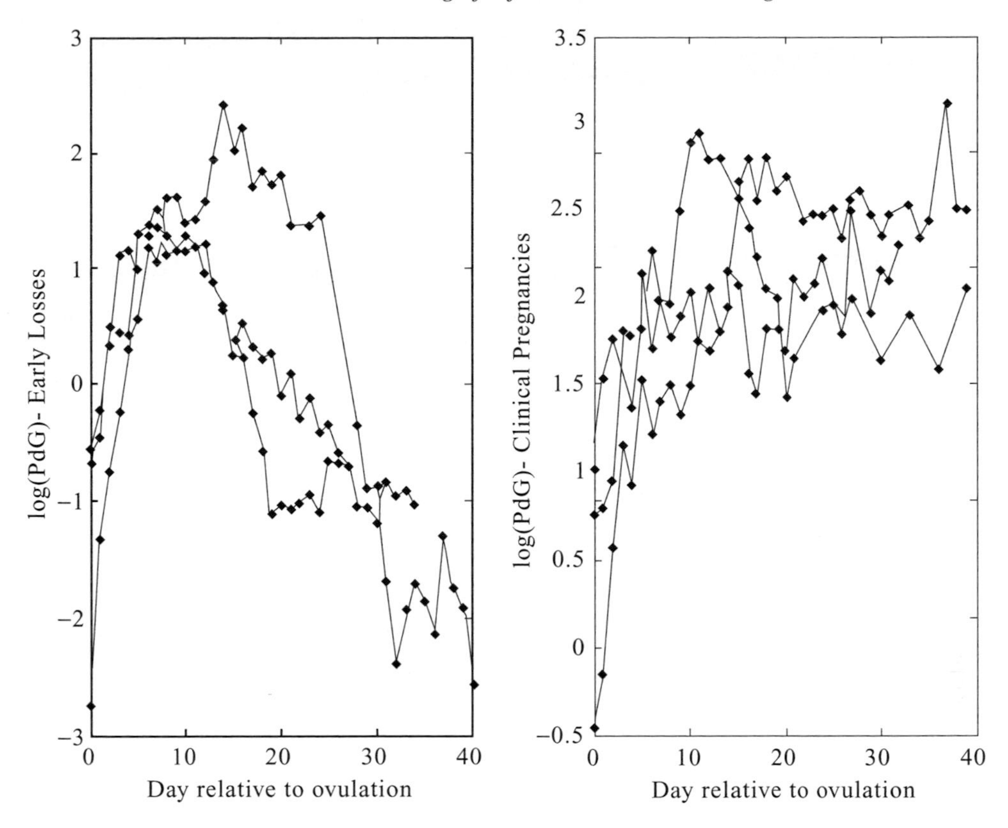

Figure 7.3 Progesterone data beginning at the estimated day of ovulation for three early losses and three clinical pregnancies.

data from Wilcox et al. (1988). Figure 7.3 shows the progesterone data for three randomly selected early losses and three clinical pregnancies. Figure 7.4 shows the progesterone data for each of the 16 identified clusters containing more than one subject, along with the number of pregnancies in the cluster and the estimated probability of early pregnancy loss. Note that it is very clear that the identified clusters match the data from these plots. In addition, the early loss probabilities varied dramatically across the clusters. This resulted in accurate out of sample prediction of impending losses soon after implantation (results not shown).

7.4 Local borrowing of information and clustering

Until this point, I have focused primarily on methods that rely on different variants of the DP for modeling of an unknown random effects distribution in a hierarchical model. As discussed above, such methods have the side effect of inducing clustering of subjects into groups, with groups defined in terms of unique random effects values

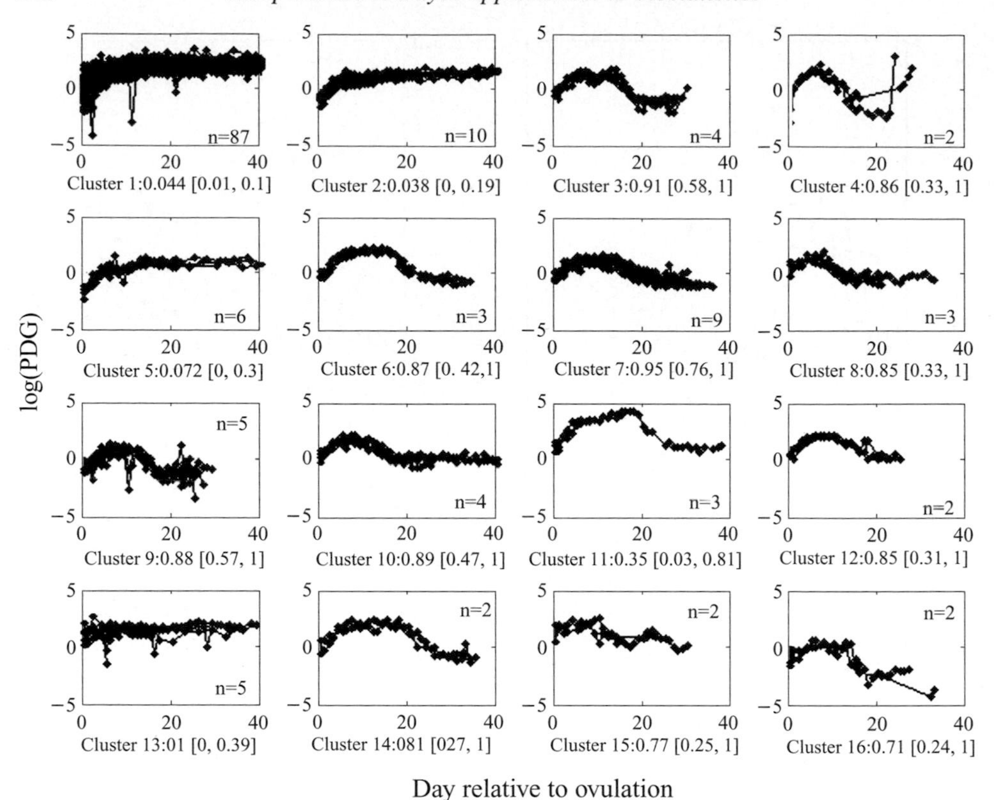

Figure 7.4 Progesterone data for pregnancies in each of the 16 clusters identified in the Wilcox et al. (1988) data. Only those clusters containing more than one pregnancy are shown. Below each plot is the estimated probability of early pregnancy loss within the cluster, along with a 95% credible interval.

or unique parameters in a parametric model. One characteristic of this type of specification is that clustering occurs globally, in that two individuals are clustered together for all their random effects or none.

For example, focus on the joint model specification in (7.15) and suppose $\boldsymbol{\beta}_i \sim P$, for $i = 1, \ldots, n$. Then, it is clear that either $\boldsymbol{\beta}_i = \boldsymbol{\beta}_{i'}$, with prior probability $1/(1+\alpha)$, or $\boldsymbol{\beta}_i$ and $\boldsymbol{\beta}_{i'}$ correspond to two independent draws from P_0, so that none of the elements of $\boldsymbol{\beta}_i$ can cluster together with the corresponding elements of $\boldsymbol{\beta}_{i'}$ (assuming nonatomic P_0). Such global clustering is quite restrictive in that two different subjects may be quite similar or even identical for most of their random effects, while having important deviations in certain components. An alternative, which allows local clustering, is to let $\boldsymbol{\beta}_{ij} \sim P_j$, with $P_j \sim \mathrm{DP}(\alpha_j P_{0j})$, independently for $j = 1, \ldots, p$. However, this approach does not allow for accrual of information about similarities between subjects. In particular, if $\boldsymbol{\beta}_{ij} = \boldsymbol{\beta}_{i'j}$ then the probability should intuitively be increased that $\boldsymbol{\beta}_{ij'} = \boldsymbol{\beta}_{i'j'}$ compared with the case in which $\boldsymbol{\beta}_{ij} \neq \boldsymbol{\beta}_{i'j}$.

Motivated by this problem, Dunson, Xue and Carin (2008) proposed a matrix stick-breaking process (MSBP), which generalizes the DP stick-breaking structure to allow row and column stick-breaking random variables to induce dependent local clustering. Posterior computation for the MSBP can proceed using a simple modification to the blocked Gibbs sampler of Ishwaran and James (2001). The MSBP has particular practical advantages over joint DP and independent DP priors in high-dimensional settings in which subjects can be very similar for most of their coefficients, while having distinct local deviations. This occurs, for example, when modeling of multiple, related functions or images using a basis representation. Often, most of the function or image is quite similar across subjects, suggesting that most of the basis coefficients are effectively identical. However, it is those local regions of heterogeneity that are most scientific interest.

Motivated by the problem of local clustering in functional data analysis, Petrone, Guindani and Gelfand (2007) proposed a hybrid functional Dirichlet process prior. The hybrid DP is based on the clever idea of introducing a collection of global species, with each individual function formulated from a patchwork of these global species. In particular, a latent Gaussian process is introduced, with the level of this latent process controlling local allocation to the global species. This results in local clustering of the functions for different individuals, with discontinuities occurring in the functions at changepoints in which the latent process crosses thresholds so that allocation switches to a different species. Petrone, Guindani and Gelfand (2007) applied this approach to an interesting brain image application. Rodriguez, Dunson and Gelfand (2009a) proposed an alternative latent stick-breaking process (LaSBP). The LaSBP is defined by introducing a fixed unknown marginal distribution, which is assigned a stick-breaking prior, and then using a latent Gaussian process copula model to control local allocation to the atoms. The LaSBP avoids problems faced by the hybrid DP in performing spatial interpolations by introducing an order constraint on the atoms. This constraint also induces skewness, which is appealing in many applications.

Friedman and Meulman (2004) defined a concept of clustering on subsets of attributes (COSA) in which two subjects can be partially clustered by having identical values for a subset of a vector of parameters. Hoff (2006) proposed a simple but clever Bayesian nonparametric version of COSA relying on a DP. In particular, if $\boldsymbol{\beta}_i = (\beta_{i1}, \ldots, \beta_{ip})'$ represents a vector of subject-specific parameters in a hierarchical model, then Hoff (2006) first lets $\boldsymbol{\beta}_i = \boldsymbol{\beta} + \mathbf{r}_i \times \boldsymbol{\delta}_i$, with $\mathbf{r}_i \in \{0, 1\}^m$, $\boldsymbol{\delta}_i \in \Re^p$ and $\times$ denoting the elementwise product. Then, letting $(\mathbf{r}_i, \boldsymbol{\delta}_i) \sim P$, with $P \sim \text{DP}(\alpha P_0)$, results in clustering of the n subjects into $k \leq n$ groups. Note that subjects in a group will deviate from the baseline in only those attributes having

With this goal in mind, Rodriguez, Dunson and Gelfand (2007) proposed the nested DP (nDP), which lets

$$P_i \sim \sum_{h=1}^{\infty} \pi_h \delta_{P_h^*}, \qquad P_h^* \overset{\text{i.i.d.}}{\sim} \text{DP}(\gamma P_0), \tag{7.19}$$

where the weights $\boldsymbol{\pi} = \{\pi_h\}_{h=1}^{\infty}$ are as defined in (7.2). Note that this specification allows P_i to be exactly equal to $P_{i'}$ with prior probability $1/(1+\alpha)$. When $P_i = P_{i'}$, the joint distribution of gad and bw in center i is identical to that for center i', and these centers are clustered together. Hence, unlike the HDP, which clusters patients within and across study centers while assuming study centers are distinct, the nDP allows clustering of both study centers and patients within centers. Such clustering may be of direct interest or may be simply a tool for flexible borrowing of information in estimating the center-specific distributions.

7.6 Flexible modeling of conditional distributions

7.6.1 Motivation

In many applications, it is of interest to study changes in the distribution of a response variable Y over time, for different spatial locations, or with variations in a vector of predictors, $\mathbf{x} = (x_1, \ldots, x_p)'$. Depending on the study design, the primary interest may be

(i) prediction of a health response Y for a new subject given demographic and clinical predictors for that subject;
(ii) inference on the impact of time, space or predictors on the conditional response distribution;
(iii) inverse regression problems involving identification of predictor values associated with an adverse health response;
(iv) clustering of subjects based on their health response while utilizing predictor information.

The methods reviewed in Section 7.5 can be used to address these interests when there is a single unordered categorical predictor, such as the study center. However, alternative methods are needed in the general case. This section reviews some approaches for modeling of predictor-dependent collections of distributions through the use of mixture models incorporating priors for collections of dependent random probability measures indexed by predictors. In particular, let

$$P_{\mathcal{X}} = \{P_{\mathbf{x}} : \mathbf{x} \in \mathcal{X}\} \sim \mathcal{P}, \tag{7.20}$$

where $P_{\mathbf{x}}$ denotes the random probability measure at location (predictor value) $x \in \mathcal{X}$, $\mathcal{X}$ is the sample space for the predictors, and $\mathcal{P}$ is the prior for the collection, $P_{\mathcal{X}}$. Here, I use the term predictor broadly to refer also to time and spatial location.

There are a wide variety of applications in which it is useful to incorporate priors for dependent collections of distributions. For example, one may be interested in modeling of conditional densities, $f(y \mid \mathbf{x})$. Revisiting the reproductive epidemiology application from Section 7.5, suppose that y_i is the gestational age at delivery for woman i and $\mathbf{x}_i = (x_{i1}, \ldots, x_{ip})'$ is a vector of predictors, including age of the woman and blood level of DDE, a persistent metabolite of the pesticide DDT. Then, it is of interest to assess how the risk of premature delivery, corresponding to the left tail of the distribution of gestational age at delivery, changes with increasing DDE exposure adjusting for age. In making such assessments, it is appealing to limit parametric assumptions, and avoid the common epidemiologic practice of reducing information on gestational age at delivery (gad) to a 0/1 indicator of gad ≤ 37 weeks.

To solve this problem, one can consider a mixture model of the form

$$f(y \mid \mathbf{x}) = \int \int g(y; \mathbf{x}, \boldsymbol{\theta}, \boldsymbol{\phi}) dP_{\mathbf{x}}(\boldsymbol{\theta}) d\pi(\boldsymbol{\phi}), \tag{7.21}$$

where $g(y; \mathbf{x}, \boldsymbol{\theta}, \boldsymbol{\phi})$ is a parametric model for the conditional density of y given $\mathbf{x}$, and (7.21) allows deviations from this parametric model through nonparametric mixing. Expression (7.21) contains both parametric and nonparametric components, with the prior distribution for the parameters $\boldsymbol{\phi}$ treated as known, while the mixture distribution for $\boldsymbol{\theta}$ is nonparametric and predictor dependent. Some possibilities for $g(y; \mathbf{x}, \boldsymbol{\theta}, \boldsymbol{\phi})$ include $\mathrm{N}(y; \mu, \sigma^2)$, with $\boldsymbol{\theta} = (\mu, \sigma^2)$, and $\mathrm{N}(y; \mathbf{x}'\boldsymbol{\beta}, \sigma^2)$, with $\boldsymbol{\theta} = (\boldsymbol{\beta}', \sigma^2)$. The advantage of incorporating a regression component in $g(y; \mathbf{x}, \boldsymbol{\theta}, \boldsymbol{\phi})$ is that a sparse structure is then favored through centering on a base parametric regression model. In this context, a sparser specification allows subjects to be allocated to few mixture components, while maintaining flexibility. In addition, we allow for interpolation across sparse data regions through the base parametric model, addressing the curse of dimensionality.

7.6.2 Dependent Dirichlet processes

One possibility for $\mathcal{P}$ in (7.20) is the dependent DP (DDP) originally proposed by MacEachern (1999, 2001; see also De Iorio, Müller, Rosner and MacEachern, 2004). In full generality, the DDP is specified as follows:

$$P_{\mathbf{x}} = \sum_{h=1}^{\infty} \pi_h(\mathbf{x}) \delta_{\Theta_h(\mathbf{x})}, \quad \Theta_h \overset{\text{i.i.d.}}{\sim} P_0, \quad \forall \mathbf{x} \in \mathcal{X}, \tag{7.22}$$

where $\pi_h(\mathbf{x}) = V_h(\mathbf{x}) \prod_{l<h} \{1 - V_l(\mathbf{x})\}$, for $h = 1, \ldots, \infty$, with the stick-breaking weights $\{V_h(\mathbf{x})\}_{h=1}^{\infty}$, at any fixed $\mathbf{x}$, consisting of independent draws from a Beta$(1, \alpha)$

distribution. In addition, Θ_h is a stochastic process over $\mathcal{X}$ generated from P_0. For example, P_0 may correspond to a Gaussian process.

Due to complications involved in allowing the weights to depend on predictors, most applications of the DDP have assumed fixed weights, resulting in the specification

$$P_{\mathbf{x}} = \sum_{h=1}^{\infty} \pi_h \delta_{\Theta_h(\mathbf{x})}, \quad \Theta_h \overset{\text{i.i.d.}}{\sim} P_0, \quad \forall \mathbf{x} \in \mathcal{X}, \tag{7.23}$$

where $\boldsymbol{\pi} = \{\pi_h\}_{h=1}^{\infty}$ are as defined in (7.2). De Iorio, Müller, Rosner and MacEachern (2004) applied this specification to develop ANOVA-type models for collection of dependent distributions. Gelfand, Kottas and MacEachern (2005) applied the DDP in spatial data analysis applications.

One can also use the fixed π DDP in (7.23) to develop a method for conditional density modeling as in (7.21) by letting

$$f(y \mid \mathbf{x}) = \sum_{h=1}^{\infty} \pi_h \mathrm{N}(y; \mu_h(\mathbf{x}), \sigma_h^2), \quad (\mu_h, \sigma_h^2) \sim P_0 = P_{01} \otimes P_{02} \tag{7.24}$$

where P_{01} is a Gaussian process over $\mathcal{X}$ and P_{02} is a probability measure on $\Re^+$ (e.g., corresponding to an inverse-gamma distribution). Note that (7.24) characterizes the conditional density using an infinite mixture of normals, with the component means varying differentially and nonlinearly with predictors. This specification is a generalization of typical Gaussian process regression models, which would correspond to letting $\pi_1 = 1$, so that the mean varies flexibly while the residual density is assumed to be constant and Gaussian. In contrast, the DDP mixture of normals in (7.24) allows multimodal residual densities that vary with $\mathbf{x}$.

It is useful to consider the gestational age at delivery application. In that case, there are likely a few dominant mixture components that are assigned most of the probability weight, with these components corresponding to early preterm birth, preterm birth and full term birth. Specification (7.24) assumes that the probability allocated to these components does not vary with age or DDE, but the locations of the component can vary. For example, the mean of the preterm birth component can shift to earlier weeks as DDE level increases to allow an increasing proportion of babies born too soon at higher exposures.

From this example, it is clear that the fixed π DDP may be overly restrictive in that biologic constraints on the timing of gestation make it more realistic and interpretable to consider models with fixed locations but weights that depend on predictors. For example, if we have three dominant components corresponding to early preterm, preterm and full term, then a varying weights model would allow the probability of early preterm birth to increase as DDE increases without necessarily changing the timing and hence the meaning of the early preterm component. In addition, a varying weights model is necessary to allow the probability that two subjects are assigned to the same cluster to depend on predictors.

Motivated by such considerations, Griffin and Steel (2006) proposed an order-based dependent DP, referred to as the π-DDP. The π-DDP allows for predictor-dependent weights in the DDP in a clever manner by allowing the ordering in the stick-breaking weights to depend on predictors. In more recent work, Griffin and Steel (2007) proposed a simplification of the π-DDP, which they used for modeling of the residual component in a flexible regression model. The resulting approach is referred to as the Dirichlet process regression smoother (DPRS).

An alternative approach that was developed for spatial applications in which it is appealing to allow the weights to vary spatially, was proposed by Duan, Guindani and Gelfand (2007). This approach places a stochastic process on the weights, which is carefully specified so that the marginal distributions at any fixed spatial location maintain the DP stick-breaking form. De la Cruz-Mesia, Quintana and Müller (2007) recently proposed an alternative extension of the DDP for classification based on longitudinal markers. Their approach incorporated dependence in the random effects distribution across groups.

7.6.3 Kernel-based approaches

As an alternative to the DDP, Dunson, Pillai and Park (2007) proposed an approach based on kernel-weighted mixtures of independent DP basis components. This approach is conceptually related to the kernel regression approach described in Section 7.3.4 and expression (7.12). However, instead of specifying a prior for a single function, the goal is to specify a prior for an uncountable collection of predictor-dependent random probability measures. To motivate the general approach, first consider the case in which there is an single continuous predictor with support on [0,1] and interest focuses on modeling the conditional density $f(y|x)$, for all $x \in [0, 1]$. In this case, a simple model would let

$$f(y|x) = \frac{1}{K(x,0) + K(x,1)}\{K(x,0)f_0^*(y) + K(x,1)f_1^*(y)\}, \tag{7.25}$$

where K is a kernel, with $K(x,x) = 1$ and $K(x,x')$ decreasing as the distance between x and x' increases, $f_0^*(y)$ is an unknown basis density located at $x = 0$, and $f_1^*(y)$ is an unknown basis density located at $x = 1$. For example, K may correspond to a Gaussian kernel, and f_0^*, f_1^* to DP mixtures of normals.

Figure 7.5 provides an example of the behavior of (7.25), letting f_0^* correspond to the standard normal density, f_1^* to a mixture of two normals, and K to the Gaussian kernel with standard deviation 0.3. As should be clear, using the mixture

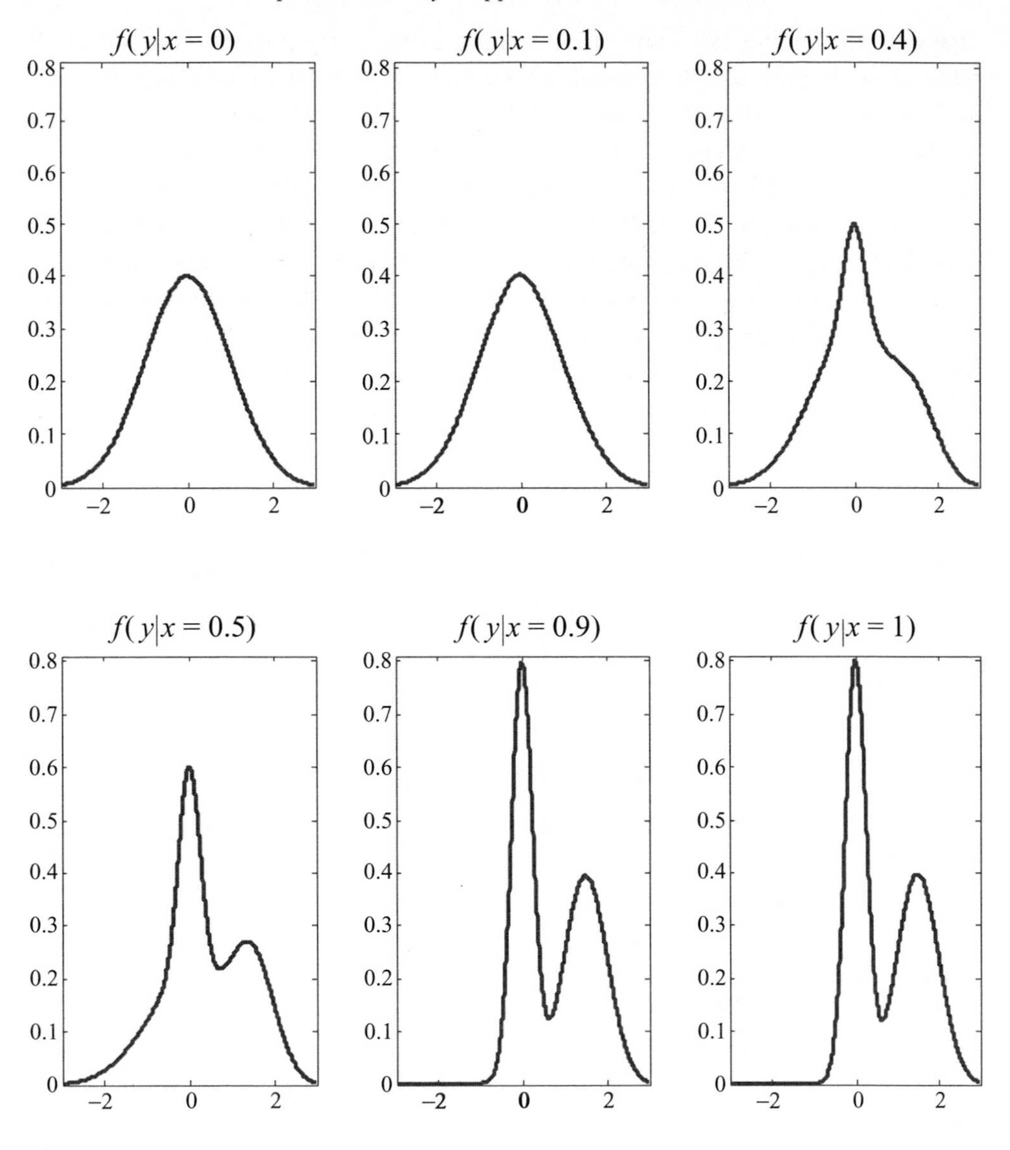

Figure 7.5 Plot illustrating the kernel mixtures approach in the simple case in which $x_i \in \{0, 1\}$ for all subjects in the sample, and a normal kernel with standard deviation 0.3 is chosen.

model (7.25) results in a smoothly morphing profile of conditional densities, with the densities very close for similar predictor values. In order to generalize the kernel mixture approach of expression (7.25) to allow multiple predictors and more flexibility, one can potentially allow unknown numbers and locations of basis densities, similarly to allowing unknown numbers and locations of knots in a spline model.

Dunson, Pillai and Park (2007) instead proposed a default approach in which DP bases were automatically placed at the sample predictor values, with each of these bases assigned a weight controlling its importance. In particular, their proposed weighted mixture of DPs (WMDP) prior had the form

$$P_{\mathbf{x}} = \sum_{i=1}^{n} \left(\frac{\gamma_i K(\mathbf{x}, \mathbf{x}_i)}{\sum_{l=1}^{n} \gamma_l K(\mathbf{x}, \mathbf{x}_l)} \right) P_i^*, \quad P_i^* \overset{\text{i.i.d.}}{\sim} \text{DP}(\alpha P_0), \quad \forall \mathbf{x} \in \mathcal{X}, \tag{7.26}$$

where $i = 1, \ldots, n$ indexes subjects in the sample, $K : \mathcal{X} \times \mathcal{X} \to 1$ is a bounded kernel (e.g., Gaussian), $\boldsymbol{\gamma} = (\gamma_1, \ldots, \gamma_n)'$ is a vector of weights, and $\{G_i^*\}_{i=1}^n$ are independent DP basis measures.

Dunson, Pillai and Park (2007) applied the WMDP approach for density regression through use of the model

$$f(y \mid \mathbf{x}) = \int \int \text{N}(y; \mathbf{x}'\boldsymbol{\beta}, \sigma^2) \mathrm{d}P_{\mathbf{x}}(\boldsymbol{\beta}) \mathrm{d}\pi(\sigma^2), \tag{7.27}$$

which is a mixture of linear regression models with predictor-dependent weights. This model is related to hierarchical mixtures-of-experts models (Jordan and Jacobs, 1994), which are widely used in the machine learning literature, but instead of assuming a finite number of experts (i.e., mixture components), the number of experts is infinite. In addition, instead of a probabilistic decision tree for the weights, a kernel model is used.

From a Bayesian perspective, (7.26) has an unappealing sample-dependence property, so that the approach is not fully Bayesian and lacks coherent updating and marginalization properties. For this reason, it is useful to consider alternative priors that borrow some of the positive characteristics of the kernel-weighted specification but without the sample dependence. One such prior is the kernel stick-breaking process (KSBP) (Dunson and Park, 2008), which modifies the DP stick-breaking specification shown in (7.2) as follows:

$$P_{\mathbf{x}} = \sum_{h=1}^{\infty} V_h K_{\psi_h}(\mathbf{x}, \Gamma_h) \prod_{l<h} \{V_l K_{\psi_l}(\mathbf{x}, \Gamma_l)\} P_h^*, \quad P_h^* \overset{\text{i.i.d.}}{\sim} \text{DP}(\alpha P_0), \tag{7.28}$$

where $V_h \overset{\text{i.i.d.}}{\sim} \text{Beta}(1, \lambda)$, K_ψ denotes a kernel having bandwidth ψ, $\{\psi_h\}_{h=1}^\infty$ is an infinite sequence of kernel bandwidths sampled from G, and $\{\Gamma_h\}_{h=1}^\infty$ is an infinite sequence of kernel locations sampled from H.

The KSBP places random basis probability measures at an infinite sequence of random locations, with the weights assigned to these bases in the formulation for $P_{\mathbf{x}}$ decreasing stochastically with the index h and the distance from $\mathbf{x}$. In using the KSBP as a prior for the mixture distributions in (7.27), one obtains a flexible, sparseness-favoring structure for conditional distribution modeling. In particular, if the linear regression model provides a good approximation, then the tendency is to assign most of the weight to few components and effectively *collapse* back to the base parametric normal linear model. This is seen quite dramatically in simulations under the normal linear model in which results obtained in the KSBP mixture analysis and the parametric model analysis are very similar. However, the KSBP mixture model does an excellent job at adaptively capturing dramatic deviations from the normal linear model even when all the subjects are assigned to a modest number of basis locations (e.g., fewer than 10).

The sparseness-favoring property is crucial in obtaining good performance, since the curse of dimensionality makes it very difficult to estimate conditional distributions reliably, even with few predictors. One does not obtain nearly as good performance using a KSBP mixture of Gaussians without the regression structure in the Gaussian kernel as in (7.27). This is partly due to the fact that the base normal linear model allows one to interpolate across sparse data regions much more reliably than a normal with mean that does not vary with predictors.

From a practical perspective, the KSBP can be implemented easily in a wide variety of settings using a simple MCMC algorithm, and produces flexible but sparse models of conditional distributions. In addition, the KSBP results in predictor-dependent clustering of subjects. In the conditional density estimation setting, one can also obtain estimates of quantile regression functions directly from the MCMC output. The KSBP has been used for classification, multitask learning and modeling of multivariate count data in unpublished work.

7.6.4 Conditional distribution modeling through DPMs

Prior to the work on DDPs and kernel-based approaches for modeling of dependent collections of random probability measures, Müller, Erkanli and West (1996) proposed a simple and clever approach for inducing a prior on $\mathrm{E}(y \mid \mathbf{x})$ through a joint DP mixture of Gaussians model for $\mathbf{z} = (y, \mathbf{x}')'$. Although they did not consider conditional density estimation, this approach also induces a prior on $f(y \mid \mathbf{x})$, for all $y \in \Re, \mathbf{x} \in \Re^p$. In recent work, Rodriguez, Dunson and Gelfand (2009b) showed that the Müller, Erkanli and West (1996) approach results in pointwise consistent estimates of $\mathrm{E}(y \mid \mathbf{x})$ under some mild conditions, and the approach can be adapted for functional data analysis.

Given that the Müller, Erkanli and West (MEW) (1996) approach can be implemented routinely through direct use of software for Bayesian multivariate density estimation using DP mixtures of Gaussians, a natural question is what is gained by using the approaches described in Sections 7.6.2 and 7.6.3 for conditional density estimation. To address this question, I start by contrasting the two types of approaches. The methods described in Sections 7.6.2 and 7.6.3 provide priors for collections of unknown distributions indexed by spatial location, time and/or predictors. The resulting specification is defined conditionally on the predictors (or time/space). In contrast, the MEW approach relies on joint modeling of the predictors and response to induce a prior on the conditional distributions. Clearly, this only makes sense if the predictors can be considered as random variables, which rules out time, space and predictors that correspond to design points. However, for many observational studies, all the predictors can be considered as random variables.

In such settings, the MEW approach is still conceptually quite different from the conditional approaches of Sections 7.6.2 and 7.6.3. All the approaches induce clustering of subjects, and the MEW approach is similar to the π-DDP, WMDP and KSBP in allowing predictor-dependent clustering. The predictor-dependent clustering of the MEW approach arises through joint clustering of the response and predictors. This implies that the MEW approach will introduce new clusters to fit better the distribution of the predictors even if such clusters are not necessary from the perspective of producing a good fit to the response data. For example, in simulations from a normal linear regression model in which the predictors have non-Gaussian continuous densities, we have observed a tendency of the MEW approach to allocate individuals to several dominant clusters to fit better the predictor distribution. In contrast, the KSBP will tend to collapse on a single dominant cluster containing all but a few subjects in such cases, as a single cluster is sufficient to fit the response distribution, and the KSBP does not model the predictor distribution.

Although the sensitivity of clustering to the predictor distribution has some unappealing characteristics, there are some potential benefits. Firstly, the approach is useful for inverse regression and calibration problems. Griffin and Holmes (2007) recently proposed a MEW-type approach for Bayesian nonparametric calibration in spatial epidemiology. Also, in providing a joint nonparametric model for predictors and response variables, the approach can automatically accommodate predictors that are missing at random through a simple step for imputing these predictors from their full conditional posterior distributions during the MCMC algorithm. Finally, instead of relying entirely on information in the response distribution, the approach tends to change the slope of the regression automatically in regions of the predictor space at which there is a change in the predictor distribution. This may be particularly appealing in semisupervised learning settings in which the responses (labels) are only available for a small subset of the subjects. In order to allow categorical responses and/or predictors, the MEW approach can be easily adapted to use an underlying normal DPM of Gaussians.

7.6.5 Reproductive epidemiology application

As an illustration, consider an application to modeling of gestational age at delivery (gad) data from a study of Longnecker et al. (2001) using the KSBP. In epidemiologic studies of premature delivery, it is standard practice to dichotomize the data on gad using 37 weeks as the cutoff for defining preterm delivery. Then, the risk of preterm birth is modeled as a function of exposures of interest and potential confounders using a logistic regression model. Similar analyses are common in studies that collect a continuous health response when the interest focuses on the effect of predictors on the risk of an adverse response. For example, in assessing factors predictive of obesity, one typically dichotomizes body mass index (bmi) to obtain a 0/1 indicator of obesity instead of analyzing the right tail of the bmi distribution. In such settings, adverse responses correspond to values in one or both of the tails of the response density, and it is certainly appealing to avoid sensitivity to arbitrary cutoffs.

Typical regression models that focus on predictor effects on the mean or median of the distribution are clearly not appropriate for characterizing changes in the tails unless the residual distribution is truly homoscedastic. In epidemiology it is often not plausible biologically for the residual distribution to satisfy such an assumption. In particular, most health conditions are multifactorial, having both genetic and environmental risk factors, with many or most of these factors unmeasured. Hence, in the presence of interactions between measured and unmeasured predictors, we would really expect the response distribution to change in shape as the values of the predictors vary. Biological constraints on values of the response also play a role.

In the gestational age at delivery setting, such biological constraints make it very unlikely that gestation continues much beyond 45 weeks, because the baby is getting so large at that point. Hence, the risk factors for premature delivery are very unlikely to result simply in a shift in the mean gad, but are more likely to impact the relative proportions of women having gads in the early preterm, preterm or full term intervals. Hence, a mixture of three normals, with the predictor-dependent weights, may be natural for modeling of the gad distribution. However, as we are not certain a priori that three components are sufficient, it is appealing to consider a nonparametric Bayes approach that allows infinitely many components in the general population, while allowing predictor-dependent weights. The KSBP mixture models proposed by Dunson and Park (2008) provide such a framework.

Note that when the focus is on predictor effects on the tails of a distribution, one can potentially apply quantile regression methods. However, most quantile regression methods allow for modeling of a single arbitrarily chosen quantile (e.g., the 10th percentile) instead or providing a general framework for modeling of all quantiles coherently. By using a nonparametric Bayes conditional distribution modeling approach, one can do inferences on shifts in the entire tail of the distribution instead of focusing narrowly on a selected quantile.

Returning to the Longnecker et al. (2007) application, there were 2313 women in the study, and we implemented the KSBP for model (7.27), with y_i = gad and $\mathbf{x}_i = (1, \text{dde}_i, \text{age}_i)'$. For the reasons discussed in detail at the end of Section 2.1, we normalized y_i, dde_i, age_i prior to analysis, and then chose the base measure in the KSBP to correspond to the unit information-type prior, $N_2(\mathbf{0}, (\mathbf{X}'\mathbf{X})^{-1}/n)$. In addition, we fixed $\alpha = 1$ and $\lambda = 1$ to favor a few clusters, and used a Gaussian kernel, with a single kernel precision, $\psi_h = \psi$, assigned a log-normal prior. The retrospective MCMC algorithm described in Dunson and Park (2008) was implemented, with 22000 iterations collected after a burn-in of 8000 iterations.

It is worth commenting on convergence assessments for nonparametric Bayes mixture models, with the KSBP providing one example. In particular, due to the label switching problem, the index on the different mixture components can change meaning across the MCMC iterates. This leads to apparently poor mixing in many cases when one examines traces plots of cluster-specific parameters. However, in the analysis of the Longnecker et al. (2001) pregnancy outcome data and in general applications in which the focus is on density estimation, conditional density estimation or inferences on unknowns that are not cluster specific, one could compellingly argue that the label-switching problem is in fact not a problem at all. It is very commonly the case that trace plots of unknowns that are not cluster specific exhibit good rates of convergence and mixing properties, while trace plots for cluster-specific parameters show high autocorrelation and fail standard convergence tests. As long as one is not interested in cluster-specific estimates and inferences, this is no cause for concern, as it just represents the cluster index identifiability problem. In KSBP applications, we monitored the value of the conditional density at different locations, and observed good mixing and rapid apparent convergence.

Figure 7.6 shows the raw data on dde and gad for the women in the Longnecker et al. (2001) study. The solid line is the posterior mean of the expected value of gad conditionally on dde. This illustrates that the KSBP mixture model induces a flexible nonlinear mean regression model, while also allowing the residual distribution to vary with predictors. The dotted lines correspond to 99% pointwise credible intervals. There is an approximately linear decrease in the mean gad, with the credible intervals becoming considerably wider for high dde values where data are sparse.

Figure 7.7 shows the estimated conditional densities of gad in days for different dde values, with posterior means shown with solid lines and 99% pointwise credible intervals shown with dashed lines. These figures suggest that the left tail corresponding to premature deliveries is increasingly fat as dde dose increases.

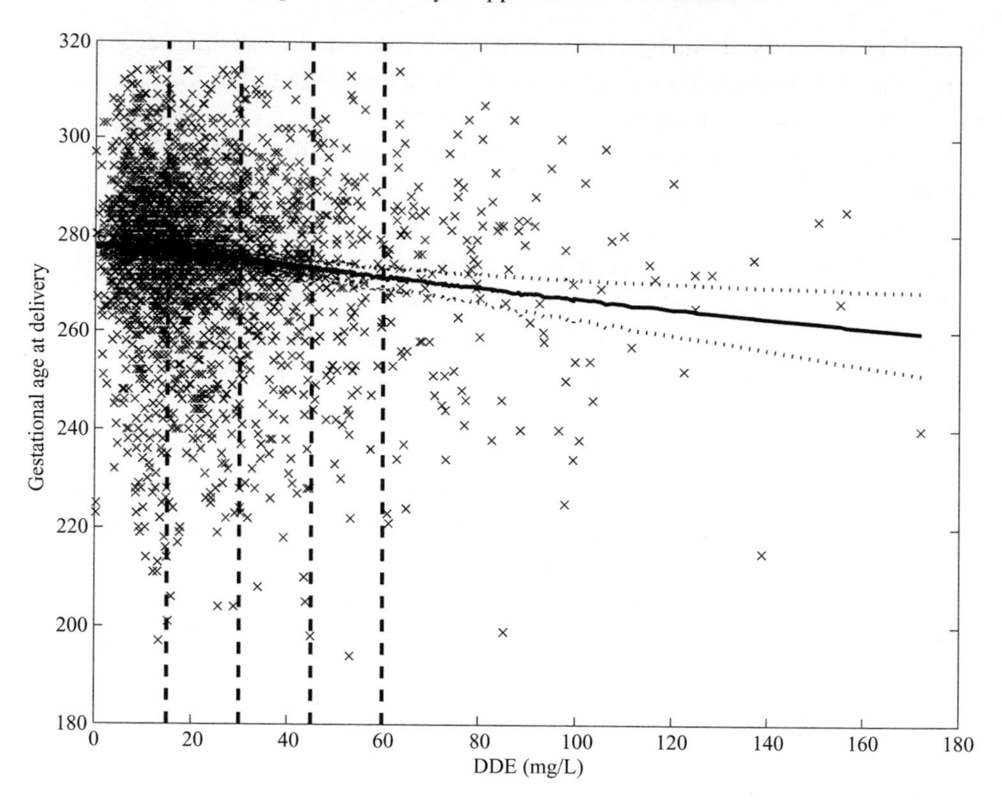

Figure 7.6 Raw data on dde and gestational age at delivery for 2313 women in the Longnecker et al. (2001) sub-study of the NCPP. Vertical dashed lines are quintiles of the empirical distribution of DDE, the solid line is the posterior mean regression curve, and the dotted lines are 99% pointwise credible intervals.

However, it is a bit hard to gauge significance of these results based on observing a sequence of conditional density estimates. Figure 7.8 provides dose response curves for the probability gad falls below different cutoffs, including (a) 33, (b) 35, (c) 37 or (d) 40 weeks. Again, solid lines are posterior means and dashed lines are 99% credible intervals. From these plots, it is clear that risk of premature delivery increases with level of DDE. It appears that DDE increases risk of early preterm birth prior to 33 weeks, which is an interesting finding given that such births represent a much more adverse response in terms of short- and long-term morbidity compared with 37 week births.

7.7 Bioinformatics

In recent years there has been a paradigm shift in biostatistics, and it is now standard to be faced with very high-dimensional data. Hence, there is clearly a

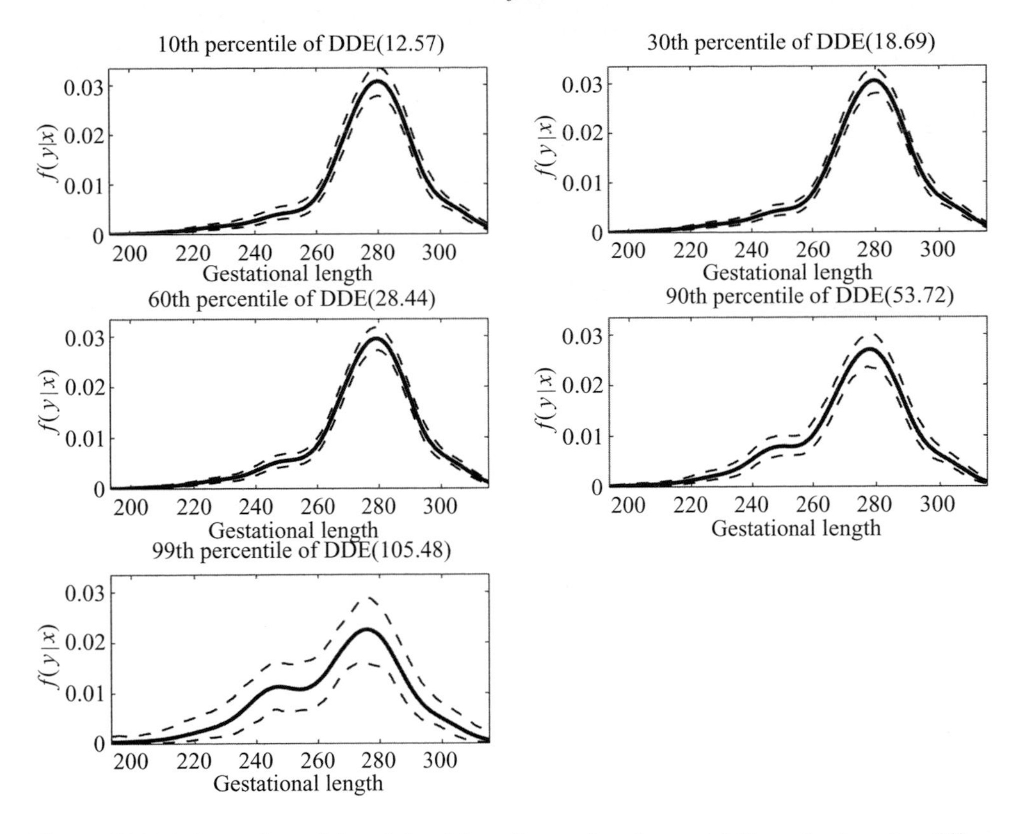

Figure 7.7 Estimated conditional densities of gestational age at delivery in days for different dde values. Solid lines are posterior means and dashed lines are 99% credible intervals.

need for automated approaches for flexible dimensionality reduction and discovery of sparse latent structure underlying very high-dimensional data. Mixture models have proven to be an extremely useful tool in such settings. For example, there is a vast literature on methods for high-dimensional variable selection using mixture priors, with one component concentrated at zero and another component being more diffuse. Nonparametric Bayes methods have proven extremely useful in this setting in providing a highly flexible framework for limiting sensitivity to arbitrary assumptions made in parametric modeling, such as knowledge of the number of mixture components. In this section, we review some of this work to give a flavor of possible applications.

7.7.1 Modeling of differential gene expression

Much of the increased interest in large p, small n problems, which started approximately a decade ago, was initiated by the development and rapidly increasing use of

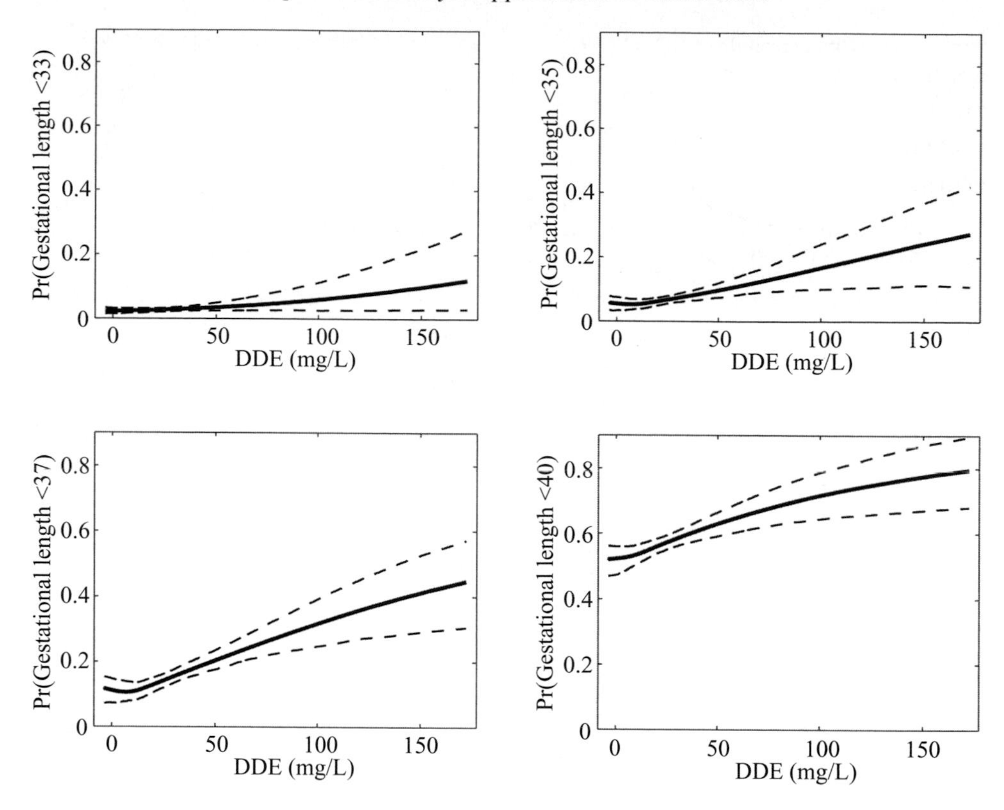

Figure 7.8 Estimated dose response curves for the probability that gestational age at delivery is less than (a) 33, (b) 35, (c) 37 or (d) 40 weeks. Solid lines are posterior means and dashed lines are 99% credible intervals.

microarray technology for measuring expression levels of large numbers of genes. There has been a particular focus on methods for clustering of gene expression profiles and for identifying genes that are differentially expressed between two groups, with these groups often representing normal and diseased or tumor tissue.

Gene expression analyses are particularly suited for nonparametric Bayes analyses due to the large sample size in terms of the number of genes and to lack of knowledge of good parametric models for approximating the joint distribution of the gene expression values. Nonparametric Bayes methods provide a flexible machinery for characterizing the complex and high-dimensional expression values, inducing a sparse latent structure through partitioning genes into clusters. However, there are some limitations relative to simpler methods in terms of ease of interpretation and computational expense.

Medvedovic and Sivaganesan (2002) proposed a Dirichlet process mixture (DPM) model for clustering genes with similar expression patterns, which they accomplished by calculating the pairwise posterior probabilities that two genes are in the same cluster from the Gibbs sampler output. Qin (2006) proposed a modified approach that relied on an iterative weighted Chinese restaurant seating scheme, designed so that the optimal number of clusters can be estimated simultaneously with assignment to clusters in an efficient manner. Medvedovic, Yeung and Bumgarner (2004) generalized the Medvedovic and Sivaganesan (2002) approach to data containing experimental replicates. Kim, Tadesse and Vannucci (2006) proposed an approach that allowed for selection of the variables to cluster upon in the DPM model. Xiang, Qin and He (2007) developed CRCView, which is a web server providing an easy to use approach for analysis and visualization of microarray gene expression data based on a DPM approach.

Do, Müller and Tang (2005) proposed a DPM of normals for the distribution of gene intensities under different conditions. Their focus was on modeling of differential gene expression between two groups, a problem which has been addressed using a variety of parametric and nonparametric empirical Bayes approaches (Newton, Noueiry, Sarkar and Ahlquist, 2004). However, Do, Müller and Tang (2005) demonstrated advantages of the fully Bayes nonparametric approach, including allowance for estimation of posterior expected false discovery rates.

7.7.2 Analyzing polymorphisms and haplotypes

In addition to gene expression data, there has been substantial interest in identifying genotypes that are predictive of an increased risk of disease. Data are now routinely collected containing the genotypes at a large number of locations (or loci) along the genome at which there is variability among individuals in the population. Such data are commonly referred to as single nucleotide polymorphisms (SNPs), with a single SNP consisting of a combination of amino acids, with one on the chromosome inherited from the mother and one from the father. Because new SNP chips allow investigators to collect data routinely from hundreds of thousands of loci, such data present quite a challenge to the statistician.

In many cases, instead of casting the net very broadly in searching for disease genes and genotypes, investigators narrow down their search to genes in a specific pathway hypothesized to play a key role in disease risk. However, even in this case, there may be many SNPs under consideration. To provide an example, Mulherin-Engel et al. (2005) related SNPs in cytokine gene regulatory regions to risk for spontaneous preterm birth. The number of loci per cytokine at which SNP data were collected ranged from 1 to 3, with 22 total across the 12 cytokines. At each loci there are 3 possible genotypes. Hence, in seeking to identify genotypes predictive of an increased risk of preterm birth, there is a very high-dimensional set of models under consideration.

Motivated by this application, Dunson, Herring and Mulherin-Engel (2008) proposed a multilevel DP prior that allowed for borrowing of information across functionally related genes, while also incorporating a variable selection component. This approach automatically grouped genotypes into null and nonnull clusters according the genotypes impact on the risk of disease. In unpublished work, we have found that this type of approach scales nicely to problems involving thousands of genotypes, providing clearly improved performance relative to parametric variable selection mixture priors. One of the reasons for this success is that most variable selection mixture priors shrink the nonnull coefficients towards zero, while the DP-based approach allows these coefficients to be shrunk towards other coefficients having similar values.

In order to address the dimensionality problem faced in searching for disease genotypes relying on high-dimensional SNP data, many articles have proposed haplotype-based analyses. Haplotypes represent a sequence of amino acids along a chromosome inherited from one parent. Due to linkage, it tends to be the case that the number of haplotypes observed in the population is substantially less than the number possible. Hence, by using haplotypes as predictors instead of the genotypes formed by a set of SNPs, one clearly obtains a reduction in dimensionality. However, the problem is that current genotyping technology does not allow SNP data to be converted into haplotypes, because the data are unphased, meaning that the amino acid pairs cannot be allocated to the chromosome for their parent of origin.

This is fundamentally a missing data problem, and a number of approaches have been proposed for imputing the missing haplotypes given the SNP data. Xing, Jordan and Sharan (2007) proposed a DPM model to address this missing data problem, with the mixture components corresponding to the pool of haplotypes in the population. Xing, Sohn, Jordan and Teh (2006) generalized this approach to the multiple population setting through the use of a hierarchical DP (Teh, Jordan, Beal and Blei, 2006). Xing and Sohn (2007) proposed an alternative approach that used a hidden Markov DP to jointly model genetic recombinations among the founders of a population and subsequent coalescence and mutation events. The goal of this approach is to identify recombination hotspots and to infer ancestral genetic patterns. An alternative approach for haplotype inference using a Bayesian hidden Markov model was independently developed by Sun, Greenwood and Neal (2007).

7.7.3 New species discovery

There has clearly been an explosion in the types of high-dimensional data generated, and nonparametric Bayes methods have seen greatly increasing use as a tool for bioinformatics. In addition to the applications presented in Sections 7.7.1 and 7.7.2, one very interesting application is to expressed sequence tag (EST) analysis. ESTs provide a useful tool for gene identification in an organism. In searching for genes using this technology, a number of interesting design questions arise. In particular, after obtaining a preliminary EST sample, scientists would like to estimate the expected number of new genes that would be detected from a further sample of a given size. Such information is critical in making decisions about the number of additional samples to sequence.

Lijoi, Mena and Prünster (2007a) addressed this problem using a nonparametric Bayes methodology for estimating the probability of discovery of a new species (Lijoi, Mena and Prünster, 2007b). In particular, Lijoi, Mena and Prünster (2007b) derived a closed form expression for a nonparametric Bayes estimator for the probability of discovery. Their approach is based on a class of priors for species sampling problems that induce Gibbs-type random partitions.

7.8 Nonparametric hypothesis testing

Most of the nonparametric Bayes literature has focused on approaches for estimation under a particular model, and there has been relatively limited consideration of hypothesis testing problems. In biomedical studies, hypothesis testing is often of primary interest. For example, in clinical trials, basic science and toxicology studies, the primary focus is typically on testing the null hypothesis of equalities in the response distribution between treatment groups against the alternative that there are differences. Estimation is often a secondary interest. In this section we review some of the work on nonparametric Bayes hypothesis testing.

Motivated by the problem of testing of equalities between groups in a study with multiple treatment groups, Gopalan and Berry (1998) proposed an approach to adjust for multiple comparisons through use of a DP prior. In particular, letting y_{hl} denote a continuous health response for the ith individual in group h, for $h = 1, \ldots, p$, let $y_{hi} \sim \mathrm{N}(\mu_h, \sigma^2)$. Then, the interest is in studying local hypotheses:

$$\mathrm{H}_{0,hl} : \mu_h = \mu_l \quad \text{versus} \quad \mathrm{H}_{1,hl} : \mu_h \neq \mu_l.$$

Clearly, the number of such hypotheses increases rapidly with p, so one faces a multiple comparisons problem. Gopalan and Berry (1998) proposed letting $\mu_h \overset{\text{i.i.d.}}{\sim} P$, with $P \sim \mathrm{DP}(\alpha P_0)$. Then, from basic properties of the DP, $\Pr(\mathrm{H}_{0,hl}) = 1/(1+\alpha)$. Using standard algorithms for posterior computation in DPM models, one can obtain estimates of the posterior hypothesis probabilities.

Motivated by epidemiologic studies having many predictors that may be highly correlated, MacLehose, Dunson, Herring and Hoppin (2007) proposed an alternative approach which assumed that the regression parameters for the different predictors in a generalized linear model, $\beta_1, \ldots, \beta_p$, were sampled from the prior:

$$\beta_j \stackrel{\text{i.i.d.}}{\sim} P = \pi_0 \delta_0 + (1 - \pi_0) P^*, \qquad P^* \sim \text{DP}(\alpha P_0), \tag{7.29}$$

where π_0 is the prior probability that the jth predictor has a zero coefficient and can thus be excluded from the model. This approach allows one to calculate posterior probabilities of $\text{H}_{0j} : \beta_j = 0$ versus $\text{H}_{0j} : \beta_j \neq 0$, while clustering the β_j for the nonnull predictors. MacLehose, Dunson, Herring and Hoppin (2007) demonstrated improved performance relative to parametric variable selection priors that replace P^* with a normal distribution.

Note that, whenever using Bayes factors and posterior probabilities as a basis for hypothesis testing, it is important to keep in mind the well-known sensitivity to the prior. This sensitivity occurs regardless of whether one is considering a parametric or nonparametric model. Using expression (7.29) as an example, note that commonly used parametric variable selection mixture priors would let $P^* \equiv P_0$, with P_0 chosen to correspond to a normal or heavier tailed density centered at zero. In this parametric special case of (7.29), choosing a very high variance P_0 will tend to lead to placing high posterior probability on small models having the predictors excluded. In the nonparametric case, this same behavior will occur for very high variance P_0. Hence, far from being noninformative, a high variance P_0 instead corresponds to a very informative prior that overly-favors small models. Motivated by applications to logistic regression selection in genetics and epidemiology, MacLehose, Dunson, Herring and Hoppin (2007) proposed to address the problem of specification of P_0 by using an informative choice, motivated by prior knowledge of plausible ranges for predictor effects. In the absence of such knowledge, one can potentially use default choices of P_0 that are used in parametric variable selection settings, though the theoretical implications of such choices remain to be fully evaluated.

The MacLehose, Dunson, Herring and Hoppin (2007) approach utilizes a DP prior for dimensionality reduction and clustering of the parameters in a parametric model, while assuming that the response distribution belongs to a parametric family. In order to compare two different nonparametric models, one can potentially calculate the marginal likelihoods for each model and then obtain a Bayes factor. For example, suppose that an experiment is run having two groups, and the focus is on testing equalities in the distributions between the two groups without assuming a parametric model for the response distribution or for the types of changes between the two groups. Then, one could fit a null model, which combines the two groups and uses a DPM of Gaussians to characterize the unknown response distribution, and an alternative model, which uses separate DPMs of Gaussians for the two groups. Using the method of Basu and Chib (2003) for estimating marginal likelihoods for DPMs, a Bayes factor can then be obtained.

When there are multiple treatment groups, such an approach faces practical difficulties, since it involves running separate MCMC algorithms for each comparison of interest. In addition, if the treatment groups correspond to increasing doses, then it is appealing to apply an approach that borrows information across the groups in characterizing the unknown dose group-specific distributions. Pennell and Dunson (2008) proposed such an approach based on a dynamic mixture of DPs, which allows adjacent dose groups to be effectively identical. Dunson and Peddada (2008) proposed an alternative approach for testing of partial stochastic ordering among multiple groups using a restricted dependent DP.

In parametric models, Bayesian hypothesis testing and model selection makes the assumption that one of the models under consideration is true. Such an assumption is often viewed as unrealistic, since it seems unlikely that any parametric model is more than a rough approximation of the truth. Walker and Gutièrrez-Peña (2007) proposed an approach that allows for coherent comparisons of parametric models, while allowing for the fact that none of the models under consideration is true by modeling the truth using a nonparametric Bayes approach. There has also been some focus in the literature on using a nonparametric Bayes alternative for goodness-of-fit testing of a parametric model. To derive Bayes factors, Carota and Parmigiani (1996) embedded the parametric model in a nonparametric alternative characterized as a mixture of Dirichlet processes. They obtained disturbing results showing that results are entirely driven by the occurrence of ties. DP mixtures or Pólya trees can be used to bypass this problem, with Berger and Guglielmi (2001) providing a Pólya tree-based approach.

7.9 Discussion

This chapter has provided a brief overview of some recent nonparametric Bayes work that is motivated by or relevant to biostatistical applications. The literature on the intersection between nonparametric Bayes and biostatistics is increasingly vast, so I have focused on only a narrow slice of this literature. In focusing on approaches that explicitly incorporate random probability measures, such as the Dirichlet process and extensions, I have ignored a rich literature on nonparametric Bayes methods for survival analysis. In addition, there are certainly many extremely interesting papers that I have overlooked, have not had space to cover, or appeared after completing this chapter.

Nonparametric Bayes is clearly a new and exciting field, with many important problems remaining to be worked out. One clear problem is computational time and complexity. For example, noting the great potential of the Bayesian approach in pattern recognition for bioinformatics, Corander (2006) also notes the limitation of current Bayesian computational strategies in high-dimensional settings, calling for the evolution of Bayesian computational strategies to meet this need. A number of fast approximations to the posterior have been proposed for DPMs and other nonparametric Bayes models, relying on variational approximations and other strategies. However, such approximations often perform poorly, and there is no general theory or methodology available for assessing approximation accuracy. Some of these issues have been covered in detail in Chapter 5.

Another issue is hyperparameter choice. In having infinitely many parameters, nonparametric Bayes methods tend to be heavily parameterized and to require specification of a number of hyperparameters. This subjective component is somewhat counter to the nonparametric spirit of avoiding assumptions, and the hyperparameters can have a subtle role, making them difficult to elicit. For this reason, it is appealing to have new approaches for subjective prior elicitation that allow for the incorporation of scientific and background knowledge. Typically, such knowledge does not take the form of guesses at parameter values, but instead may be represented as functional constraints or knowledge of plausible values for functions of the parameters. It is also appealing to have default priors approaches available for routine use.

There is also a pressing need for new methods and ways of thinking about model selection in nonparametric Bayes. For example, for a given problem (e.g., conditional distribution estimation), one can write down many different nonparametric Bayes models, so how does the applied statistician go about choosing between the different possibilities? Often such a choice may be pragmatic, motivated by the available software for implementation and the plausibility of the results that are produced. However, it would be appealing to have formal methods for routinely comparing competing approaches in terms of goodness-of-fit versus parsimony, asymptotic efficiency and other criteria. It may be that different formulations are primarily distinguished by sparseness-favoring properties, with "sparseness-favoring" in this context meaning that the approach favors using relatively few parameters in characterizing the distributions and functions of interest.

References

Basu, S. and Chib, S. (2003). Marginal likelihood and Bayes factors for Dirichlet process mixture models. *Journal of the American Statistical Association*, **98**, 224–35.

Berger, J. O. and Guglielmi, A. (2001). Bayesian and conditional frequentist testing of a parametric model versus nonparametric alternatives. *Journal of the American Statistical Association*, **96**, 174–84.

Bigelow, J. L. and Dunson, D. B. (2009). Bayesian semiparametric joint models for functional predictors. *Journal of the American Statistical Association*, **104**, 26–36.

Blackwell, D. and MacQueen, J. B. (1973) Ferguson distributions via Pólya urn schemes. *Annals of Statistics*, **1**, 353–55.

Blei, D. M. and Jordan, M. I. (2006). Variational inference for Dirichlet process mixtures. *Bayesian Analysis*, **1**, 121–44.

Brumback, B. A. and Rice, J. A. (1998). Smoothing spline models for the analysis of nested and crossed samples of curves. *Journal of the American Statistical Association*, **93**, 961–76.

Burr, D. and Doss, H. (2005). A Bayesian semiparametric model for random-effects meta analysis. *Journal of the American Statistical Association*, **100**, 242–51.

Bush, C. A. and MacEachern, S. N. (1996). A semiparametric Bayesian model for randomised block designs. *Biometrika*, **83**, 275–85.

Carota, C. (2006). Some faults of the Bayes factor in nonparametric model selection. *Statistical Methods and Application*, **15**, 37–42.

Carota, C. and Parmigiani, G. (1996). On Bayes factors for nonparametric alternatives. In *Bayesian Statistics 5*, ed. J. M. Bernardo et al., 507–11. Oxford: Oxford University Press.

Chakraborty, S., Ghosh, M. and Mallick, B. (2005). Bayesian non-linear regression for large *p*, small *n* problems. *Journal of the American Statistical Association*, under revision.

Corander, J. (2006). Is there a real Bayesian revolution in pattern recognition for bioinformatics? *Current Bioinformatics*, **1**, 161–65.

Crandell, J. L. and Dunson, D. B. (2009). Posterior simulation across nonparametric models for functional clustering. *Sankhya*, to appear.

Dahl, D. B. (2006). Model-based clustering for expression data via a Dirichlet process mixture model. In *Bayesian Inference for Gene Expression and Proteomics*, ed. K.-A. Do, P. Müller and M. Vannucci, 201–18. Cambridge: Cambridge University Press.

Dahl, D. B. (2007). Sequentially-allocated merge-split sampler for conjugate and nonconjugate Dirichlet process mixture models. *Journal of Computational and Graphical Statistics*, under revision.

De Iorio, M., Müller, P., Rosner, G. L. and MacEachern, S. N. (2004). An ANOVA model for dependent random measures. *Journal of the American Statistical Association*, **99**, 205–15.

De la Cruz-Mesia, R., Quintana, F. A. and Müller, P. (2007). Semiparametric Bayesian classification with longitudinal markers. *Applied Statistics*, **56**, 119–37.

Do, K.-A., Müller, P. and Tang, F. (2005). A Bayesian mixture model for differential gene expression. *Applied Statistics*, **54**, 627–44.

Duan, J., Guindani, M. and Gelfand, A. E. (2007). Generalized spatial Dirichlet process models. *Biometrika*, **94**, 809–25.

Dunson, D. B. (2006). Bayesian dynamic modeling of latent trait distributions. *Biostatistics*, **7**, 551–68.

Dunson, D. B., Herring, A. H. and Mulherin-Engel, S. M. (2008). Bayesian selection and clustering of polymorphisms in functionally related genes. *Journal of the American Statistical Association*, **103**, 534–46.

Dunson, D. B. and Park, J.-H. (2008). Kernel stick-breaking processes. *Biometrika*, **95**, 307–23.

Dunson, D. B. and Peddada, S. D. (2008). Bayesian nonparametric inference on stochastic ordering. *Biometrika*, **95**, 859–74.

Dunson, D. B., Pillai, N. and Park, J.-H. (2007). Bayesian density regression. *Journal of the Royal Statistical Society B*, **69**, 163–83.

Dunson, D. B., Xue, Y. and Carin, L. (2008). The matrix stick-breaking process: flexible Bayes meta analysis. *Journal of the American Statistical Association*, **103**, 317–27.

Dunson, D. B., Yang, M. and Baird, D. D. (2007). Semiparametric Bayes hierarchical models with mean and variance constraints. *Discussion Paper*, 2007–8, Department of Statistical Science, Duke University.

Escobar, M. D. and West, M. (1995). Bayesian density estimation and inference using mixtures. *Journal of the American Statistical Association*, **90**, 577–88.

Ferguson, T. S. (1973). A Bayesian analysis of some nonparametric problems. *Annals of Statistics*, **1**, 209–30.

Ferguson, T. S. (1974). Prior distributions on spaces of probability measures. *Annals of Statistics*, **2**, 615–29.

Friedman, J. H. and Meulman, J. J. (2004). Clustering objects on subsets of attributes (with discussion). *Journal of the Royal Statistical Society*, **66**, 815–49.

Gelfand, A. E., Kottas, A. and MacEachern, S. N. (2005). Bayesian nonparametric spatial modeling with Dirichlet process mixing. *Journal of the American Statistical Association*, **100**, 1021–35.

Gopalan, R. and Berry, D. A. (1998). Bayesian multiple comparisons using Dirichlet process priors. *Journal of the American Statistical Association*, **93**, 1130–9.

Green, P. J. and Richardson, S. (2001). Modelling heterogeneity with and without the Dirichlet process. *Scandinavian Journal of Statistics*, **28**, 355–75.

Griffin, J. E. and Holmes, C. C. (2007). Bayesian nonparametric calibration with applications in spatial epidemiology. *Technical Report*, Institute of Mathematics, Statistics and Actuarial Science, University of Kent.

Griffin, J. E. and Steel, M. F. J. (2006). Order-based dependent Dirichlet process. *Journal of the American Statistical Association*, **101**, 179–94.

Griffin, J. E. and Steel, M. F. J. (2007). Bayesian nonparametric modelling with the Dirichlet process regression smoother. *Technical Report*, Institute of Mathematics, Statistics and Actuarial Science, University of Kent. *Statistica Sinica*, to appear.

Hanson, T. (2006). Inference for mixtures of finite Polya tree models. *Journal of the American Statistical Association*, **101**, 1548–65.

Hoff, P. D. (2006). Model-based subspace clustering. *Bayesian Analysis*, **1**, 321–44.

Ishwaran, H. and James, L. F. (2001). Gibbs sampling methods for stick-breaking priors. *Journal of the American Statistical Association*, **101**, 179–94.

Ishwaran, H. and Takahara, G. (2002). Independent and identically distributed Monte Carlo algorithms for semiparametric linear mixed models. *Journal of the American Statistical Association*, **97**, 1154–66.

Jain, S. and Neal, R. M. (2004). A split-merge Markov chain Monte Carlo procedure for the Dirichlet process mixture model. *Journal of Computational and Graphical Statistics*, **13**, 158–82.

Jara, A. (2007). Applied Bayesian non- and semi-parametric inference using DPpackage. *Technical Report*, Biostatistical Center, Catholic University of Leuven.

Jasra, A., Holmes, C. C. and Stephens, D. A. (2005). Markov chain Monte Carlo methods and the label switching problem in Bayesian mixture modeling. *Statistical Science*, **20**, 50–67.

Jefferys, W. and Berger, J. (1992). Ockham's razor and Bayesian analysis. *American Statistician*, **80**, 64–72.

Jones, B. L., Nagin, D. S. and Roeder, K. (2001). A SAS procedure based on mixture models for estimating developmental trajectories. *Sociological Methods & Research*, **29**, 374–93.

Jordan, M. I. and Jacobs, R. A. (1994). Hierarchical mixtures of experts and the EM algorithm. *Neural Computation*, **6**, 181–214.

Kim, S., Tadesse, M. G. and Vannucci, M. (2006), Variable selection in clustering via Dirichlet process mixture models, *Biometrika*, **93**, 877–93.

Kimeldorf, G. S. and Wahba, G. (1971). A correspondence between Bayesian estimation on stochastic processes and smoothing by splines. *Annals of Mathematical Statistics*, **41**, 495–502.

Kleinman, K. P. and Ibrahim, J. G. (1998a). A semiparametric Bayesian approach to the random effects model. *Biometrics*, **54**, 921–38.

Kleinman, K. P. and Ibrahim, J. G. (1998b). A semi-parametric Bayesian approach to generalized linear mixed models. *Statistics in Medicine*, **17**, 2579–96.

Kurihara, K., Welling, M. and Teh, Y. W. (2007). Collapsed variational Dirichlet process mixture models. *Twentieth International Joint Conference on Artificial Intelligence*, 2796–801.

Kurihara, K., Welling, M. and Vlassis, N. (2007). Accelerated variational Dirichlet mixture models. *Advances in Neural Information Processing Systems, 19*, 761–8.

Laird, N. M. and Ware, J. H. (1982). Random-effects models for longitudinal data. *Biometrics*, **38**, 963–74.

Lau, J. W. and Green, P. J. (2007). Bayesian model-based clustering procedures. *Journal of Computational and Graphical Statistics*, **16**, 526–58.

Lee, K. J. and Thompson, S. G. (2008). Flexible parametric models for random-effects distributions. *Statistics in Medicine*, **27**, 418–34.

Li, Y., Lin, X. and Müller, P. (2009). Bayesian inference in semiparametric mixed models for longitudinal data. *Department of Biostatistics Working Paper Series*, University of Texas M. D. Anderson Cancer Center. *Biometrika*, to appear.

Liang, F., Liao, M., Mao, K., Mukherjee, S. and West, M. (2007). Non-parametric Bayesian kernel models. *Discussion Paper*, 2007–10, Department of Statistical Science, Duke University, Durham, NC.

Lijoi, A., Mena, R. H. and Prünster, I. (2007a). A Bayesian nonparametric method for prediction in EST analysis. *BMC Bioinformatics*, **8**, to appear.

Lijoi, A., Mena, R. H. and Prünster, I. (2007b). Bayesian nonparametric estimation of the probability of discovering new species. *Biometrika*, **94**, 769–86.

Lo, A. Y. (1984). On a class of Bayesian nonparametric estimates. 1. Density estimates. *Annals of Statistics*, **12**, 351–7.

Longnecker, M. P., Klebanoff, M. A., Zhou, H. and Brock, J. W. (2001). Association between maternal serum concentration of the DDT metabolite DDE and preterm and small-for-gestational-age babies at birth. *Lancet*, **358**, 110–14.

MacEachern, S. N. (1994). Estimating normal means with a conjugate style Dirichlet process prior. *Communications in Statistics: Simulation and Computation*, **23**, 727–41.

MacEachern, S. N. (1999). Dependent nonparametric processes. *ASA Proceedings of the Section on Bayesian Statistical Science*, 50–5. Alexandria, Va.: American Statistical Association.

MacEachern, S. N., Clyde, M. and Liu, J. S. (1999). Sequential importance sampling for nonparametric Bayes models: The next generation. *Canadian Journal of Statistics*, **27**, 251–67.

MacLehose, R. F. and Dunson, D. B. (2009). Nonparametric Bayes kernel-based priors for functional data analysis. *Statistica Sinica*, **19**, 611–29.

MacLehose, R. F., Dunson, D. B., Herring, A. H. and Hoppin, J. A. (2007). Bayesian methods for highly correlated exposure data. *Epidemiology*, **18**, 199–207.

McAuliffe, J. D., Blei, D. M. and Jordan, M. I. (2006). Nonparametric empirical Bayes for the Dirichlet process mixture model. *Statistics and Computing*, **16**, 5–14.

Medvedovic, M. and Sivaganesan, S. (2002). Bayesian infinite mixture model based clustering of gene expression profiles. *Bioinformatics*, **18**, 1194–1206.

Medvedovic, M., Yeung, K. Y. and Bumgarner, R. E. (2004). Bayesian mixture model based clustering of replicated microarray data. *Bioinformatics*, **20**, 1222–32.

van der Merwe, A. J. and Pretorius, A. L. (2003). Bayesian estimation in animal breeding using the Dirichlet process prior for correlated random effects. *Genetics Selection Evolution*, **35**, 137–58.

Mukhopadhyay, S. and Gelfand, A. E. (1997). Dirichlet process mixed generalized linear models. *Journal of the American Statistical Association*, **92**, 633–9.

Mulherin-Engel, S. A., Eriksen, H. C., Savitz, D. A., Thorp, J., Chanock, S. J. and Olshan, A. F. (2005). Risk of spontaneous preterm birth is associated with common proinflammatory cytokine polymorphisms. *Epidemiology*, **16**, 469–77.

Muliere, P. and Tardella, L. (1998). Approximating distributions of random functionals of Ferguson–Dirichlet priors. *Canadian Journal of Statistics*, **26**, 283–97.

Müller, P., Erkanli, A. and West, M. (1996). Bayesian curve fitting using multivariate normal mixtures. *Biometrika*, **83**, 67–79.

Müller, P., Quintana, F. and Rosner, G. (2004). A method for combining inference across related nonparametric Bayesian models. *Journal of the Royal Statistical Society B*, **66**, 735–49.

Müller, P., Quintana, F. and Rosner, G. L. (2007). Semiparametric Bayesian inference for multilevel repeated measurement data. *Biometrics*, **63**, 280–9.

Müller, P. and Rosner, G. L. (1997). A Bayesian population model with hierarchical mixture priors applied blood count data. *Journal of the American Statistical Association*, **92**, 1279–92.

Muthén, B. and Shedden, K. (1999). Finite mixture modeling with mixture outcomes using the EM algorithm. *Biometrics*, **55**, 463–9.

Newton, M. A., Noueiry, A., Sarkar, A. and Ahlquist, P. (2004). Detecting differential gene expression with a semiparametric hierarchical mixture method. *Biostatistics*, **5**, 155–76.

Newton, M. A. and Zhang, Y. (1999). A recursive algorithm for nonparametric analysis with missing data. *Biometrika*, **86**, 15–26.

Ohlssen, D. I., Sharples, L. D. and Spiegelhalter, D. J. (2007). Flexible random-effects models using Bayesian semi-parametric models: Applications to institutional comparisons. *Statistics in Medicine*, **26**, 2088–112.

Papaspiliopoulos, O. and Roberts, G. (2008). Retrospective Markov chain Monte Carlo methods for Dirichlet process hierarchical models. *Biometrika*, **95**, 169–86.

Pennell, M. L. and Dunson, D. B. (2008). Nonparametric Bayes testing of changes in a response distribution with an ordinal predictor. *Biometrics*, **64**, 413–23.

Petrone, S., Guindani, M. and Gelfand, A. E. (2007). Hybrid Dirichlet processes for functional data. *Technical Report*, Bocconi University, Milan, Italy.

Petrone, S. and Raftery, A. E. (1997). A note on the Dirichlet process prior in Bayesian nonparametric inference with partial exchangeability. *Statistics and Probability Letters*, **36**, 69–83.

Pillai, N. S, Wu, Q., Liang, F., Mukherjee, S. and Wolpert, R. L. (2007). Characterizing the function space for Bayesian kernel models. *Journal of Machine Learning Research*, **8**, 1769–97.

Qin, Z. S. (2006). Clustering microarray gene expression data using weighted Chinese restaurant process. *Bioinformatics*, **22**, 1988–97.

Ramsay, J. O. and Silverman, B. W. (1997). *Functional Data Analysis*, Springer.

Rasmussen, C. E. and Williams, C. K. I. (2006). *Gaussian Processes for Machine Learning*. Cambridge, Mass.: MIT Press.

Ray, S. and Mallick, B. (2006). Functional clustering by Bayesian wavelet methods. *Journal of the Royal Statistical Society B*, **68**, 305–32.

Rodriguez, A., Dunson, D. B. and Gelfand, A. E. (2009a). Latent stick-breaking processes. *Journal of the American Statistical Association*, to appear.

Rodriguez, A., Dunson, D. B. and Gelfand, A. E. (2009b). Nonparametric functional data analysis through Bayesian density estimation. *Biometrika*, **96**, 149–62.

Rodriguez, A., Dunson, D. B. and Gelfand, A. E. (2007). The nested Dirichlet process (with discussion). *Journal of the American Statistical Association*, to appear.

Sethuraman, J. (1994). A constructive definition of Dirichlet priors. *Statistica Sinica*, **4**, 639–50.

Sollich, P. (2002). Bayesian methods for support vector machines: evidence and predictive class probabilities. *Machine Learning*, **46**, 21–52.

Stephens, M. (2000). Dealing with label switching in mixture models. *Journal of the Royal Statistical Society B*, **62**, 795–809.

Sun, S., Greenwood, C. M. T. and Neal, R. M. (2007). Haplotype inference using a Bayesian hidden Markov model. *Genetic Epidemiology*, **31**, 937–48.

Teh, Y. W., Jordan, M. I., Beal, M. J. and Blei, D. M. (2006). Hierarchical Dirichlet processes. *Journal of the American Statistical Association*, **101**, 1566–81.

Tipping, M. E. (2001). Sparse Bayesian learning and the relevance vector machine. *Journal of Machine Learning Research*, **1**, 211–44.

Tomlinson, G. (1998). Analysis of densities. *Unpublished Dissertation*, University of Toronto.

Walker, S. G. (2007). Sampling the Dirichlet mixture model with slices. *Communications in Statistics: Simulation and Computation*, **36**, 45–54.

Walker, S. G. and Gutièrrez-Peña, E. (2007). Bayesian parametric inference in a nonparametric framework. *TEST*, **16**, 188–97.

Walker, S. G. and Mallick, B. K. (1997). Hierarchical generalized linear models and frailty models with Bayesian nonparametric mixing. *Journal of the Royal Statistical Society B*, **59**, 845–60.

West, M., Müller, P. and Escobar, M. D. (1994). Hierarchical priors and mixture models, with application in regression and density estimation. In *Aspects of Uncertainty: A Tribute to D.V. Lindley*, ed. P. R. Freeman and A. F. Smith, 363–86. Wiley.

Wilcox, A. J., Weinberg, C. R., O'Connor, J. F., Baird, D. D., Schlatterer, J. P., Canfield, R. E., Armstrong, E. G. and Nisula, B. C. (1988). Incidence of early loss of pregnancy. *New England Journal of Medicine*, **319**, 189–94.

Xiang, Z. S., Qin, Z. H. S. and He, Y. Q. (2007). CRCView: a web server for analyzing and visualizing microarray gene expression data using model-based clustering. *Bioinformatics*, **23**, 1843–5.

Xing, E. P., Jordan, M. I. and Sharan, R. (2007). Bayesian haplotype inference via the Dirichlet process. *Journal of Computational Biology*, **14**, 267–84.

Xing, E. P. and Sohn, K. (2007). Hidden Markov Dirichlet process: modeling genetic recombination in open ancestral space. *Bayesian Analysis*, **2**, 501–28.

Xing, E. P., Sohn, K., Jordan, M. I. and Teh, Y. W. (2006). Bayesian multi-population haplotype inference via a hierarchical Dirichlet process mixture. *Proceedings of the 23rd International Conference on Machine Learning*. ACM International Conference Proceeding Series, Volume 148, 1049–56. New York: ACM Press.

8

More nonparametric Bayesian models for biostatistics

Peter Müller and Fernando Quintana

In this companion to Chapter 7 we discuss and extend some of the models and inference approaches introduced there. We elaborate on the discussion of random partition priors implied by the Dirichlet process. We review some additional variations of dependent Dirichlet process models and we review in more detail the Pólya tree prior used briefly in Chapter 7. Finally, we review variations of Dirichlet process models for data formats beyond continuous responses.

8.1 Introduction

In Chapter 7, Dunson introduced many interesting applications of nonparametric priors for inference in biomedical problems. The focus of the discussion was on Dirichlet process (DP) priors and variations. While the DP prior defines a probability model for a (discrete) random probability distribution G, the primary objective of inference in many recent applications is not inference on G. Instead many applications of the DP prior exploit the random partition of the Pólya urn scheme that is implied by the configuration of ties among the random draws from a discrete measure with DP prior. When the emphasis is on inference for the clustering, it is helpful to recognize the DP as a special case of more general clustering models. In particular we will review the product partition (PPM) and species sampling models (SSM). We discuss these models in Section 8.2. A definition and discussion of the SSM as a random probability measure also appears in Section 3.3.4. Another useful characterization of the DP is as a special case of the Pólya tree (PT) prior. A particularly attractive feature of PT priors is the possibility to model absolutely continuous distributions. In Section 8.3 we will define PT priors and discuss specific algorithms to implement inference. For more discussion of the PT prior see also Section 3.4.2. In Section 8.4 we discuss more variations of the dependent DP (DDP) models. In Section 8.5 we review some examples of DP priors for biostatistical applications that involve non-continuous data. Finally, in Section 8.6 we discuss implementation details. We show some example R code using the popular R package `DPpackage` to implement nonparametric Bayesian inference.

8.2 Random partitions

Let $[n] = \{1, \ldots, n\}$ denote a set of experimental units. A partition is a family of subsets $\rho_n = \{S_1, \ldots, S_k\}$ with $\bigcup S_j = [n]$ and $S_j \cap S_\ell = \emptyset$ for all $j \neq \ell$. The partitioning subsets S_j define clusters of experimental units. Often it is convenient to describe a partition by cluster membership indicators $s_i = j$ if $i \in S_j$. We use notation $n_{nj} = |S_j|$ and $\boldsymbol{n}_n = (n_1, \ldots, n_k)$ to denote the cluster sizes and $k_n = |\rho_n|$ to denote the number of clusters. When the sample size n is understood from the context we drop the subindex n in $\rho, n_j, \boldsymbol{n}$ and k.

Several probability models $p(\rho_n)$ are introduced in the literature. For an extensive recent review of probability models and Bayesian inference for clustering, see, for example, Quintana (2006). Popular models include the product partition models (PPM), species sampling models (SSM), model based clustering (MBC) and Voronoi tessellations. The PPM (Hartigan, 1990; Barry and Hartigan, 1993) requires a cohesion function $c(S_j) \geq 0$ (see an example below). A PPM for a partition ρ and data y is defined as

$$p(\rho) \propto \prod c(S_i) \qquad \text{and} \qquad p(y \mid \rho) = \prod_{j=1}^{k} p_j(y_{S_j}). \tag{8.1}$$

Here, p_j is any sampling model for the observations in the jth cluster. Model (8.1) is conjugate. The posterior $p(\theta \mid y)$ is again in the same product form.

Most recent applications of such models in biomedical applications use the special case of DP priors (Ferguson, 1973; Antoniak, 1974). The DP implicitly defines a probability model on $p(\rho_n)$ by defining a discrete random probability measure G. An i.i.d. sample $x_i \sim G$, $i = 1, \ldots, n$, includes with positive probability ties among the x_i. Let $x_j^\star$, $j = 1, \ldots, k \leq n$ denote the unique values of x_i and define clusters $S_j = \{i : x_i = x_j^\star\}$. By defining the probability of ties the DP prior has implicitly defined $p(\rho_n)$. The probability model is known as the Pólya urn and is a special case of the PPM with cohesions $c(A) = M \times (|A| - 1)!$ (Dahl, 2003; Quintana and Iglesias, 2003).

A typical recent use of the DP random partition model in biostatistics applications appears in Dahl and Newton (2007) who use a clustering model to improve power for multiple hypothesis tests by pooling tests within clusters. Dahl (2006) describes a similar model with focus on inference for the clustering only. Tadesse, Sha and Vannucci (2005) combine inference on clustering, again based on the DP prior, with variable selection to identify a subset of genes whose sampling model is defined by the clustering.

Another class of random partition models are the species sampling models (SSM) (Pitman, 1996). An SSM defines a probability model $p(\rho)$ that depends on ρ only indirectly through the cardinality of the partitioning subsets, $p(\rho) = p(|S_1|, \ldots, |S_k|)$. The SSM can be alternatively characterized by a sequence of predictive probability functions (PPF) that describe how individuals are sequentially assigned either to already formed clusters or to start new ones. Let $n_j = |S_j|$ and $\boldsymbol{n}_n = (n_1, \ldots, n_k)$. The PPFs are the probabilities $p_{nj}(\boldsymbol{n}_n) = \Pr(s_{n+1} = j \mid \rho_n)$, $j = 1, \ldots, k_n + 1$. Compare Theorem 3.30 in Chapter 3. The opposite is not true. Not every sequence of PPFs characterizes an SSM. Pitman (1996) states the conditions. Let $\boldsymbol{n}^{j+}$ denote $\boldsymbol{n}$ with n_j incremented by one. Essentially the PPFs have to arise as $p_{nj}(\boldsymbol{n}_n) = p(\boldsymbol{n}^{j+}_{n+1})/p(\boldsymbol{n}_n)$, where $p(\boldsymbol{n}_n)$ is a probability measure on $\bigcup_n \{\boldsymbol{n}_n\}$ that is symmetric in its arguments. An important implication is that $p(\boldsymbol{n}_n)$ has to arise as the marginal of $p(\boldsymbol{n}_{n+1})$, i.e., $p(\boldsymbol{n}_n) = \sum_{j=1}^{k_n+1} p(\boldsymbol{n}^{j+}_{n+1})$. The probability model $p(\boldsymbol{n}_n)$ is known as the exchangeable partition probability function. See Section 3.3.2 for more discussion. Again, the random clustering implied by the DP model is a special case of an SSM, i.e., the random partition model implied by the DP is a special case of both PPM and SSM.

Model based clustering (Banfield and Raftery, 1993; Dasgupta and Raftery, 1998) implicitly defines a probability model on clustering by assuming a mixture model

$$p(y_i \mid \eta, k) = \sum_{j=1}^{k} \tau_j \, f_j(y_i \mid \theta_j),$$

where $\eta = (\theta_1, \ldots, \theta_k, \tau_1, \ldots, \tau_k)$ are the parameters of a size k mixture model. Together with a prior $p(k)$ on k, the mixture implicitly defines a probability model on clustering. Consider the equivalent hierarchical model

$$p(y_i \mid s_i = j, k, \eta) = f_j(y_i \mid \theta_j) \qquad \text{and} \qquad \Pr(s_i = j \mid k, \eta) = \tau_j. \tag{8.2}$$

The implied posterior distribution on $(s_1, \ldots, s_n)$ and k implicitly defines a probability model on ρ_n. Richardson and Green (1997) develop posterior simulation strategies for mixture of normal models. Green and Richardson (1999) discuss the relationship to DP mixture models.

Heard, Holmes and Stephens (2006) model gene expression profiles with a hierarchical clustering model. The prior model on the random partition is an example of model based clustering with a mixture of regression models, a uniform prior on the number of clusters and a Dirichlet prior for cluster membership probabilities.

8.3 Pólya trees

Lavine (1992, 1994) proposed Pólya trees (PT) as a useful nonparametric Bayesian prior for random probability measures. In contrast to the DP, an appropriate choice of the PT parameters allows the analyst to specify absolutely continuous distributions. In the following discussion we briefly review the definition of the PT model, and give explicit implementation details for posterior inference under the PT. See also Section 3.4.2 for more discussion of the definition. The definition starts with a nested sequence $\Pi = \{\pi_m,\ m = 1, 2, \ldots\}$ of partitions of the sample space Ω. Without loss of generality, we assume that the partitions are binary. We start with a partition $\pi_1 = \{B_0, B_1\}$ of the sample space, $\Omega = B_0 \cup B_1$, and continue with nested partitions defined by $B_0 = B_{00} \cup B_{01}$, $B_1 = B_{10} \cup B_{11}$, etc. Thus the partition at level m is $\pi_m = \{B_\epsilon,\ \epsilon = \epsilon_1 \ldots \epsilon_m\}$, where ϵ are all binary sequences of length m. A PT prior for a random probability measure G is defined by beta-distributed random branching probabilities. Let $Y_{\epsilon 0} \equiv G(B_{\epsilon 0} \mid B_\epsilon)$, and let $\mathcal{A} \equiv \{\alpha_\epsilon\}$ denote a sequence of nonnegative numbers, one for each partitioning subset. If $Y_{\epsilon 0} \sim \text{Beta}(\alpha_{\epsilon 0}, \alpha_{\epsilon 1})$ then we say that G has a PT prior, $G \sim \text{PT}(\Pi, \mathcal{A})$.

The parameters α_ϵ are usually chosen as $\alpha_\epsilon = cm^r$ for level m subsets. For $r = 2$ the random probability measure G is a.s. absolutely continuous. With $r = -1/2$ the PT reduces to the DP as a special case. The partitioning subsets B_ϵ can be chosen to achieve a desired prior mean $G^\star$. Let $q_{mk} = G^{\star -1}(k/2^m)$, $k = 0, \ldots, 2^m$, denote the inverse c.d.f. under $G^\star$ evaluated at dyadic fractions. If $\alpha_{\epsilon 0} = \alpha_{\epsilon 1}$, for example $\alpha_\epsilon = cm^r$, and the dyadic quantile sets $[q_{mk}, q_{m,k+1})$ are used as the partitioning subsets B_ϵ, then $E(G) = G^\star$. Alternatively the prior mean can be fixed to $G^\star$ by choosing $\alpha_{\epsilon 0}/(\alpha_{\epsilon 0} + \alpha_{\epsilon 1}) = G^\star(B_{\epsilon 0} \mid B_\epsilon)$ for any choice of the nested partitions Π.

The main attraction of PT models for nonparametric Bayesian inference is the simplicity of posterior updating. Assume $x_i \sim G$, i.i.d., $i = 1, \ldots, n$, and $G \sim \text{PT}(\Pi, \mathcal{A})$. Consider first $n = 1$, i.e., a single sample from the unknown distribution G. The posterior $p(G \mid x)$ is again a Pólya tree, $p(G \mid x) = \mathcal{P}(\Pi, \mathcal{A}')$ with the beta parameters in $\mathcal{A}'$ defined as

$$\alpha'_\epsilon = \begin{cases} \alpha_\epsilon & \text{if } x_1 \notin B_\epsilon \\ \alpha_\epsilon + 1 & \text{if } x_1 \in B_\epsilon. \end{cases} \tag{8.3}$$

The general case with a sample of size $n > 1$ follows by induction.

The result (8.3) can be used to implement exact posterior predictive simulation, i.e., simulation from $p(x_{n+1} \mid x_1, \ldots, x_n)$. In words, we "drop" a ball down (well, really up) the Pólya tree. Starting with (B_0, B_1) at the root we generate the random probabilities $(Y_{\epsilon 0}, Y_{\epsilon 1} = 1 - Y_{\epsilon 0})$ for picking the two nested partitions $B_{\epsilon 0}$ and $B_{\epsilon 1}$ at the next level. Recall that $Y_{\epsilon 0} = P(x \in B_{\epsilon 0} \mid x \in B_\epsilon)$ and $Y_{\epsilon 0} \sim \text{Beta}(\alpha'_{\epsilon 0}, \alpha'_{\epsilon 1})$.

Going down the tree we run into some good luck. At some level m we will drop the ball into a subset B_ϵ, $\epsilon = \epsilon_1\epsilon_2 \ldots \epsilon_m$, that does not contain any data point. From level m onwards, dropping the ball proceeds as if we had no data observed. Thus we can generate the posterior predictive draw from the base measure $G^\star$, restricted to B_ϵ. The following algorithm summarizes posterior predictive simulation of $x_{n+1} \sim p(x_{n+1} \mid x_1 \ldots x_n)$.

Algorithm 1

1. *Initialize:* $\epsilon = \emptyset$ (nil).
2. *Iteration:* Loop over $m = 1, 2, \ldots$:

   (a) *Posterior PT parameters:* Find $n_{\epsilon 0} = \sum I(x_i \in B_{\epsilon 0})$ and $n_{\epsilon 1} = \sum I(x_i \in B_{\epsilon 1})$, the number of data points in the two partitioning subsets for $B_\epsilon = B_{\epsilon 0} \cup B_{\epsilon 1}$. Let $\alpha'_{\epsilon 0} = \alpha_{\epsilon 0} + n_{\epsilon 0}$, and same for $\alpha'_{\epsilon 1}$.

   (b) *Generate random branching probability:* Generate $Y_{\epsilon 0} \sim \text{Beta}(\alpha'_{\epsilon 0}, \alpha'_{\epsilon 1})$, and set $\epsilon_m \sim \text{Bernoulli}(1 - Y_{\epsilon 0})$.

3. *Stop the recursion:* Stop the iteration over m for the smallest m^* such that $n_\epsilon = 0$ at $m = m^*$.
4. *Generate x_{n+1}:* Draw $x_{n+1} \sim G^\star(x_{n+1} \mid B_\epsilon)$.

We can use the same algorithm to generate a prior predictive, i.e., marginal samples $x_i \sim G$, i.i.d., with $G \sim \text{PT}(\Pi, \mathcal{A})$.

1. Generate $x_1 \sim G^\star$
2. Iterate over $i = 2, \ldots, n$:
   Use algorithm 1 to generate $x_i \sim p(x_i \mid x_1, \ldots, x_{i-1})$.

A minor variation of the algorithm computes the posterior mean $E(G \mid x_1, \ldots, x_n)$. Let $x = x_1, \ldots, x_n$ denote the data. Consider some maximum level M, say $M = 10$. For all levels $m = 1, \ldots, M$ compute $\overline{Y}_{\epsilon_1\epsilon_2\cdots\epsilon_{m-1}0} = E(Y_{\epsilon_1\epsilon_2\cdots\epsilon_{m-1}0} \mid x) = \alpha'_{\epsilon_1\epsilon_2\cdots 0}/(\alpha'_{\epsilon_1\epsilon_2\cdots 0} + \alpha'_{\epsilon_1\epsilon_2\cdots 1})$. Record $\overline{Y}_{\epsilon_1\epsilon_2\cdots\epsilon_{m-1}1} = 1 - \overline{Y}_{\epsilon_1\epsilon_2\cdots\epsilon_{m-1}0}$ as the complement to one. Computing $\overline{Y}_\epsilon$ is most elegantly implemented as a recursion. Let $\boldsymbol{\epsilon} = \epsilon_1 \ldots \epsilon_M$ denote a dyadic number, $\boldsymbol{\epsilon} \in [0, 1]$. We find

$$E(G(\boldsymbol{\epsilon}) \mid x_1, \ldots, x_n) \approx \prod_{m=1}^{M} \overline{Y}_{\epsilon_1\epsilon_2\cdots\epsilon_m} \equiv \overline{G}(\boldsymbol{\epsilon}).$$

Here $G(\cdot)$ is the p.d.f. for the random measure G.

To generate random draws $G \sim p(G \mid x_1, \ldots, x_n)$ proceed as above replacing $\overline{Y}_\epsilon$ by $Y_{\epsilon 0} \sim \text{Beta}(\alpha'_{\epsilon 0}, \alpha'_{\epsilon 1})$. Let $G(\boldsymbol{\epsilon}) = \prod_{m=1}^{M} Y_{\epsilon_1 \epsilon_2 \cdots \epsilon_m}$. Plotting G against $\boldsymbol{\epsilon}$ shows a random posterior draw of G. Plotting multiple draws G_j, $j = 1, 2, \ldots, J$ in the same figure illustrates posterior uncertainty on the random measure.

An important simplification for applications is achieved by restricting the PT prior to finitely many levels, e.g. $m \leq 10$. Actual inference differs little, but the required computational effort is greatly reduced. Posterior predictive draws and posterior estimated densities can still be carried out exactly, as outlined above. Nonparametric Bayesian inference under the PT prior and the finite PT prior for some important models is implemented in the public domain R package `DPpackage`. See Section 8.6 for an example.

Applications of PT models in biomedical problems are less common than the widely used DP. The limited use of PT priors for nonparametric Bayesian data analysis is due in part to the awkward sensitivity of posterior inference to the choice of the partitioning subsets. Consider a model for density estimation, $x_i \sim G$, i.i.d., $i = 1, \ldots, n$, with a PT prior for the unknown distribution, $G \sim \text{PT}(\mathcal{A}, \mathcal{P})$. The posterior estimate $\overline{G} = E(G \mid x)$ shows characteristic discontinuities at the boundaries of the partitioning subsets. An exception is the special case of the DP, i.e., when $\alpha_\epsilon = \alpha_{\epsilon 0} + \alpha_{\epsilon 1}$. In that case the probability model $p(G)$ remains invariant under any change of the partitioning sequence. Several authors have proposed variations of the PT prior model to mitigate this undesirable feature of posterior inference. Hanson and Johnson (2002) and Hanson (2006) defined mixture of Pólya trees, with the mixture being with respect to the centering measure $G^\star$. Paddock, Ruggeri, Lavine and West (2003) introduced an explicit perturbation of the partition boundaries. The posterior dependence on the partition boundaries is less of an issue when the focus of the inference is not on the density itself. For example, Branscum, Johnson, Hanson and Gardner (2008) develop inference for ROC curves with a PT prior for the distribution of the recorded measurements for true positives and negatives. Hanson and Yang (2007) use PT priors for survival data.

8.4 More DDP models

Many applications in biostatistics involve hierarchical models across different subpopulations and naturally lead to models that include several random probability measures. Appropriate nonparametric probability models require modeling of dependent families of random distributions. One of the most commonly used models to achieve this aim are variations of dependent DP (DDP) models. Since the first proposal of DDP models in MacEachern (1999) many authors have developed variations and implementations for specific problems. Some are discussed in

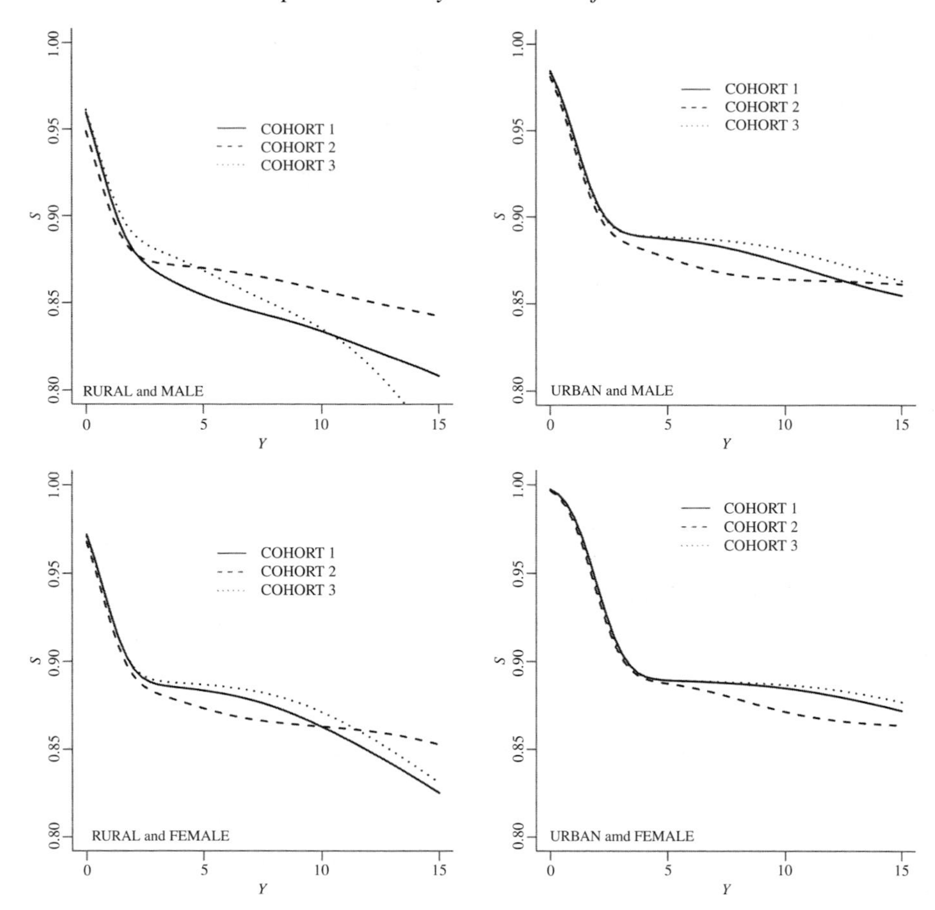

Figure 8.2 Colombian children data. Posterior survivor functions for children from the three birth cohorts, arranged by rural versus urban and male versus female. The solid line corresponds to children in the first birth cohort, the dashed line represents a child in the second birth cohort and the dotted line refers to children in the third birth cohort.

$i = n + 1$ conditional observed responses y_{n+1}, i.e.,

$$p(x_{n+1} \mid y_{n+1}, y_1, x_1, \ldots, y_n, x_n).$$

Let $p(y_i \mid x_i = x) = \int p(y_i \mid \theta_i)\, \mathrm{d}G_x(\theta_i)$ be a semiparametric sampling model for outcomes in group x, see for example De la Cruz-Mesía, Quintana and Müller (2007) for details of the model for the pregnancy data. Marginally, for each $x \in X$ we use a DP mixture model, i.e., we assume a DP prior for G_x. The submodels for all x are combined into one encompassing hierarchical model by linking the marginal DP priors through an ANOVA DDP across x. Finally the model is completed by assuming a marginal distribution $p(x_i)$ for x_i, for example $x_i \sim \mathrm{Dir}(\alpha_1, \ldots, \alpha_k)$.

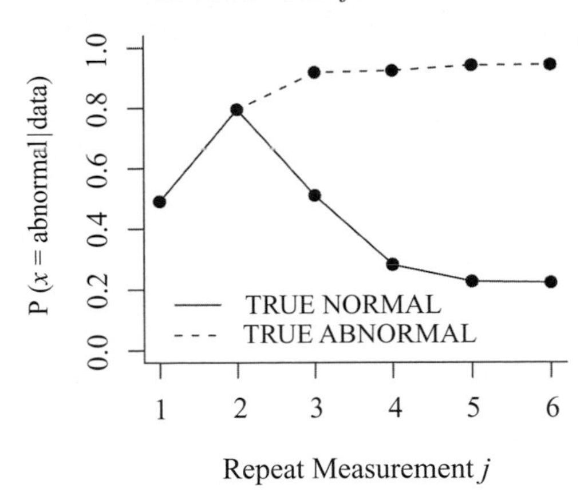

Figure 8.3 Pregnancy data: sequentially updated classification probabilities $p(x_{n+1} = 1 \mid y_{n+1,1}, \ldots, y_{n+1,j}, \text{data})$ for a future case. The probabilities are plotted against j. The solid (dashed) line is for a future case with normal (abnormal) pregnancy.

The ANOVA DDP model defines $p(y_1, \ldots, y_{n+1} \mid x_1, \ldots, x_{n+1})$. Together with the marginal model $p(x_1, \ldots, x_{n+1}) = \prod p(x_i)$ we can use Bayes theorem to derive the desired $p(x_{n+1} \mid y_{n+1}, x_1, y_1, \ldots, x_n, y_n)$. One of the attractions of this principled model-based approach is the possibility for coherent sequential updating. Assume $y_i = (y_{i1}, \ldots, y_{im_i})$ are repeated measurement data as in the pregnancy example. The classification probability $p(x_{n+1} \mid y_{n+1,1}, \ldots, y_{n+1,j}, y_1, x_1, \ldots, y_n, x_n)$ can be sequentially updated and reported as increasingly more data become available for $j = 1, \ldots, m_{n+1}$. Figure 8.3 shows sequentially updated classification probabilities $p(x_{n+1} = 1 \mid y_{n+1,1}, \ldots, y_{n+1,j}, \; y_1, x_1, \ldots, y_n, x_n)$ for the pregnancy example. Probabilities are plotted against j for two hypothetical future patients. The first patient (solid line) is a woman with a truly normal pregnancy. The second case (dashed line) is a truly abnormal pregnancy. See De la Cruz-Mesía, Quintana and Müller (2007) for details of the simulation. Note how the two classification probabilities start to diverge from the third repeat measurement onwards.

8.5 Other data formats

Many discussions of DP models, including most of the discussion in Chapter 7, focus on continuous outcomes. But many biomedical data analysis problems involve other data formats. We briefly review two applications of nonparametric Bayesian inference to categorical and binary data.

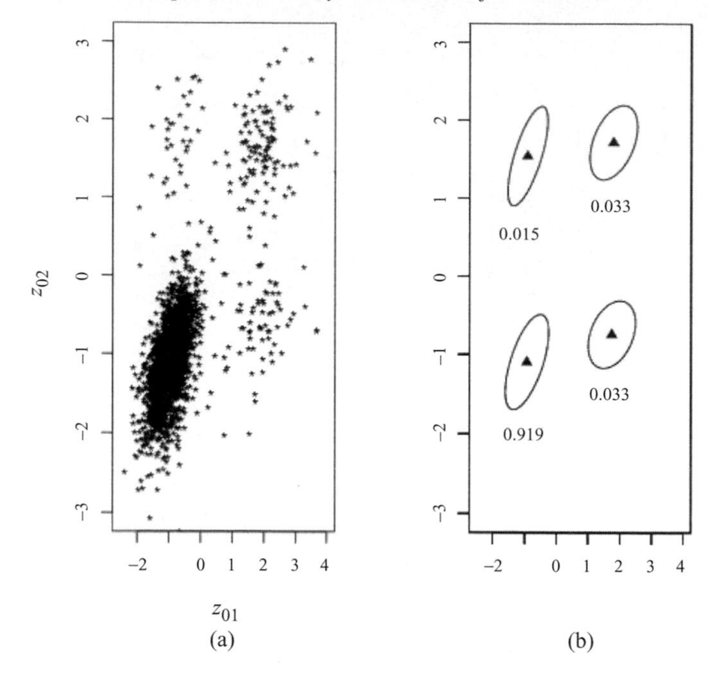

Figure 8.4 Interrater agreement data. Panel (a) plots draws from the latent probit scores (with a DP mixture of normal prior). Panel (b) shows one posterior draw for the moments of the mixture of normal terms (location and scale are shown as triangle and ellipse), together with the corresponding weights (number below the ellipse). Notice the varying degree of polychoric correlation across scores.

Kottas, Müller and Quintana (2005) propose nonparametric Bayesian inference for multivariate ordinal data, for example the rating of the extent of tumor invasion by two raters. Tumor invasion is coded on an ordinal scale from none to extensive invasion. A feature of the data example reported in Kottas, Müller and Quintana (2005) is that the raters tend to agree on extreme cases, but less so on intermediate cases. This makes it inappropriate to use a bivariate ordinal probit model, as described for example in Johnson and Albert (1999). Instead Kottas, Müller and Quintana (2005) propose an ordinal probit model based on a DP mixture of normal distributions for the latent variable. The mixture of normal model allows us to formalize the notion of varying degrees of interrater agreement across scores. Let Σ_j denote the variance–covariance matrix of the jth term in the mixture of normal model for the latent variables. The correlation of the latent scores is known as polychoric correlation. We refer to the correlation that is implied by Σ_j for each term of the mixture as *local polychoric correlation.* The use of different Σ_j for each term in the mixture allows for varying degrees of interrater agreement, as desired. Figure 8.4 shows imputed ordinal probit scores and summaries of the mixture

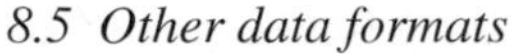

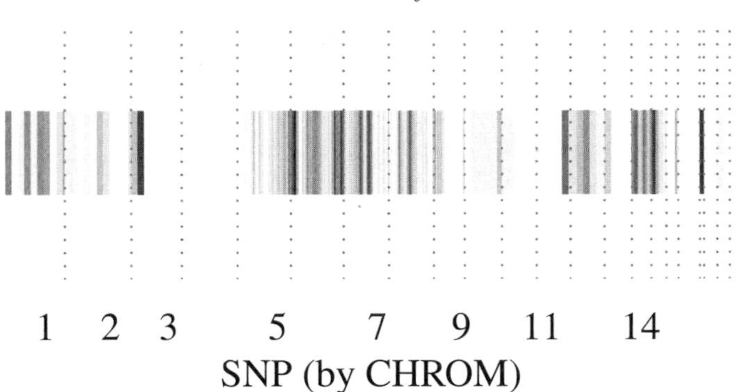

Figure 8.5 LOH data: gray shades show the probability of increased LOH for a given sample. SNPs are arranged by chromosomes indicated by vertical dashed lines. Darker gray shade indicates higher probability of increased LOH. The underlying model makes minimal assumptions beyond order ℓ exchangeability.

of normal model for the interrater agreement example. The plotted variables z_{01} and z_{02} are the latent ordinal probit scores. The observed scores are defined by thresholds on the latent scores; see Kottas, Müller and Quintana (2005) for details. For normal cases (with low scores) the two raters are in strong agreement, i.e., high polychoric correlation. For extreme cases the strength of agreement is considerably less.

Quintana, Müller, Rosner and Relling (2008) discusses semiparametric Bayesian inference for binary sequences of indicators of loss of heterozygosity (LOH). Nonparametric inference for sequences of binary indicators subject to a partial exchangeability assumptions are defined in Quintana and Müller (2004). The exchangeability assumption is order ℓ exchangeability, i.e., the assumption that the probability model is invariant with respect to any permutation of the sequence that leaves the order ℓ transition counts unchanged. Quintana and Newton (2000) show that any such distribution can be represented as a mixture of order ℓ Markov chains. The mixture is with respect to the transition probabilities of the Markov chain. In the application to the LOH data we implemented this by a DP prior on the mixture measure of the transition probabilities. We allow different transition probabilities for each region of the chromosome. Regions are defined as sequences of 55 to 835 SNPs (single nucleotide polymorphism); see Quintana, Müller, Rosner and Relling (2008) for details. The transition probabilities of the binary Markov chain with states { no LOH, LOH } imply a limiting probability of LOH. Let π_{cj} denote this limiting probability for region j in chromosome c. We can map the posterior distribution on the transition probabilities into $p(\pi_{cj} \mid \text{data})$. Figure 8.5 shows $I_{cj} \equiv p(\pi_{cj} > 0.01 \mid \text{data})$ by region j, arranged by chromosome c.

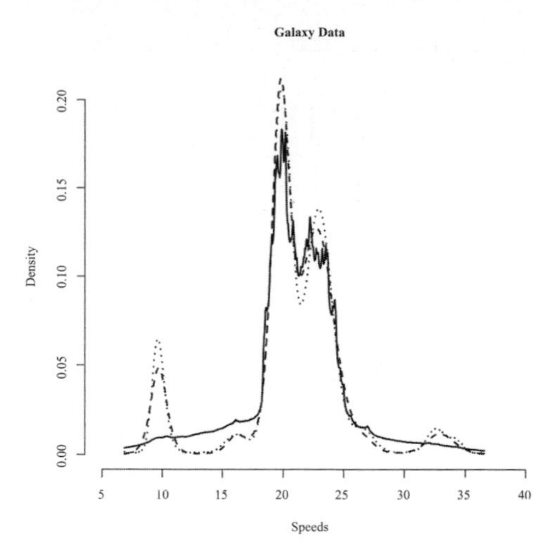

Figure 8.6 Galaxy data. The histogram shows the data with $n = 82$ observations. The curves show the density estimate using a DP mixture model (*dashed line*), a PT prior (*solid line*) and a random Bernstein polynomial (*dotted line*).

8.6 An R package for nonparametric Bayesian inference

An important impediment to the wider use of nonparametric Bayesian models was, until recently, the limited availability of reliable public domain software. This is true for biomedical applications in particular, where the research focus is often on the application, and only limited resources are available for the development of problem-specific software. Also reproducibility is an increasingly important issue. The impact of research publications proposing new methods and approaches remains limited unless readers can reproduce inference and implement the proposed methods for their problems.

A popular software platform for biomedical research, in particular for bioinformatics applications, is the public domain R language (R Development Core Team, 2008). The R package `DPpackage` (Jara, 2007) implements inference for some of the models discussed in Chapter 7, including DP mixture density estimation, PT priors for density estimation, nonparametric random effects models including generalized linear models. `DPpackage` is available from the package repository CRAN. We show a simple example.

Buta (1987) reports velocities (y_i) and radial positions (x_i) of galaxy NGC7531 at 323 different locations. We use the first 82 velocity measurements. The data are available as `galaxy` data set in `DPpackage`. The following R code implements density estimation using PT priors, a DP mixture model and random Bernstein polynomials (Petrone, 1999a, 1999b). See Section 3.4.1 for a discussion of random Bernstein polynomials. Figure 8.6 shows a histogram of the galaxy data and the three density estimates based on DP mixture model, the PT prior and the random Bernstein polynomials.

```
library("DPpackage")

data(galaxy)              # Data
galaxy <- data.frame(galaxy,speeds=galaxy$speed/1000)
attach(galaxy)

state <- NULL             # MCMC parameters
nburn <- 10000; nsave <- 1000; nskip <- 50; ndisplay <- 10
mcmc <- list(nburn=nburn,nsave=nsave,nskip=nskip,ndisplay=ndisplay,
             tune1=0.15,tune2=1.1,tune3=1.1)
             ## tune parameters only needed for PTdensity()

## POLYA TREE
   prior<-list(alpha=1,M=6) # Prior information
                            # Fitting the model
   fit1 <- PTdensity(y=speeds,ngrid=1500,prior=prior,mcmc=mcmc,
                          state=state,status=TRUE)


## DIRICHLET PROCESS
                            # Prior information
   prior <- list(a0=2,b0=4,m2=rep(20,1),s2=diag(100000,1),
            psiinv2=solve(diag(0.5,1)), nu1=4,nu2=4,tau1=1,tau2=100)
                            # Fitting the model
   fit2 <- DPdensity(y=speeds,ngrid=1500,prior=prior,mcmc=mcmc,
                 state=state,status=TRUE)

## BERNSTEIN DIRICHLET PROCESS
                            # Prior information
   prior <- list(aa0=2.01,ab0=1.01,kmax=1000,a0=1,b0=1)
                            # Fitting the model
   fit3 <- BDPdensity(y=speeds,ngrid=1500,prior=prior,mcmc=mcmc,
                 state=state,status=TRUE)

   rg <- range(c(fit1$dens,fit2$dens,fit3$fun))        ## Plots
   hist(galaxy$speeds,xlim=c(5,40),nclass=30,ylim=rg)
   lines(fit1$x1,fit1$dens,lty=1,lwd=2)
   lines(fit2$x1,fit2$dens,lty=2,lwd=2)
   lines(fit3$grid,fit3$fun,lty=3,lwd=2)
```

Inference for the ANOVA DDP is implemented in the `LDDPdensity()` function of `DPpackage`. For the following example we generate $n = 500$ observations each from an assumed simulation truth $F_0^o = N(3, 0.8)$ and $F_1^o = 0.6\,N(1.5, 0.8) +$

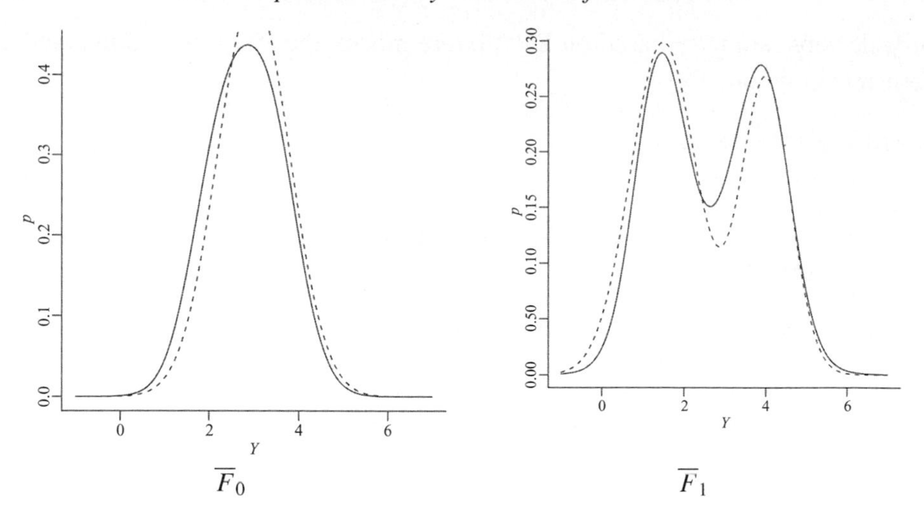

Figure 8.7 Simulated mixture of normal data. The figures show the simulation truth (*dashed line*) and the estimated distributions $\overline{F}_x = E(F_x \mid \text{data})$ (*solid lines*). The posterior means are equal to the posterior predictive distribution for a future observation $\overline{F}_x = p(y_{n+1} \mid x_{n+1} = x, \text{data})$.

$0.4N(4, 0.6)$. Based on the simulated data we estimate $\{F_0, F_1\}$ under a DDP ANOVA prior for $\{F_0, F_1\}$. We assume $F_x(y) = \int N(y; m, s)\,\mathrm{d}G_x$ with $\{G_0, G_1\} \sim$ DDP ANOVA. The DDP ANOVA model is based on an ANOVA model $\mu_{xh} = M_h + xA_h$, i.e., a main effect for $x = 1$ and a common intercept M_h. Let $d = (1, x)$ denote a design vector and let $\beta_h = (M_h, A_h)$ denote the vector of ANOVA parameters. The DDP ANOVA model for $\{F_0, F_1\}$ can be written as a mixture of DP model

$$y \mid x \sim \int N(y;\, \beta' d, \sigma_x)\,\mathrm{d}G(\beta) \quad \text{with} \quad G \sim \mathrm{DP}(G^\star, \alpha).$$

See De Iorio, Müller, Rosner and MacEachern (2004) for more details of the model. The data structure *prior* in the R code fragment below sets the parameters for the base measure $G^\star$, a hyperprior on the residual variance σ_x and the total mass parameter α. We assume $1/\sigma_x^2 \sim Ga(\tau_1/2, \tau_2/2)$ and $\alpha \sim Ga(a_0, b_0)$. Let $B \sim IW(\nu, A)$ denote an inverse Wishart random $(q \times q)$ matrix B with expectation $E(B) = A^{-1}/(\nu - q - 1)$. The base measure is $G^\star(\beta) = N(\mu_b, \Sigma_b)$ with conditionally conjugate hyperpriors $\mu_b \sim N(m, S)$ and $\Sigma_b \sim IW(\nu, \psi)$. Figure 8.7 shows the density estimates $\overline{F}_x = E(F_x \mid \text{data})$, $x = 0, 1$. The estimates are produced by the R code below.

```
  library(DPpackage)

## prepare simulated data: mixture of two normals
  nobs <- 50;                            y1   <-rnorm(nobs, 3,.8)
  y21 <- rnorm(nobs,1.5, 0.8);           y22 <- rnorm(nobs,4.0, 0.6)
  y2 <- ifelse(runif(nobs)<0.6,y21,y22); y <- c(y1,y2)

  trt <- c(rep(0,nobs),rep(1,nobs))  # design matrix with a
                                       single factor
  xpred <- rbind(c(1,0),c(1,1))      # design matrix for posterior
                                       predictive
  m <- rep(0,2);  psiinv <- diag(1,2); s <- diag(100,2)       # prior
  prior <- list(a0=1,b0=1/5,nu=4,m=m,s=s,psiinv=psiinv,
              tau1=0.01,tau2=0.01)

## Fit the DDP ANOVA model
  mcmc <- list(nburn=100, nsave=500, nskip=5, ndisplay=100)
  fit <- LDDPdensity(y~trt,prior=prior,mcmc=mcmc,state=NULL,status=TRUE,
                    grid=seq(-1,7,length=200),xpred=xpred)

## Estimated densities F_x (posterior predictive distributions)
  plot(fit$grid,fit$dens[1,],type="l")
  lines(fit$grid, dnorm(fit$grid, 3.0, 0.8), lty=2) # simulation truth
                                     # ... and x0=(1,1)
  plot(fit$grid,fit$dens[2,],type="l",xlab="Y",ylab="p",bty="l")
  p2 <- 0.6*dnorm(fit$grid, 1.5, 0.8) + 0.4*dnorm(fit$grid, 4.0, 0.6)
  lines(fit$grid,p2, lty=2)
```

The implementation of DDP ANOVA in DPpackage is based on the R package ddpanova which can be downloaded from http://odin.mdacc.tmc.edu/~pm. The ddpanova package has additional options, including options for censored event time data.

8.7 Discussion

We have discussed some extensions and elaborations of the models introduced in Chapter 7. The discussion is by no means an exhaustive list of Bayesian nonparametric models. Keeping to the theme of Chapter 7 we have focused on models that find applications in biostatistics. This focus excluded, for example, interesting recent applications with spatial and spatiotemporal data.

Many more nonparametric Bayesian models and methods are reviewed in other chapters of this volume, including among many others, the beta process, and the Indian buffet process. As an alternative to the R package discussed, DPpackage, the package bayesm (Rossi and McCulloch, 2008) also implements many of the DP-based models.

References

Antoniak, C. E. (1974). Mixtures of Dirichlet processes with applications to Bayesian nonparametric problems. *Annals of Statistics*, **2**, 1152–74.

Banfield, J. D. and Raftery, A. E. (1993). Model-based Gaussian and non-Gaussian clustering. *Biometrics*, **49**, 803–21.

Barry, D. and Hartigan, J. A. (1993). A Bayesian analysis for change point problems. *Journal of the American Statistical Association*, **88**, 309–19.

Branscum, A. J., Johnson, W. O., Hanson, T. E. and Gardner, I. A. (2008). Bayesian semiparametric roc curve estimation and disease diagnosis. *Statistics in Medicine*, **27**, 2474–96.

Buta, R. (1987). The structure and dynamics of ringed galaxies, iii. *Astrophysical Journal, Supplement Series*, **64**, 1–37.

Dahl, D. B. (2003). Modal clustering in a univariate class of product partition models. *Technical Report* 1085, Department of Statistics, University of Wisconsin.

Dahl, D. B. (2006). Model-based clustering for expression data via a Dirichlet process mixture model. In *Bayesian Inference for Gene Expression and Proteomics*, ed. K.-A. Do, P. Müller and M. Vannucci, 201–18. Cambridge: Cambridge University Press.

Dahl, D. B. and Newton, M. A. (2007). Multiple hypothesis testing by clustering treatment effects. *Journal of the American Statistical Association*, **102**, 517–26.

Dasgupta, A. and Raftery, A. E. (1998). Detecting features in spatial point processes with clutter via model-based clustering. *Journal of the American Statistical Association*, **93**, 294–302.

De Iorio, M., Johnson, W., Müller, P. and Rosner, G. (2008). A ddp model for survival regression. *Technical Report*, Texas University M. D. Anderson Cancer Center.

De Iorio, M., Müller, P., Rosner, G. L. and MacEachern, S. N. (2004). An anova model for dependent random measures. *Journal of the American Statistical Association*, **99**, 205–15.

De la Cruz-Mesía, R., Quintana, F. and Müller, P. (2007). Semiparametric Bayesian classification with longitudinal markers, *Applied Statistics*, **56**, 119–37.

Ferguson, T. S. (1973). A Bayesian analysis of some nonparametric problems. *Annals of Statistics*, **1**, 209–30.

Green, P. J. and Richardson, S. (1999). Modelling heterogeneity with and without the Dirichlet process. *Technical Report*, Department of Mathematics, University of Bristol.

Hanson, T. E. (2006). Inference for mixtures of finite Polya tree models. *Journal of the American Statistical Association*, **101**, 1548–64.

Hanson, T. and Johnson, W. (2002). Modeling regression error with a mixture of polya trees. *Journal of the American Statistical Association*, **97**, 1020–33.

Hanson, T. and Yang, M. (2007). Bayesian semiparametric proportional odds models. *Biometrics*, **63**, 88–95.

Hartigan, J. A. (1990). Partition models. *Communications in Statistics, Part A – Theory and Methods*, **19**, 2745–56.

Heard, N. A., Holmes, C. C. and Stephens, D. A. (2006). A quantitative study of gene regulation involved in the immune response of Anopheline mosquitoes: an application of Bayesian hierarchical clustering of curves. *Journal of the American Statistical Association*, **101**, 18–29.

Jara, A. (2007). Applied Bayesian non- and semi-parametric inference using dppackage. *Rnews*, 17–26.

Johnson, V. E. and Albert, J. H. (1999). *Ordinal Data Modeling*. New York: Springer.

Kottas, A., Müller, P. and Quintana, F. (2005). Nonparametric Bayesian modeling for multivariate ordinal data. *Journal of Computational and Graphical Statistics*, **14**, 610–25.

Lavine, M. (1992). Some aspects of Polya tree distributions for statistical modelling. *Annals of Statistics*, **20**, 1222–35.

Lavine, M. (1994). More aspects of Polya tree distributions for statistical modelling. *Annals of Statistics*, **22**, 1161–76.

MacEachern, S. (1999). Dependent nonparametric processes. In *ASA Proceedings of the Section on Bayesian Statistical Science*, 50–5. Alexandria, Va.: American Statistical Association.

Paddock, S., Ruggeri, F., Lavine, M. and West, M. (2003). Randomised Polya tree models for nonparametric Bayesian inference. *Statistica Sinica*, **13**, 443–60.

Petrone, S. (1999a). Bayesian density estimation using Bernstein polynomials. *Canadian Journal of Statistics*, **27**, 105–26.

Petrone, S. (1999b). Random Bernstein polynomials. *Scandinavian Journal of Statistics*, **26**, 373–93.

Pitman, J. (1996). Some developments of the Blackwell–MacQueen urn scheme. In *Statistics, Probability and Game Theory. Papers in Honor of David Blackwell*, ed. T. S. Ferguson, L. S. Shapeley and J. B. MacQueen, IMS Lecture Notes/Monographs, 245–68. Hayward, Calif.: Institute of Mathematical Statistics.

Quintana, F. A. (2006). A predictive view of Bayesian clustering. *Journal of Statistical Planning and Inference*, **136**, 2407–29.

Quintana, F. A. and Iglesias, P. L. (2003). Bayesian clustering and product partition models. *Journal of the Royal Statistical Society, Series B*, **65**, 557–74.

Quintana, F. A. and Newton, M. A. (2000). Computational aspects of nonparametric Bayesian analysis with applications to the modeling of multiple binary sequences. *Journal of Computational and Graphical Statistics*, **9**, 711–37.

Quintana, F. and Müller, P. (2004). Nonparametric Bayesian assessment of the order of dependence for binary sequences. *Journal of Computational and Graphical Statistics*, **13**, 213–31.

Quintana, F., Müller, P., Rosner, G. and Relling, M. (2008). A semiparametric Bayesian model for repeatedly repeated binary outcomes. *Journal of the Royal Statistical Society, Series C*, **57**, 419–31.

R Development Core Team (2008). *R: A Language and Environment for Statistical Computing*. Vienna: R Foundation for Statistical Computing. `http://www.R-project.org`

Richardson, S. and Green, P. J. (1997). On Bayesian analysis of mixtures with an unknown number of components (with discussion). *Journal of the Royal Statistical Society, Series B*, **59**, 731–92.

Rossi, P. and McCulloch, R. (2008). `bayesm`: Bayesian Inference for Marketing/Microeconometrics. R package version 2.2-2. `http://faculty.chicagogsb.edu/peter.rossi/research/bsm.html`

Somoza, J. L. (1980). Illustrative analysis: infant and child mortality in Colombia. *World Fertility Survey Scientific Reports 10*. The Hague: International Statistical Institute.

Tadesse, M. G., Sha, N. and Vannucci, M. (2005). Bayesian variable selection in clustering high-dimensional data. *Journal of the American Statistical Association*, **100**, 602–17.

Author index

Subject index